Forensic Science, 5th Edition

An Introduction to Scientific and Investigative Techniques

Forensic Science, 5th Edition

An Introduction to Scientific and Investigative Techniques

By
Suzanne Bell

CRC Press
Taylor & Francis Group
Boca Raton London New York

CRC Press is an imprint of the
Taylor & Francis Group, an **informa** business

CRC Press
Taylor & Francis Group
6000 Broken Sound Parkway NW, Suite 300
Boca Raton, FL 33487-2742

© 2019 by Taylor & Francis Group, LLC
CRC Press is an imprint of Taylor & Francis Group, an Informa business

No claim to original U.S. Government works

Printed on acid-free paper

International Standard Book Number-13: 978-1-138-04812-6 (Hardback)

Visit the Taylor & Francis Web site at
http://www.taylorandfrancis.com

and the CRC Press Web site at
http://www.crcpress.com

To all students of forensic science past and present and for all the forensic scientists working to ensure that justice is done.

Contents

Notes to the Instructor

This edition of *Forensic Science: An Introduction to Scientific and Investigative Techniques* includes several new features designed to assist you in teaching this class at a variety of levels, from high school through freshman and sophomores in college. It is not meant as a definitive textbook in any one area of forensic science, but rather as an overview of the most commonly known and widely practiced areas. I hope this treatment will help students engage in the material and develop an interest in the underlying science. For a typical semester-long introductory course, the core chapters are Chapters 1–6 and 9–15. Chapters 7 and 8 (forensic anthropology and entomology) are shorter and slightly more advanced, but students typically find these two topics of great interest.

Chapters 17 and 19 are a bit more advanced and invoke some mathematics (algebra). Students sometimes struggle with the core concepts of microscopy but understanding these concepts is critical given that microscopy and trace evidence are inseparable. The remaining chapters are designed to stand on their own to the extent possible. Digital evidence and forensic computing are emerging as a critical forensic discipline, and Chapter 20 could easily round out a semester class.

Two new features have been added. First is a section of advanced questions and exercises. These are designed for advanced students or students who require an honors add-in component. For the more ambitious, these could be extra credit assignments or group activities. Second, a complete crime scene scenario has been added to Chapter 3, and subsequent chapters refer to it either explicitly in the last section of the chapter or as part of the advanced questions and exercises. The scenario is meant to help integrate what might otherwise seem to be disconnected units, particularly later in the book, as student memories of earlier material may have faded. Finally, inserts entitled "Myths of Forensic Science" appear in most chapters and are meant to counter false impressions that many students have regarding the practice of forensic science and how it has evolved in the last few years.

It is critical to note that this is not meant to be a definitive treatment of any one forensic discipline, nor is it intended as a primary reference in a judicial setting. It is designed to provide students an overview of the field and examples of practice, but it is not a manual.

About the Editor

Dr. Suzanne Bell earned a BS (chemistry and criminal justice majors) from Northern Arizona State University and an MS in forensic science from the University of New Haven. She joined the New Mexico State Police in 1983 and worked as a forensic chemist and crime scene processor. She worked at the Los Alamos National Laboratory starting in 1985 as a technical staff member. During this time, she obtained a PhD from New Mexico State University. She started her career in academia in 1994 at Eastern Washington University, where she taught an undergraduate chemistry course and helped the university and the Washington State Patrol develop a forensic chemistry major. In 2003 she joined the Chemistry Department at West Virginia University.

Dr. Bell is a past member of the Scientific Working Group for Seized Drug Analysis (SWGDRUG) and a current member of the Scientific Working Group for Gunshot Residue (SWGGSR) and the gunshot residue subcommittee of the National Institutes of Standards and Technology's (NIST) Organization of Scientific Area Committees (OSAC). She is also a member of the National Commission on Forensic Science (NCFS) and the Forensic Education Programs Accreditation Commission (FEPAC).

Currently, Dr. Bell is the Chair of the department of Forensic and Investigative Science at WVU, where she mentors chemistry and forensic chemistry students from the BS to the postdoctoral level. Her areas of research interest are forensic chemistry, forensic toxicology, ion mobility spectrometry, chemometrics, energetics and explosives, and gunshot residue. In addition to numerous research articles, she has authored and edited many textbooks and reference books, including *Forensic Chemistry* and the 4th edition of *Forensic Science: An Introduction to Scientific and Investigative Techniques.*

Justice and Science

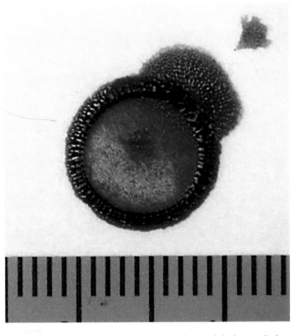

Criminalistics (forensic science) is concerned with the unlikely and the unusual. Other sciences are concerned primarily with the likely and the usual. The derivation of equations, formulas, and generalizations summarizing the normal behavior of any system in the universe is a major goal of the established sciences. It is not normal to be murdered, and most persons never experience this unlikely event. Yet, when a murder occurs, some combination of circumstances suddenly alters the situation from unlikely to certain.

Paul Kirk (1902–1970)
American forensic scientist

1.1 THE ROLE OF FORENSIC SCIENCE

One thing that all forensic disciplines share is an application of a tool, method, or technique to some aspect of the judicial system. Often this is law enforcement and criminal prosecution, but not exclusively. Biology becomes forensic biology with DNA typing, for example, and chemistry becomes forensic chemistry when chemical methods are used to analyze samples like plant matter or pills. The results of drug and DNA testing are used in legal matters, and so we describe these sciences as *forensic* science. Just as biology can be further categorized as cellular biology or microbiology, so too can forensic science be categorized into disciplines such as toxicology, death investigation, or trace evidence. In this book, we examine the most common forensic disciplines, but by no means all of them.

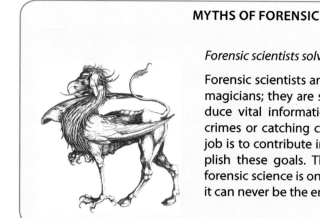

MYTHS OF FORENSIC SCIENCE

Forensic scientists solve cases.

Forensic scientists are not crime scene heroes or courtroom magicians; they are scientists that analyze evidence to produce vital information. They are not tasked with solving crimes or catching criminals. Rather, the forensic scientist's job is to contribute information that can be used to accomplish these goals. The data and information provided by forensic science is only part of a case—often a vital part, but it can never be the entire story.

The quote from Dr. Paul Kirk at the beginning of the chapter describes how forensic science deals with unique and rare events. The phrase "every case is different" has become a cliché because it is true. No two murders, drug samples, firearms, and so forth, are the same, but how and why they are analyzed based on similar underlying concepts. Every crime scene is different, but physics can explain how blood moves and how it behaves when it strikes a surface. When you analyze many bloodstains at a crime scene and put the information together, the data can demonstrate what actions could have produced those stains or, just as importantly, what could *not* have happened. We will learn more about bloodstain patterns in Chapter 4, but all forensic analyses share this core characteristic of being useful in supporting or refuting possibilities or hypotheses regarding events that took place in the recent past.

Events that have occurred in the distant past are studied and evaluated using techniques of archaeology and the study of archaeological artifacts. As such, the discipline shares common elements with forensic science. Imagine that an archaeologist has found a promising site in a cave. Her hypothesis (a possible explanation or interpretation) is that early humans lived in this cave in the distant past (Figure 1.1). The next task is to gather data and information from the site to evaluate this idea. Next would come a lengthy excavation of the site during which artifacts and evidence would be recovered. If none were recovered, the scientist would have to revise her hypothesis or abandon it, but she would do so based on data and information. In this example, assume she did recover artifacts that included small animal bones with striation marks, sharp rocks, charcoal, and a few human bones piled together neatly at the far back of the cave. These items are physical evidence. Taken alone, this evidence does not tell us what happened in the cave tens of thousands of years ago, but it does support the hypothesis that people once inhabited the cave.

There is still much work to be done with the evidence. For example, the archaeologist could send the bone fragments to an expert to see if the striation marks on the animal bones were made

FIGURE 1.1   An archaeological excavation. Image courtesy of the National Parks Service.

by some form of primitive knife or axe. Another expert could look at the sharp rock and determine if it was artificially shaped by chipping to make it into a tool. Specialized laboratories could be used to estimate the date when the wood was burned. All this information becomes data that the archaeologist would use to evaluate her hypothesis.

Suppose the rock was determined to be a tool, the marks on the animal bones were from a primitive knife, and the charcoal dated back tens of thousands of years. Add this data to finding human bones in the cave, and all the evidence supports the archaeologist's original hypothesis—humans did occupy this site in the distant past. On the other hand, imagine that the marks on the animal bones were found to be nothing more than damage accumulating over the years and the sharp rock was just that and not a tool. Knowing that, the archaeologist must reconsider her ideas. Perhaps the evidence is there and has yet to be found, or maybe it was there but long since washed away. New data means that interpretations must be reexamined. The data always shapes conclusions—that is the essence of the *science* of forensic science.

1.2 FORENSIC SCIENCE AND COMPETING STORIES

Data generated by forensic scientists are often used to support or refute different versions of recent events rather than events from the distant past. In the legal context, this could be a suspect's version of an event compared with an investigator's or prosecutor's version of that same event. Suppose police are called to the scene of a fatal shooting. They find a relative of the victim and his clothes are heavily stained with what appears to be blood. This person is identified as the victim's brother. Police call the crime scene unit to process the scene, collect evidence, and deliver it to laboratory. The information provided will be integrated into the investigation and used to determine what could have happened and, just as importantly, what could not have happened.

The suspect (the victim's brother) might claim that he was upstairs, heard a shot, and ran down to find his brother mortally wounded. He called 911 and comforted him until help arrived. This is the suspect's story. On the other hand, investigators might initially believe that the suspect purposely shot and killed his brother after a heated argument and the blood on his clothes shows that he was standing close to the victim when the shot was fired. Both stories are initially plausible, and the investigation will begin. The role of the crime scene investigator would be to document and collect evidence at the scene and deliver it to the appropriate forensic scientists for examination. In turn, the scientists will examine the evidence and generate reports that will become part of the body of evidence in the investigation. Table 1.1 illustrates the type of information and findings that could be produced in the example shooting case.

TABLE 1.1   Example Findings

Finding	Determined by	Story Supported
The brother called 911	Cell phone records	Suspect
The stains on the suspect's clothing were blood	Forensic laboratory analysis	Both
DNA results showed the blood to be that of the victim	Forensic laboratory analysis	Both
A gun was found at the scene	Crime scene investigator	Both
One set of fingerprints on the gun; not the suspect's	Forensic analysis	Suspect
A small smudge of what looks like red nail polish was found in the gun	Investigator	Neither
Autopsy indicates that the weapon was held close to the victim when fired	Medical examiner or coroner	Neither
No residues from shooting a gun found on the suspect's hands or clothing	Forensic laboratory analysis	Suspect
Faint bloody shoeprints were found on the floor leading away from the victim and toward the front door	Crime scene investigator	Suspect
The shoes that made the prints were at least two sizes smaller than those of the suspect	Forensic analysis	Suspect
The footwear impressions were from a woman's athletic shoe	Forensic analysis	Suspect

The first piece of evidence produced is the cell phone records that prove the suspect made the 911 call as he claimed, which supports his version of the story, but by themselves prove nothing. Moving down the list, finding out that the stains on the suspect's clothing were blood and that the source of the blood was the victim could support either story. The suspect said that he held his brother before he died, and therefore the transfer of blood to his clothing is expected. Thus, this information doesn't lend additional weight to either story. Moving down the list, findings of bloody shoeprints and the lack of fingerprints from the suspect on the gun support the suspect's version of events and point to the possibility that someone else committed the crime.

This is a simplified example that makes a crucial point. Isolated pieces of evidence and information are always part of the larger investigation. For a story to be valid, the evidence must be **internally consistent**; *every* finding must support it. If a reliable finding is inconsistent with that version of events, then that story cannot be correct.

Even simple cases often come down to using data to sort out different versions of events, with forensic science and forensic scientists contributing information that can be used to establish the likelihood of one version of events versus another. These two versions are not limited to "the suspect" versus "investigators" or prosecution versus defense. For example, suppose a toxicologist detects ethanol (alcohol) in a blood sample that has been stored for several weeks. Several explanations fit the result. The person who was the source of the blood might have been drinking before the sample was taken, but it is also conceivable that the ethanol was a by-product of improper storage and decomposition. Additional data would be needed to determine which explanation (which "story") is most likely.

MYTHS OF FORENSIC SCIENCE

Forensic science and criminology are the same thing.

Criminology is a social science that focuses on the social and human aspects of crime, and as such, criminology shares many characteristics of sociology. Criminology is not the same thing as criminal justice, which focuses on law enforcement and the criminal justice system. Forensic science incorporates the natural sciences and requires an understanding of the basics of biology, chemistry, and physics. If you are interested in a career as a law enforcement officer, criminal justice is a reasonable choice for a college major. If you are interested in understanding street gangs and helping to counsel former gang members, criminology is a fit. If you want to pursue forensic science, you will need a solid foundation in math, statistics, biology, chemistry, and physics.

1.3 SCIENCE

"Data! Data! Data!" he cried impatiently. "I can't make bricks without clay."

Sherlock Holmes quote, The Adventure of the Copper Beeches

The process of gathering information to sort out different versions of events shares elements of what is referred to as the scientific method. However, the practice of science is better thought of as a collection of procedures and approaches founded in the principles of science. At its core, science is based on evidence and data. This data can come from observations, tests, and experiments. If a scientist develops a new theory or modifies an existing one, the validity of the new concept must be supported by data. This data should be obtainable by anyone, and anyone who repeats the experiments should obtain comparable results. This is the concept of **reproducibility**. Scientific findings must also be **falsifiable**, meaning that they are presented in such a way that if incorrect, other scientists could clearly demonstrate that they are false.

Related to this are the properties of explanation and prediction. For example, centuries ago Isaac Newton proposed what is now called the second law of motion, which links force needed to accelerate an object in terms of its mass: $F = ma$. This equation is falsifiable because it can easily be tested for accuracy, as it has been over centuries. The equation is also predictive because if we know the mass of an object and the magnitude of the acceleration, force is simple to calculate. A scientific claim or result does not have to be in the form of an equation, but it must still retain the elements of falsifiability and reproducibility.

Ideally, science is transparent. If a claim is made, data and methods used in formulating that claim should be clearly stated so that others can evaluate it. One method of evaluating scientific claims and data is **peer review**. If a physicist claims that Newton's second law must be modified, typically he or she would prepare a paper and submit it for publication in a scientific journal. The paper would be sent to other physicists for review and discussion before a decision to publish is made. Once published, other scientists will attempt to reproduce the results as part of their own work, which is another form of review. Ideally, science is self-correcting and dynamic. It advances over time as innovative ideas, new data, and new observations are proposed, tested, reviewed, and accepted. The system of peer review is not flawless, but it contributes to a process of **self-correction** that is also a core principle of science.

These same concepts apply to forensic science. When an analyst in a toxicology laboratory measures the concentration of alcohol in blood, he or she writes a report that states how the measurement was done and what the result was, such as 0.09% ± 0.01%. This is testable and falsifiable because another analyst can obtain a portion of the same blood and repeat the test. Using the term *falsifiable* does not mean that the results are inherently suspected of being wrong; it just means that there is a mechanism to uncover that should they be false. In addition, the toxicologist uses a validated method that has undergone peer review, as has the report. Anyone who wanted to test the repeatability of the results would have access to all the information necessary to do so.

1.3.1 BIAS IN SCIENCE

As human beings, scientists have ideas and notions that could influence how they look at data and results. The concern over such **bias** has become an important topic of discussion in the forensic science community. While scientists strive for objectivity, bias can arise despite the best of intentions. Because bias arises unconsciously, it can be difficult to avoid and eliminate. There are several potential sources of bias that can impact forensic science. As an example, suppose you were working in a fingerprint laboratory and your supervisor presented you two fingerprints— one a reference print from a database and one recovered from a crime scene. He tells you that the reference print is from a suspect with three prior arrests. This information, called **contextual information**, could impart bias. Because you have been told that the person has a criminal record, you might, consciously or unconsciously, assume the likelihood of guilt. As a result, you may be predisposed to determine that the reference print and the crime scene evidence probably came from the same person.

One way to limit this source of bias is **sequential unmasking**, in which information is provided or obtained in a stepwise process. Suppose your supervisor simply presented you with the latent fingerprint recovered from the crime scene and asked you to perform the database search. Your search might produce several possible sources, of which the convicted felon is one. You would then do further analysis on the prints without knowing that one of these is from a convicted felon, thus eliminating that source of bias. You can't be biased by something you don't know.

Another potential source of bias arises from where forensic science is practiced. Suppose a forensic DNA analyst works for a laboratory that is run by the police department of the city in which he lives and works. Does that cause him to be biased toward the police officers that submit the evidence? It is not hard to imagine such an analyst feeling a sense of loyalty to his colleagues in the department, which could conceivably lead to **prosecutorial bias**, which like all bias can be unconscious. Forensic scientists must be constantly aware of their primary responsibility, which is to perform scientific analyses and report their results regardless of how the results might help or hurt either side in a case. It is not the forensic scientist's job to identify the guilty party and make sure that they are punished. Rather, their job is to supply scientific data and information that can be used to find the truth in any given case, regardless of who or what their data supports, incriminates, or exonerates.

1.3.2 JUNK SCIENCE

Sometimes claims are made that are presented as scientific even when none of the elements of science are involved. As an example, astrology purports that that the position of the stars on the day you were born makes it possible to make sweeping predictions regarding your personality. As fun as it might be to read the daily horoscope, astrology is not a science. It is not testable, nor is it falsifiable. For example, suppose your daily horoscope says, "Something interesting will happen to you today." The term *interesting* is vague and subjective; what is interesting to you may not be interesting to someone else. Also, there is no way to test whether this experience is directly related to, and caused by, the position of the stars at your birth. Astrology passes none of the tests of science. Astronomy *is* a science in that it is based on the scientific principles we discussed above. Astronomers make observations, publish and present findings, formulate innovative ideas based on this data, and revise as new data is uncovered and verified.

If all we use horoscopes for is entertainment, no harm is done. However, suppose an expert was hired in a murder case who testified that the defendant committed a crime and is thus guilty of this crime because of the position of the stars on the day they were born. Astrology in this context would fall under the category of **junk science** (or **pseudoscience**) because it is not a science at all. As such, it has no place in a courtroom or any kind of legal setting in which the outcome has profound consequences for a person and for society. Defining what type of evidence can be used in a law enforcement setting is determined through the process of admissibility, which we will discuss in detail in Chapter 2. The goal of determining admissibility is to ensure that junk science or unreliable or unproven information is not allowed to be used in deciding the outcome of a case.

1.4 THE RECENT PAST OF FORENSIC SCIENCE

This history of death investigation is surprisingly long. Consider the following commonsense advice quoted from a forensic text:

> The similarities between those who jump into wells, those who are thrown in, and those who lose their footing and fall are very great. The differences are slight.... If the victim was thrown in or fell in accidentally, the hands will be open and the eyes slightly open, and about the person he may have money or other valuables. But, if he was committing suicide, then his eyes will be shut and the hands clenched. There will be no valuables on the body. Generally, when someone deliberately jumps into a well, they enter feet first. If a body is found to have gone in head first, it is probable that the victim was being chased or was throw in by others. If he lost his footing and fell in, you must check the point where his feet slipped to see if the ground has been disturbed.

You might assume that this passage was taken from a Sherlock Holmes story from around 1900 or even from a modern crime thriller or television episode. It was written in 1247 and was printed in a book in China entitled (translated) *The Washing Away of Wrongs*. While the investigation of death was only one motivating force in the development of forensic science, it certainly was one of the most important. Besides the obvious issue of crime, the nature and circumstances of death have long had legal and social consequences, such as inheritance and taxes, making death more than a personal or family issue.

ALFRED LUCAS (1867–1945), ARCHAEOLOGIST AND FORENSIC CHEMIST

Alfred Lucas worked as both a forensic chemist and an archaeologist. In 1897, he left England to work in Egypt in hopes that the dry climate would help him recover from tuberculosis. He headed a British government laboratory and worked on cases involving handwriting, poisons, and firearms. Lucas authored the first known forensic chemistry textbook in 1912. He is probably best known for working with Howard Carter during the excavation and study of the tomb of Tutankhamen from 1927 to 1933. He worked to restore the throne recovered from the tomb shown here.

Over the centuries, what we could call forensic science developed in parallel with the emergence of science, but a recognizably modern form of the discipline coalesced in the 1800s in Europe. Many famous cases date from that era forward and are highlighted in textboxes throughout this book. In the United States, medical examiners and forensic laboratories were organized in the early part of the 20th century, with notable developments such as the formation of the New York City Medical Examiner's Office in 1918 and the Federal Bureau of Investigation (FBI) Laboratory in 1932. Government entities such as cities, counties, and states established laboratories and laboratory systems that were, for the most part, associated with law enforcement agencies, such as state police, city police, or sheriff's offices. According to the "Census of Publicly Funded Forensic Crime Laboratories" published by the U.S. Department of Justice, there were 411 such laboratories in the United States in 2009, the latest date for which data are available.

A milestone in the practice of forensic science in the United States came in 2009 with the publication of a report created by the National Research Council and the National Academy of Sciences (NAS) entitled *Forensic Science in the United States: A Path Forward*. The report was critical of many forensic disciplines, particularly those working with pattern evidence, such as fingerprints, tire and shoe treads, and tool marks. This report was instrumental in launching entities such as the Organization of Scientific Area Committees (OSAC) managed by the National Institute of Standards and Technology (NIST). The purpose of the OSAC is to develop and approve standard methods for use in a broad range of forensic disciplines. The documents produced are expected to have a significant impact on forensic science through improving and standardizing procedures used by forensic scientists and forensic laboratories.

Another critical recent development is the increasing attention being paid to **accreditation**. Many kinds of laboratories are accredited; for example, clinical laboratories in hospitals that perform blood tests are accredited. Accreditation means that a laboratory has agreed to operate per a professional or industry standard and has proven that it can and does operate this way. The process of obtaining accreditation is often long and complex and involves planning, documentation, record keeping, training, and site visits by the accrediting agency. Once a laboratory is accredited, it must be reaccredited on a set schedule, such as every 5 years.

Accreditation is increasingly being demanded of forensic laboratories. As noted in the 2009 NAS report, *"Laboratory accreditation and individual certification of forensic science professionals should be mandatory.... All laboratories and facilities (public or private) should be accredited."* Currently, the largest forensic accrediting body is the American Association of Crime Laboratory Directors/Laboratory Accreditation Board (ASCLD/LAB). As of February 2017, 396 laboratories in the United States and abroad were accredited by ASCLD/LAB.

Accreditation is associated with systems of **quality assurance/quality control** (QA/QC). Collectively, these are procedures implemented on the flow of evidence from when it enters the laboratory through to reporting and final disposition. Elements of QA/QC include the use of special samples that are known to be positive (**positive control** or **known**) or negative (**blanks** or **negative control**), **blind samples**, **double blinds**, peer review, quality review, and **standard operating procedures** (SOPs). For example, suppose a blood sample is submitted to a laboratory and is to be tested for the presence of an illegal drug such as cocaine. The analyst will use the laboratory's standard analytical method to prepare the blood sample, place it into the correct instrument, and complete the analysis (the SOP). If for some reason the analyst must modify the process, he would note that in the case documentation. To be certain that any drug he finds is really from the sample and not a contaminant, he would also analyze a known clean blood sample using the exact same procedure (the blank), as well as a blood sample to which a known amount of that same drug has been purposely added (positive control). If he finds the drug in the submitted sample and in the positive control but not in the blank, he can be confident that the drug was present and that all the procedures are working properly. In other words, the quality is ensured. He would write up his report and hand it over to the QA/QC reviewer, whose job it is to check the file and all the data to be sure everything is in order and the results are reliable and complete.

Other steps would also be taken to control and ensure quality. Every few weeks, the analyst would be assigned a sample by his supervisor to test; he would not be told what is in the sample, but the supervisor would know. This is an example of a blind sample, and its purpose is to ensure that the analyst's work is acceptable. In some cases, the laboratory would receive samples from a testing agency that would be given to the analyst for the same reason. The difference is that no one in the laboratory knows what is in this sample, including the supervisor. This is an example of a double-blind test. These samples are designed to insure the quality of the laboratory's work, not just that of the analyst. Any accredited laboratory must have a robust QA/QC system in place for all sections of the laboratory.

While laboratories can be accredited, analysts in some forensic disciplines can become certified. **Certification** means that a forensic scientist has completed a written test covering his or her discipline and that the analyst participates in proficiency testing to ensure that their laboratory methods and techniques are sound. Because forensic science is such a broad discipline, accreditation is not currently the responsibility of a single organization. For example, forensic toxicologists are certified by the American Board of Forensic Toxicology (http://www.abft.org/), an organization that also accredits certain types of forensic toxicology laboratories. The National Association of Medical Examiners (NAME) accredits medical examiner's offices, while the American Board of Pathology (ABP) certifies physicians in forensic pathology. The International Association of Identification certifies several forensic disciplines. The American Board of Criminalistics (ABC) covers the most diverse set of forensic disciplines, from chemistry to tool marks. The certification process begins by passing a multiple-choice test. Forensic scientists can then elect to be further certified in a specialty area. Currently, these areas are molecular biology, drug analysis, fire debris analysis, and two types of trace evidence (hairs/fibers and paints/polymers). This level of certification requires the successful annual completion of proficiency tests.

With all the advances in forensic science since 1990, such as the widespread adoption of DNA testing, the NAS report, the OSAC, accreditation, and certification, forensic science has entered a time of change and advancement not seen since its inception nearly two centuries ago.

1.5 THE ODD COUPLE

Lawyers and forensic scientists enjoy a close, yet often uneasy, relationship. In this sense, at least, forensic scientists and lawyers speak different languages with different objectives, unfortunately using many of the same words. The words *truth, fact, certainty, possible, probable*, and even *error* can mean very different things in law and in science. Lawyers work in an **adversarial system** where the objective is winning a favorable decision for one's client through knowledge of the law. The opposing side works toward the same outcome for their client. A key goal of the legal system is to settle disputes. In criminal cases, the dispute is between some government entity (the state of California, the City of New York, etc., "the people" or the prosecution) and an individual or group of individuals (the accused or the defendant). In civil cases, the dispute can be between individuals, companies, etc. Each side presents an argument before the decision-maker and whomever makes the best argument wins. A critical consideration is legal **precedent**, or what earlier courts

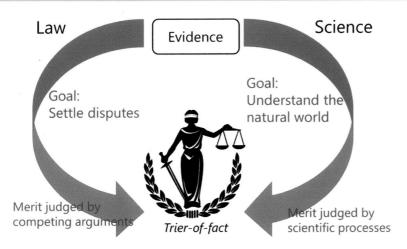

FIGURE 1.2   The different goals and processes of science and the law.

have decided in similar cases. Past decisions are important, and courts strive for consistency. This makes sense; a case today should be treated the same as yesterday's case unless something has changed (laws, rules, etc.) (Figure 1.2)

The test of merit in the legal system is the argument—whichever side makes the best case for their position will "win." Scientific analysis of evidence and the reports and testimony of forensic scientists are integrated into the argument, but such information is always part of a larger whole. In contrast, the goal of science, broadly speaking, is to improve our understanding of the natural world. This goal has nothing to do with justice directly; science produces information used by the justice system.

The two models—the scientific method and the adversarial system—make for an odd couple and are even *at* odds in some ways. The legal system relies on and respects precedent (what has gone before) and as such can be slow to change. Science, on the other hand, strives to move forward and prizes advancement and the change it is built on. At its best, science is dynamic, ever improving, and self-correcting. Scientists are motivated to change and move forward, while the law values the past and precedent.

Unfortunately, rapid change and evolution can cause problems in the legal setting. As an example, suppose a person was accused of a crime in the early 1980s. The analysis of blood evidence would be limited to the ABO blood type, and perhaps a few other markers, and would require a stain the size of a dime for the analysis. At that time, this was the state of the art and there would be no scientific problems at the time with that approach. Today, the size of the stain needed would be about the size of the period at the end of this sentence, and depending on the condition of the evidence, a DNA type and sex of the source of the stain could be determined. ABO blood typing, good as it was at the time, is an archaic forensic technique less than 40 years later. In another 40 years (probably much less), the analysis of blood will be based on sophisticated genetic sequencing, and our current methods may seem as archaic as ABO typing does to us today. Thus, when older evidence is reexamined, findings can change. When it does, there can be a suspicion that the older data was somehow wrong or mistaken. That is rarely the case. New data is better in the scientific sense, but obtaining it does not mean that analysts in the 1980s were doing anything wrong; they were just working with the scientific understanding of the time. Science values change, law values precedent, and sometimes those two perspectives see things very differently.

1.6 TYPES OF LEGAL PROCEEDINGS

A key part of most forensic science jobs is interacting with the justice system through different procedures and mechanisms. Sometimes, all that is required is a signed report and the forensic scientist's role in the case ends. At other times, days spent in grueling testimony are required before a judge, jury, and even national television audiences. We'll discuss the most common ways, but certainly not the only ways, in which a forensic scientist interacts with the judicial system.

Broadly speaking, the legal system in the United States can be divided into the realm of **civil law** and **criminal law**. To simplify, civil cases are cases between individuals or parties—the common meaning of the terms *lawsuit* or *being sued*. Another example is a patent infringement suit in which one company has accused another of violating a patent. On the other hand, criminal cases involve violation of criminal laws and involve the government as the body that is charging an individual, individuals, or companies with violation of criminal laws. Typically, the party that files criminal charges is called the **prosecution** (**plaintiff** in civil actions) and the accused is the **defendant**. Forensic scientists can testify in either type of legal action and for either party, the accused or the defendant.

A **jurisdiction** is a region or geographical area over which law enforcement or legal entity can exercise authority. In the United States, different legal rules and procedures often apply depending on the jurisdiction in which the procedure is conducted. For example, the rules that apply in the City of New York may differ from those in Chicago, which may differ from the rules that apply in the states of New York and Illinois, which in turn may differ from rules used in federal courts. As an example, a person who sells a sample of an illegal drug in a state may be tried at a local or state level. If that same person smuggles drugs across state lines, then federal jurisdiction applies. The legal procedures may be similar, but they are not necessarily the same. We will discuss this in the context of admissibility in detail in Chapter 2.

It is important to know to whom evidence is presented and who will make the decision based on the evidence presented. This entity is defined as the **trier of fact**, which can be a jury of some type or a judge. In a criminal proceeding, the prosecutor bears the **burden of proof**, meaning the prosecutor must prove a defendant guilty beyond a reasonable doubt, a burden that never shifts. However, in a civil case, the plaintiff need only prove his case by a preponderance of the evidence, a much easier standard to meet. Informally, beyond a reasonable doubt might be thought of as 99% certainty while preponderance of evidence would be 51%. These are not hard numbers but examples to put the difference between civil and criminal burdens in perspective. Criminal cases are divided into the more serious (**felony**) versus less serious (**misdemeanor**), a division that is important in determining the severity of the punishment handed out to a defendant found guilty.

Most people assume that when someone is arrested and charged with a crime, they will end up in a courtroom in front of a jury. That outcome is rare. As shown in Figure 1.3, less than 5% of people charged with a crime are tried before a jury. Instead, most cases are **adjudicated** (settled) by **plea bargaining**. Forensic science may or may not play a role in such cases. As we will see in Chapter 11, a method for screening for the presence of a drug is the use of simple chemical tests that produce color changes when drugs are present. In the lab, these screening tests were once routinely used as a first step in analyzing a sample suspected to contain

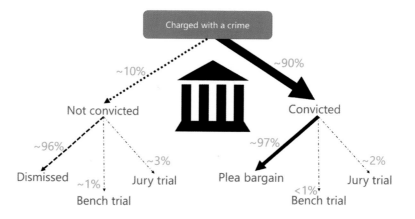

FIGURE 1.3   The fate of cases in which charges have been filed. Data current to 2015.

illegal drugs. The results were not considered conclusive and were used to direct laboratory work to follow. The same tests were adapted for use in the field by police officers and other law enforcement officials as a quick means of checking for the possible presence of an illegal drug. A positive test meant that there could be **probable cause** to investigate further and even arrest the person who had the sample in their possession. However, a color test is far from sufficient to definitively identify a substance as containing a drug. Unfortunately, the link between scientific use and interpretation of these tests can be broken when this type of result is used to spur plea bargaining. There is nothing wrong with color testing when used and interpreted properly and the limitations of the findings are clearly stated and understood by all involved. However, a positive test does not mean with 100% certainty that a drug is present, while a negative does not mean that there is no drug present. A screening test is just that—not a definitive result.

1.7 ETHICS

ETHICS AND ACCREDITATION

In 2012, Annie Dookhan was arrested in Boston. She was suspended from a Boston-area forensic laboratory in 2011 on suspicion of faking case results and other actions that put into question nearly 34,000 defendants. She served time in jail and in 2017, the district attorney announced that all but about 1000 of those convictions would be removed from the record. Her conduct was clearly unethical but had the laboratory been accredited, it is likely that problems would have been uncovered long before these thousands of cases were settled.

The importance of ethics as part of forensic science is obvious. As with most professions, there are formal written codes of ethics that members of professional societies agree to and are expected to follow. However, because there are so many different subdisciplines in forensic science, there is no one code of ethics that covers all forensic scientists, although one has been proposed by the National Commission on Forensic Science. The American Academy of Forensic Sciences (AAFS) has a code of ethics and an ethics committee that deals with ethics issues and challenges. Other societies and groups have separate codes. Examples include the Society of Forensic Toxicology, the American Society of Questioned Document Examiners, and the organizations discussed in the section describing accreditation. However, as we all know, having a code of ethics does not always prevent ethical lapses from occurring. In a profession such as forensic science, these ethical lapses can have serious and long-term effects.

The fundamentals of forensic ethics are simple to define in a broad sense. First is ethics, which we can think of as a set of rules that govern the conduct of a professional working in each field. These rules define what is proper, acceptable, and honest behavior within that discipline. Here, a forensic scientist is expected to do the best work that they can on every case that is submitted, provide a complete and honest report of their work, and testify as to their scientific opinion based on that report and their knowledge in the field. This does not imply that disagreements between two experts automatically mean that one of the scientists is being unethical—far from it. In complex cases that involve opinion evidence such as crime scene reconstruction, death investigations, and toxicological testing and data interpretation, different experts often have different opinions that are derived from the same data. Testimony provided by accepted scientific experts is still opinion evidence. The degree to which expert opinions can be reasonably expected to disagree usually depends on the complexity of a case. You would not expect two experienced forensic chemists to disagree over the identification of a white powder as cocaine, for example, but you can reasonably expect two blood stain pattern experts to disagree over stains found in a large and complex crime scene. It is up to the experts to present their opinion and defend it, but in the end, it is the trier of fact that will decide how that testimony is integrated into their decision in the case.

1.8 REVIEW MATERIAL

1.8.1 KEY TERMS AND CONCEPTS

Accreditation
Adjudicated
Adversarial system
Bias
Blank
Blind sample
Burden of proof
Certification
Civil law
Contextual information
Criminal law
Defendant
Double-blind sample
Falsifiability
Felony
Internally consistent
Junk science
Jurisdiction
Known
Misdemeanor
Negative control
Peer review
Plaintiff
Plea bargain
Positive control
Precedent
Probable cause
Prosecution
Prosecutorial bias
Pseudoscience
Quality assurance/quality control
Reproducibility
Self-correction
Sequential unmasking
Standard operating procedure
Trier of fact

1.8.2 REVIEW QUESTIONS

1. If a police officer suspects a driver of driving under the influence, he can pull a car over and request that the driver take a breath test. The results of this test can provide probable cause. Explain what this means.
2. Suppose that the results of this breath test were positive, and the driver is arrested and a blood sample collected and sent to the laboratory. The laboratory assigns the case to an analyst but does not provide her with any other information. What practice is this an example of?
3. Graphology practitioners claim that they can identify personality traits from your handwriting. Is this a science? Why or why not?
4. When your doctor orders a blood test for you, the sample is sent to an accredited laboratory. Why?
5. What is the difference between accreditation and certification?

6. In the United States, a person is supposed to be presumed innocent until proven guilty. How does this relate to the concept of the burden of proof?
7. The chapter discussed prosecutorial bias. Can there be defense bias? If so, provide an example.
8. What ethical codes exist for students at your school?

1.8.3 ADVANCED QUESTIONS AND EXERCISES

1. Suppose you work in a laboratory and notice that your coworker is consistently completing more cases than you think is possible. What do you think your ethical responsibilities are in this situation?
2. A forensic analyst receives a bloody shirt as evidence. The stained area is large, and she collects five small samples from the stained area. She completes the analysis and reports the DNA type, which is the same in all five stains. How is this data repeatable and falsifiable?
3. The following article is freely available online by open access:
   Amy Jeanguenat and Itel Dror, Human Factors Effecting Forensic Decision Making: Workplace Stress and Well-Being, *Journal of Forensic Sciences*, Vol. 63, Issue 1, 2018.
   Find it online and answer the following questions:
   a. Why and how would irrelevant case information cause pressure or stress for a forensic examiner?
   b. How is the adversarial system stressful for forensic scientists?
   c. Which one of the sources of stress was most surprising to you and why?
4. Does the scientific method have elements of the adversarial system in it? Does the legal system have elements of the scientific method in it?
5. The Annie Dookan case probably would not have been as catastrophic if a few processes and procedures were in place. Discuss how accreditation, quality assurance, and ethics could have limited the damage in this and other cases involving rogue analysts.

BIBLIOGRAPHY AND FURTHER READING

BOOKS, BOOK SECTIONS, AND REPORTS

Bowen, R. T. *Ethics and the Practice of Forensic Science.* 2nd ed. Boca Raton, FL: CRC Press, 2017. ISBN: 978-0-309-13135-3.
Committee on Identifying the Needs of the Forensic Sciences Community, National Research Council. *Strengthening Forensic Science in the United States: A Path Forward.* Washington, DC: National Academies Press, 2009. ISBN: 978-1-498-77713-1.

JOURNAL ARTICLES

Archer, M. S., and Wallman, J. F. Context Effects in Forensic Entomology and Use of Sequential Unmasking in Casework. *Journal of Forensic Sciences* 61, no. 5 (Sep 2016): 1270–1277.
Canter, D., Hammond, L., and Youngs, D. Cognitive Bias in Line-Up Identifications: The Impact of Administrator Knowledge. *Science and Justice* 53, no. 2 (Jun 2013): 83–88.
Dror, I. E. Cognitive Neuroscience in Forensic Science: Understanding and Utilizing the Human Element. *Philosophical Transactions of the Royal Society B—Biological Sciences* 370, no. 1674 (Aug 2015).
Eastwood, J., and Caldwell, J. Educating Jurors about Forensic Evidence: Using an Expert Witness and Judicial Instructions to Mitigate the Impact of Invalid Forensic Science Testimony. *Journal of Forensic Sciences* 60, no. 6 (Nov 2015): 1523–1528.
Edmond, G., Tangen, J. M., Searston, R. A., and Dror, I. E. Contextual Bias and Cross-Contamination in the Forensic Sciences: The Corrosive Implications for Investigations, Plea Bargains, Trials and Appeals. *Law Probability & Risk* 14, no. 1 (Mar 2015): 1–25.

Edmond, G., Towler, A., Growns, B., Ribeiro, G., Found, B., White, D., Ballantyne, K., et al. Thinking Forensics: Cognitive Science for Forensic Practitioners. *Science and Justice* 57, no. 2 (Mar 2017): 144–154.

Kafadar, K. Statistical Issues in Assessing Forensic Evidence. *International Statistical Review* 83, no. 1 (Apr 2015): 111–134.

Kelty, S. F., Julian, R., and Ross, A. Dismantling the Justice Silos: Avoiding the Pitfalls and Reaping the Benefits of Information-Sharing between Forensic Science, Medicine and Law. *Forensic Science International* 230, no. 1–3 (Jul 2013): 8–15.

Koehler, J. J., Schweitzer, N. J., Saks, M. J., and McQuiston, D. E. Science, Technology, or the Expert Witness: What Influences Jurors' Judgments about Forensic Science Testimony? *Psychology Public Policy and Law* 22, no. 4 (Nov 2016): 401–413.

Laurin, J. E. Remapping the Path Forward: Toward a Systemic View of Forensic Science Reform and Oversight. *Texas Law Review* 91, no. 5 (Apr 2013): 1051–1118.

Murrie, D. C., Boccaccini, M. T., Guarnera, L. A., and Rufino, K. A. Are Forensic Experts Biased by the Side That Retained Them? *Psychological Science* 24, no. 10 (Oct 2013): 1889–1897.

Nakhaeizadeh, S., Dror, I. E., and Morgan, R. M. Cognitive Bias in Forensic Anthropology: Visual Assessment of Skeletal Remains Is Susceptible to Confirmation Bias. *Science and Justice* 54, no. 3 (May 2014): 208–214.

Osborne, N. K. P., Taylor, M. C., and Zajac, R. Exploring the Role of Contextual Information in Bloodstain Pattern Analysis: A Qualitative Approach. *Forensic Science International* 260 (Mar 2016): 1–8.

Wechsler, H. J., Kehn, A., Wise, R. A., and Cramer, R. J. Attorney Beliefs Concerning Scientific Evidence and Expert Witness Credibility. *International Journal of Law and Psychiatry* 41 (Jul–Aug 2015): 58–66.

WEBSITES

Link	Description
www.aafs.org	American Academy of Forensic Sciences
www.criminalistics.com	American Board of Criminalistics
www.abpath.org	American Board of Pathology
http://www.forensicsciencesimplified.org/	Overview of many forensic science disciplines with many linked resources
http://www.ascld.org/	American Association of Crime Laboratory Directors
http://www.ascld-lab.org/	American Association of Crime Laboratory Directors/ Laboratory Accreditation Board
www.asqde.org	American Society of Questioned Document Examiners
http://asq.org/learn-about-quality/quality-assurance-quality-control/overview/overview.html	American Society of Questioned Document Examiners
www.fbi.gov	Federal Bureau of Investigation
https://www.fbi.gov/services/laboratory	Federal Bureau of Investigation Laboratory
https://www.nist.gov/topics/forensic-science	Forensic Science at NIST
http://thefsab.org/	Forensic Specialties Accreditation Board
www.theiai.org	International Association for Identification
www.thename.org	National Association of Medical Examiners
http://www.nasonline.org	National Academy of Sciences
http://www.ndaa.org/	National District Attorneys Association
www.nist.gov	National Institute of Standards and Technology
https://www.nist.gov/forensics/organization-scientific-area-committees-forensic-science	Organization of Scientific Area Committees
www.soft-tox.org	Society of Forensic Toxicologists
www.swgdrug.org	Scientific Working Group for the Analysis of Seized Drugs
https://www.innocenceproject.org/	Innocence Project
http://thelawdictionary.org/	The Law Dictionary online
http://www.publicdefenders.us/	National Association for Public Defense

Evidence

Origins, Types, and Admissibility

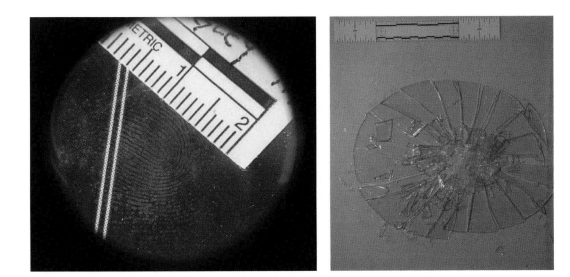

2.1 EVIDENCE

Evidence is the "stuff" of forensic science, and forensic scientists, need to understand how it is generated, what it can be used for, how it is analyzed, and how the legal system will decide to admit and use it. Evidence can be tangible things, such as a bloody shirt, a tiny fiber on a baby blanket, a blood sample, or a fingerprint. These are examples of physical evidence, and we will spend most of the rest of this book talking about diverse types of physical evidence and the scientific techniques used to evaluate it. Testimony can also be a type of evidence; a forensic scientist offers his or her opinion (**opinion evidence**) as part of the body of evidence that is considered in any given case.

In Chapter 1, we introduced an example of an archaeologist exploring a cave and using the evidence she found to understand more about what happened in that cave in the ancient past. Now we will update that example to a crime scene to illustrate how evidence is created and how forensic science and the law interact relative to that evidence. You may have heard some version of the phrase "every contact leaves a trace," which is attributed to Edmund Locard and called **Locard's exchange principle**. This concept explains how evidence is created and how it can be used to recreate an event or sort out different versions of an event. According to this concept, contact or interaction between people or between a person and a place inevitably produces some type of exchange of material between them that is evidence of the interaction. This type of evidence is sometimes referred to as **transfer evidence** because of how it is created.

Suppose two people meet on a baseball field, shake hands, and walk away. This meeting is the event in question at the scene. Locard's principle tells us that transfers have occurred between the two people but also between each person and the scene. Transfer evidence is the tangible link between the two people and each person and the baseball field. Since they shook hands, DNA has been exchanged, as has sweat. Each person has grass and dirt wedged in their shoes from the baseball field. In turn, each person left footwear impressions on the field. If we were to stop one of these people, collect the shoes, and recover the grass and dirt, this is the type of evidence that can link the person to the scene. There is no magic test that will tell us that the dirt and grass came from that specific field and nowhere else, but finding it does support a theory that the person was at the field at some point. It helps to link that person to the scene. If that person's shoeprint is found at the field, this is also links the person to the scene. Each finding is one more piece of supporting evidence for the idea that the person was at the baseball field.

EDMUND LOCARD (1877–1966)

The namesake of the exchange principle was trained in both law and medicine. In 1910, he established a forensic laboratory in Lyon, France. The lab was primitively equipped, but Locard was able to establish a reputation and increase the visibility of forensic science in Europe. Locard was interested in microscopic and trace evidence, particularly dust, and believed that such trace evidence was crucial in linking people to places. The success of his laboratory and methods encouraged other European nations to form forensic science laboratories after the conclusion of World War I. In Lyon, he founded and directed the Institute of Criminalistics located at the University of Lyon, and he remained a dominant presence in forensic science into the 1940s. In 2016, France issued a postage stamp in his honor.

Evidence can be described as **inclusive**, **exclusive**, or **indeterminate**. The dirt and grass on the shoes is inclusive because if we are trying to determine if the person was at the baseball field, the finding supports that and includes them in a list of people that could have been at the baseball field. If beach sand was found on the shoes, the evidence is indeterminate because it doesn't help answer the question related to being at the baseball field. Exclusive evidence would, as the name suggests, tend to exclude a possibility. Suppose that neither person lives in the city where the baseball field is located; that would be exclusive evidence.

The same concepts apply in crime investigation. Figure 2.1 illustrates how an event, in this example a fatal shooting, creates evidence and how it can be used to link a person to a scene or an event. In this case, the perpetrator of the shooting enters a home and kills the victim with a gun, which he takes with him when he flees the scene. Per the exchange principle, contact between the perpetrator and the scene has created physical evidence. The scene will retain traces of the perpetrator and the perpetrator will carry with him traces from the scene. If this scene is processed,

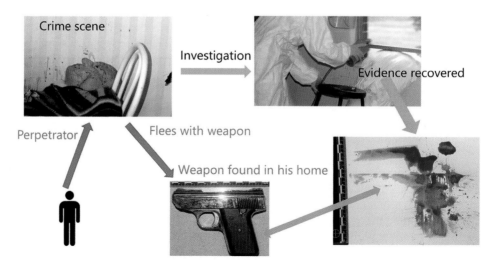

FIGURE 2.1   When one person (here the perpetrator) comes in contact with the victim at his home, evidence is created and transferred to and from people to and from the scene as per Locard's exchange principle.

many types of evidence may be recovered, such as fingerprints (a trace left by the perpetrator) and bloodstain patterns. One of those patterns could be the outline of the gun in blood found on a counter, for example. The investigation may then identify a suspect, in this case the perpetrator. Police might go to his house and find the weapon, which can then be compared to the blood pattern found at the scene. Fingerprints on the gun could link it to the perpetrator, while the bloody pattern links the gun to the crime scene. The perpetrator is thus linked to the gun, which is linked to the scene through evidence that arose from contact. The linkage is made by transfer evidence.

MYTHS OF FORENSIC SCIENCE

Eyewitness testimony is the best evidence.

Eyewitness testimony can be invaluable in an investigation, but such evidence should always be considered and scrutinized the same way scientific evidence would be. In the worst cases, witnesses may lie, but even those with the best of intentions can be mistaken; memories can even be false memories, such as when a child "remembers" being abused by a parent even when there was no abuse. Eyewitnesses can miss or omit critical details or even create or embellish upon what happened without realizing it.

Evidence in this context is useful in linking a person to a place and a place to a person. We can expand on this to define types of evidence that forge this link. Again, assume that a shooting occurs (Figure 2.2). There are several types of evidence that can be created during such an incident. First, people other than the suspect and the victim may be present and can act as witnesses. This would become **testimonial evidence** or eyewitness evidence. The study of such evidence is a fascinating subject, but beyond the scope of this book. This leaves **physical**

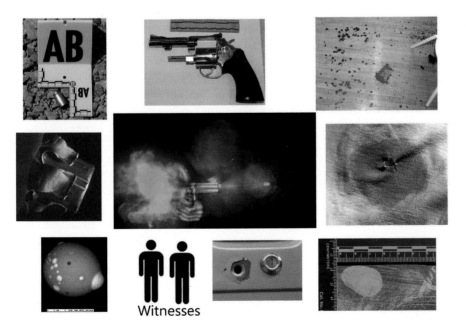

FIGURE 2.2 A shooting (center frame) creates a wealth of evidence in this example. Tiny particulates of gunshot residue (lower left), the bullet (middle left), and cartridge case (left corner) are examples, as is the gun itself. Witnesses can provide testimonial evidence; bullet holes in metal (lower center), fingerprints (lower right), a bullet hole in clothing (center right), and bloodstains may also be produced.

evidence, which is the raw material needed by forensic scientists. When the gun was fired, residues are deposited on the hands of the shooter, his clothing, and any surface nearby; this is gunshot residue (Chapter 17). The bullet travels down the barrel of the gun and picks up markings (firearms and tool mark evidence), and where the bullet strikes will also have residues from the propellant. The closer the gun to the target, the greater the amount of residue deposited on it. The gun itself is another form of physical evidence, as is the cartridge case in which the bullet was supplied. If a person was injured or killed by the shot, patterns of blood (bloodstain patterns; Chapter 4) can be produced. If the bullet passes through objects such as a car door, more physical evidence is produced. Finally, fingerprints (Chapter 14) may be found on the weapon or at the scene. Such evidence may be documented and collected at the crime scene (Chapter 3) or by other means and analyzed in a laboratory to generate reports, a form of testimonial evidence.

2.2 RULES OF EVIDENCE

Evidence is analyzed by forensic scientists, but to be used in court, it must still be admitted as evidence. If not, it will never be seen by those who are making critical decisions. A court must admit evidence before it can be considered in the case. The **admissibility** or inadmissibility of trial information—whether eyewitness testimony, photographs, physical objects, or scientifically generated information such as DNA—is determined by the trial court's application of the **rules of evidence**. Such rules are designed to ensure that only reliable and relevant scientific information is considered by the court. In practice, this is much more difficult than it sounds, and the pace of scientific advancement increases the challenge.

The person or entity that decides if scientific evidence is admissible is not the scientist, but the court. The ones that make these key decisions (judges) usually do not have a scientific background, yet they are charged with deciding what scientific information can be admitted. The decision is based on competing arguments (adversarial system) and not strictly on scientific merits. This means that in the worst cases, junk science may be admitted because a more convincing argument is made to do so than to exclude it. This does not mean that judges purposely admit junk science, but given that judges aren't scientists, it can happen. As a result, the rules of evidence have evolved over the past century to assist judges in making critical admissibility decisions.

2.3 ADMISSIBILITY OF EVIDENCE

How does a judge decide if scientific evidence can be admitted? Remember that the law values precedent, and thus this is one of the first considerations. For example, chemical analysis of blood for detection of drugs and poisons (forensic toxicology; Chapter 6) has been used in legal proceedings for decades and admissibility of such evidence is rarely an issue. For many forensic techniques, this holds true. However, slight changes, modernizations, or development of new scientific knowledge can lead to revisiting admissibility criteria, even for long-established methods and techniques.

2.3.1 ADMISSIBILITY HEARINGS

Before any scientific evidence is presented before a court, it must be determined to be admissible. Admissible evidence must be reliable and relevant to the case at hand, and for scientific analysis, the court must be assured that the methods used are scientifically acceptable and reliable. The intent of admissibility proceedings is to prevent the introduction of results obtained using poor science or pseudoscience, or admission of evidence that has no bearing on the case. Admissibility hearings provide a way for new scientific test methods to be introduced and accepted as viable tools in forensic science. If required, these hearings are held separate from the case presentation.

The standards that courts use to determine admissibility of evidence vary among the jurisdictions, and we will discuss the most important ones. These standards are based on past cases (precedent again) and are referred to by that case name.

Those jurisdictions following the *"Frye standard"* (*Frye v. United States*) require that scientific methods be generally acceptable to a significant proportion of the scientific discipline to which they belong. For example, new chemical tests would have to be generally accepted as reliable among most analytical chemists. Jurisdictions that follow the **Federal Rules of Evidence** and the ***Daubert* standard** (*Daubert v. Merrell Dow Pharmaceuticals*) decision use more flexible guidelines. Essentially, under *Daubert*, the trial judge is responsible for determining if the scientific evidence is relevant and that the expert presenting it is qualified to discuss the results and offer an opinion. The judge must also determine if the testing method rests on a reliable and reasonable scientific foundation. Such hearings are referred to as *"Daubert* hearings." As of 2016, 16% of states relied on the *Frye* standard, 76% on *Daubert*, with the rest using other admissibility standards.

2.3.2 THE FRYE DECISION AND FRYE STANDARD

In 1923, the D.C. Circuit Court of Appeals (293 F. 1013 [D.C. Cir 1923]) handed down the first ruling that applied to the modern era of forensic science. The ruling was a rejection of the validity of the polygraph (lie detector) test, which stated in part:

> Just when a scientific principle or discovery crosses the line between the experimental and demonstrable stages is difficult to define. Somewhere in the twilight zone the evidential force of the principle must be recognized, and while courts will go a long way in admitting expert testimony deduced from a well-recognized scientific principle or discovery, the thing from which the deduction is made must be sufficiently established to have gained general acceptance in the particular field in which it belongs.

This ruling led to criteria referred to as **general acceptance** that governed the admissibility of scientific evidence in many jurisdictions. However, the ruling became more problematic as scientific advances continued and scientific disciplines became more specialized and compartmentalized. As this occurred, the idea of general acceptance within a scientific field became more difficult to obtain or define. Furthermore, the criteria set forth in *Frye* can limit acceptance of innovative techniques that might be known to only a small group of scientists within any given discipline. This was the case when DNA evidence was first used; a comparatively small number of scientists worked in that area and understood how DNA typing methods work. Thus, while the *Frye* standard was a step forward in 1923, rapid scientific advances have made the standard more problematic (Figure 2.3).

2.3.3 FEDERAL RULES OF EVIDENCE

From 1923 until the late 1960s, there was little change relative to the admissibility of scientific evidence. However, it became obvious that the *Frye* standard was not sufficient in an era of incredible scientific advancement. In 1969, a draft of what would become the **Federal Rules of Evidence** was put forth by a committee formed at the request of the U.S. Supreme Court. The rules were enacted by Congress in 1975, 52 years after the *Frye* decision. Rule 702, "Testimony by Experts," states:

> If scientific, technical, or other specialized knowledge will assist the trier of fact to understand the evidence or to determine a fact in issue, a witness qualified as an expert by knowledge, skill, experience, training, or education, may testify thereto in the form of an opinion or otherwise.

While a step forward, the rules did not discuss how a witness was to be qualified as an expert and what specifically the court should weigh in determining qualifications of an expert

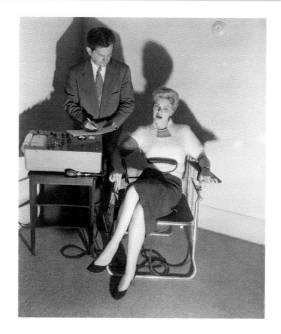

FIGURE 2.3   Fictional depiction of a lie detector test, the technique at question in the *Frye* decision.

and the validity of the science or technology in question. The Federal Rules of Evidence apply only to federal jurisdictions, but many other jurisdictions adopted these or similar rules. It would be nearly 20 years before the Supreme Court would directly address admissibility in a general way.

2.3.4 *DAUBERT* DECISION (1993)

Daubert v. Merrell Dow Pharmaceuticals (113 S. Ct. 2786 [1993]) was a landmark case regarding admissibility. This was a civil case in which the plaintiffs were parents of child born with birth defects. The parents believed that the birth defects were the result of a morning sickness medication taken by the mother during her pregnancy. In these types of cases, scientific evidence is often the critical information in reaching a decision. In their ruling, the Supreme Court stated that under the Federal Rules of Evidence, Rule 702, general acceptance (from the *Frye* decision) is not an absolute requirement for determining admissibility. Rather, it is the responsibility of the trial judge to determine if scientific evidence is relevant and reliable. This role assigned to the judge is often referred to as the **gatekeeper**, and the court offered suggestions for making that determination, while leaving flexibility to the judges. General acceptance was one criterion, as were peer review of the technique, standards for the method, validation of the method, potential errors, and testable (and thus falsifiable) theories. All these concepts provided a toolbox to judges (nonscientists) to assist them in evaluating the utility and reliability of scientific evidence.

2.3.5 THE TRILOGY

The *Daubert* decision was one of three in the 1990s that significantly impacted the way in which many jurisdictions addressed the admissibility of evidence. The other two cases (*G.E. v. Joiner* and *Kumho Tire v. Carmichael*), along with the *Daubert* decision, constitute what is often referred to as the ***Daubert* trilogy**. All three cases in the trilogy are civil cases, but the rulings are at the heart of the admissibility of forensic science in criminal cases. The *Daubert* standard set the

FIGURE 2.4 Damaged tires were at issue in the *Kumho* case.

baseline while the other two cases built upon it while retaining the concept of the judge as the gatekeeper.

The *Joiner* case was a complex civil suit in which an employee sued General Electric, claiming that his cancer was directly attributable to his being exposed to chemicals while working for the company. The plaintiff (Joiner) asked to have scientific studies on animals admitted supporting his case. The court rejected this on the basis that the study in question was not directly applicable to the issue. The study was focused on different chemicals and different exposures that were experienced by Mr. Joiner and thus not applicable to the case. In other words, the court decided that the scientific study was not relevant to the case. This isn't a criticism of the science in the study, just about its relevancy to the case. Thus, admissibility under the trilogy model is judged on both **reliability** and **utility**. The science behind evidence may be Nobel Prize worthy, but if it is not relevant to the case, it should not be admitted.

The last case in the trilogy was the *Kumho* decision, centered on the responsibility of the Kumho Tire Company in a fatality that resulted from an accident attributed to tire failure. One of the expert witnesses presented was an engineer, and the court ruled that the testimony offered by the engineer fell under the umbrella of scientific expert. Thus, *Kumho* extended the *Daubert* ruling to all experts, not just scientists such as biologists or chemists (Figure 2.4).

2.4 CATEGORIES OF EVIDENCE

We have already discussed transfer evidence and evidence as inclusive, exclusive, or indeterminate. There are many other ways to categorize forensic evidence. For example, a piece of evidence such as a fingerprint recovered from a murder weapon can be **inculpatory** or **exculpatory**, meaning that its presence tends to include or exclude a person as a source and thus tends to incriminate or exonerate him or her. The concept is similar to inclusive versus exclusive. Evidence can also be **direct** or **circumstantial**. Direct evidence is information that establishes directly, without the need for further inference, the fact for which the information is offered. Eyewitness testimony can fall in this category, as can a photograph, or a forensic identification of a powder and an illegal drug. Direct evidence speaks directly about a fact in question and does not require any additional inference.

MYTHS OF FORENSIC SCIENCE

The evidence is all circumstantial!

There is a misconception that circumstantial evidence is somehow weak or imperfect. Most evidence is circumstantial—it speaks to a circumstance or version of events but directly to a fact. In other word, it helps sort out different versions of stories to narrow down what could have happened and what could not have happened. For example, finding a fingerprint on a gun means that the person with that fingerprint touched the gun. It doesn't tell us when or why, just that at some point he or she did touch it. Look at Figure 2.1 again. Finding a gun in the suspect's home is circumstantial evidence, as is finding a bloody impression of the gun at the scene. When considered together, these two pieces of circumstantial evidence become convincing evidence that the perpetrator was at the scene at the time there was blood. It is still not proof, but this circumstantial data is internally consistent with the theory that the suspect is the perpetrator.

Circumstantial evidence can be the basis from which additional information can be inferred. For example, suppose that a defendant's blood is found at a crime scene and linked to her by DNA results. From this information, you can infer that she was at the crime scene. This is Locard's principle in action; evidence links a person to a place. However, taken alone, the DNA results cannot speak to guilt or innocence in a murder case. More information is required for such a leap; the DNA results are just one part of a much larger story. If this woman lives in the house, finding her DNA there is no surprise and carries little meaning.

2.5 METHODS OF INTERPRETATION

How often have I said to you that if you eliminate the impossible, whatever remains, however improbable, "Improbable" must be the truth?

Sherlock Holmes in
The Sign of Four

We have talked a lot in this and the last chapter about how evidence is used to understand a past event and to sort through different versions of events to determine what could have happened and what could not have happened. We noted the importance of internal consistency when evaluating ideas and theories regarding what happened and that multiple pieces of evidence are essential in any investigation. Forensic evidence becomes part of the body of evidence.

When interpreting a collection of evidence, there are three types of reasoning that are used—**deductive, inductive,** and **abductive** reasoning. Deductive reasoning is sometimes compared to algebra and mathematical reasoning, such as $x = 2$ and $y = 2$; therefore, $x = y$. A forensic example can come from DNA typing, the topic of Chapter 10. Females have amelogenin type of X, X; the bloodstain tested in the laboratory had an amelogenin type of X, X; therefore, the stain came from a woman. A key to deductive reasoning is that the core concepts must be true—here, it must be true that all women are type X, X and the type obtained from the forensic analysis must also be reliable. Deductive thinking is also referred to as **inference**; you infer that the stain is from a woman because of what you know about DNA typing. In other words, you draw a logical conclusion based on facts.

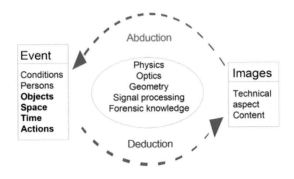

FIGURE 2.5   How abductive reasoning can be applied to an image or photograph. (Reproduced from Milliet, Q., et al., A Forensic Science Perspective on the Role of Images in Crime Investigation and Reconstruction, *Science and Justice* 54, no. 6 [Dec 2014]: 470–480. With permission.)

Inductive reasoning uses existing data to make predictions and generalizations. As an example, most serial killers are men, so an investigation into a series of murders might start based on the theory that the killer is male. Inductive reasoning is useful for proposing ideas or hypotheses that can be tested and checked using deductive reasoning. Inductive reasoning can also be based on tendencies, trends, or expectations developed from data. Suppose a laboratory seizes a collection of 1000 white tablets that all have the same general appearance. A forensic chemist tests 100 of them and all contain the same drug. Inductive reasoning is that all the tablets are the same. This can be rephrased as a hypothesis that can then be tested quantitatively.

Abductive reasoning (Figure 2.5) is probably the most commonly used in criminal investigations. Abductive reasoning is based on gathering what is known and using this information to come up with the simplest and most likely explanation. This is how different pieces of evidence are integrated into the most likely theory regarding how something happened. Note that there is an element of probability that differs from deductive reasoning. Abductive reasoning differs from inductive thinking in that inductive reasoning starts from data and then generalizes it. Abduction is integrative and offers the most reasonable explanation based on what is known.

Figure 2.5 illustrates how abductive and deductive reasoning can be applied to a photograph such as a crime scene photo. Suppose an image is taken such as the one in the bottom right of Figure 2.1. Starting with the technical aspect and content, an investigator would know that this is an image made by the transfer of blood from something to the surface. The blood is still wet, so the imprint was made recently. There is also a pattern in the imprint. Using abduction, he could draw the most likely conclusion that this imprint occurred during the crime being investigated and that the blood is from the victim. Later, when the gun is collected as evidence (lower middle of Figure 2.1), it could be tested to see if it produced a similar impression. If it did, the investigator could then deduce that the imprint at the crime scene was made by that weapon.

2.6 REVIEW MATERIAL

2.6.1 KEY TERMS AND CONCEPTS

Abductive reasoning
Admissibility
Circumstantial evidence
Daubert standard
Daubert trilogy
Deductive reasoning
Direct evidence
Exclusive evidence
Exculpatory

Federal Rules of Evidence
Frye standard
Inclusive evidence
Inculpatory evidence
Inductive reasoning
Locard's exchange principle
Opinion evidence
Physical evidence
Rules of evidence
Testimonial evidence
Transfer evidence

2.6.2 REVIEW QUESTIONS

1. Early in the chapter there was an example of two people meeting on a baseball field. If we find the second person's DNA on the first person's hand, does this link the people to the scene (the baseball field)? Why or why not?
2. The Federal Rules of Evidence are exclusionary. Explain what this means.
3. What are the key differences between the *Frye* admissibility standards and the *Daubert* admissibility standards?
4. Summarize the key components of each decision in the *Daubert* trilogy.
5. Figure 2.1 shows a bloody outline of a gun found at the home of a suspect in a hypothetical crime scene. Categorize this evidence in as many ways as possible.
6. Is bloodstain pattern evidence as shown in Figure 2.2 direct or circumstantial evidence? Explain.
7. When you go to the doctor, by what process of reasoning does she employ to diagnose what is wrong with you?

2.6.3 ADVANCED QUESTIONS AND EXERCISES

1. All three cases in the *Daubert* trilogy were civil cases and not criminal cases. Why do you suppose the admissibility of scientific evidence in these cases is so critical and how does it differ from admissibility in a criminal case?
2. How would scientific ethics play a role in determining the admissibility of evidence using the *Daubert* method?
3. Explain Locard's principle in terms of abductive, inductive, and deductive reasoning. Create an example to illustrate.

BIBLIOGRAPHY AND FURTHER READING

BOOKS, BOOK SECTIONS, AND REPORTS

Bowers, C. M. *Forensic Testimony: Science, Law, and Expert Evidence.* San Diego: Academic Press, 2014. ISBN: 978-0123970053.
Bronstein, D. A. *Law for the Expert Witness.* 4th ed. Boca Raton, FL: CRC Press, 2012. ISBN: 978-1439851562.
Kiely, T. F. *Forensic Evidence: Science and the Criminal Law.* 2nd ed. Boca Raton, FL: CRC Press, 2006. ISBN: 978-0849328589.

ARTICLES

Milliet, Q., Delemont, O., and Margot, P. A Forensic Science Perspective on the Role of Images in Crime Investigation and Reconstruction. *Science and Justice* 54, no. 6 (Dec 2014): 470–480.

Milliet, Q., Delemont, O., Sapin, E., and Margot, P. A Methodology to Event Reconstruction from Trace Images. *Science and Justice* 55, no. 2 (Mar 2015): 107–117.

Strom, K. J., Hickman, M. J., McDonald, H. M. S., Ropero-Miller, J., and Stout, P. R. Crime Laboratory Personnel as Criminal Justice Decision Makers: A Study of Controlled Substance Case Processing in Ten Jurisdictions. *Forensic Science Policy and Management: An International Journal* 2, no. 2 (2011): 57–69.

WEBSITES

Link	Description
https://www.mccrone.com/mm/	Modern microscopy, hosted by the McCrone Group; historical information, trace evidence, and microscopy
https://www.livescience.com/21569-deduction-vs-induction.html	Good examples of inductive and deductive thinking
http://www.forensicsciencesimplified.org/legal/index.htm	Discussion of admissibility in the context of forensic science

Crime Scene Investigation

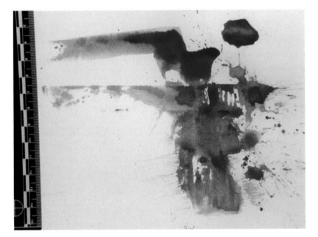

3

In the previous section, we established the foundations and fundamentals of forensic science as a discipline and how science operates within the legal system. We discussed how evidence is created, how it can link people to places, and how it is (or is not) admitted into court proceedings. This leads us next into specific types of evidence and where it is collected and documented. We will start with crime scenes, a primary but certainly not the only source of many kinds of evidence. Not all physical evidence arises from crime scenes, but all crime scenes create physical evidence. The topics of crime scenes and death investigation are inextricably linked, and we will delve more deeply into that subject in the next section (Section 3). Not all crime scenes involve a death. Crime scenes can be as simple (relatively speaking) as a burglary, as dangerous as a clandestine drug laboratory, or as complex as a multiple murder that takes place in several locations. While each scene is unique, the principles that underlie its investigation are consistent. In this section, we focus on these common elements.

3.1 A CRIME SCENE AND ITS STORY

Any place can become a crime scene—a car, a home, or even underwater. Each crime scene is unique because the chain of events that create it are unique, but there are unifying themes that dictate how a crime scene is studied and processed. Crime scenes are also fragile and start to change and decay the moment they are created. One of the biggest challenges in crime scene processing is to ensure that evidence collected reflects what happened at the time of interest (the criminal act) and is not an artifact of something that happened before the crime or afterwards.

MYTHS OF FORENSIC SCIENCE

The CSI does everything in the field and the lab.

Thanks to recent television shows, many people believe that forensic scientists go to a scene, document and collect evidence, take it back to the lab, test it, arrest the suspect, and testify. While laboratory-based forensic scientists or medical examiners may on occasion go to a crime scene, this is not standard practice. Crime scene investigators (CSIs) may be police officers trained in crime scene work or civilians trained as CSIs. The primary job of the CSI is to document, collect, preserve, and deliver evidence to the appropriate location. Conversely, the job of laboratory-based forensic scientists is to analyze evidence delivered to them and generate reports.

A crime scene presents one of the most obvious examples of Locard's exchange principle in action, which was discussed in detail in Chapter 2 and is illustrated in Figure 3.1. The interaction of the victim, suspect, objects at the scene, and the scene itself creates physical evidence of those contacts on all things and people involved. Crime scene processing is geared toward documenting and recovering as much of that evidence as possible without collecting items that are irrelevant to the crime. This is a difficult assignment in the best of circumstances.

If someone is a victim of a violent crime, the first person to stumble upon the scene will likely rush to the victim's side to help. A call to emergency services is made and the **first responders** (police, fire services, and ambulances) arrive. Their responsibility is to provide medical assistance to the victim, which starts with checking for vital signs. If the victim is deceased, these efforts will end, but the scene has already been changed significantly by their intervention. Sorting through the evidence of rescue efforts from evidence associated with the crime can be challenging. Suppose the crime scene analyst arrives and finds a pair of gloves discarded by the front door. These could have been shed by the killer on his way out or they might have been dropped by the paramedics as they left. This evidence must be documented and collected, but later investigation and analysis will be needed to determine which explanation is correct. In this chapter, we focus on methods of documenting a crime scene to preserve the information for later study.

The insert on the next pages will serve as an example crime scene and scenario that we will use in this chapter. We will also revisit it in later chapters to illustrate how different forensic disciplines interact with crime scenes and evidence collected there. The goal of the processing of any scene is to document it and collect sufficient evidence to uncover the most likely sequence of events that led to its creation. Although it may never be possible to tell the story in every detail, the evidence should be sufficient to support or refute ideas or versions of events offered as explanations. In some cases, a formal **crime scene reconstruction** is done to test competing hypotheses proposed to explain the physical evidence. As discussed in Chapter 2, the evidence and any reconstruction performed must be internally consistent. This is yet another reason why it is so critical that evidence that is collected is relevant to the incident in question; irrelevant evidence can cause internal inconsistencies or disguise them, making it all the more difficult to generate the best reconstruction of events.

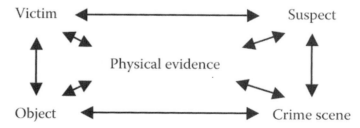

FIGURE 3.1 Locard's principle and the interactions at a crime scene.

VIGNETTE 3.1   Crime Scene Scenario

This scenario is a fictional but realistic crime scene scenario that we will reference in this and later chapters. Bookmark this page to return to it as needed. Remember that although you, as the reader, know how the crime was committed and how the evidence was created, investigators do not; they must collect information and evidence to integrate it into theories about what happened. The correct theory must be internally consistent.

The fictional crime (Figure B.1) is a homicide with one perpetrator and one victim, husband and wife, respectively. The crime occurs after a domestic dispute at the home. The neighbors hear yelling and what sounds like fighting, but the couple frequently gets into loud arguments, so no one takes notice. The noises and shouting end within a few minutes after one neighbor reports hearing a loud sharp sound that might have been a gunshot. However, none of the other neighbors report hearing that sound. About an hour later, one of the neighbors sees the husband's red truck leaving the house at a high rate of speed, which is also typical behavior for this couple. Although no one realizes it at this point, the husband has wrapped his wife's body in a wool blanket and placed it in the bed of his pickup truck. He covered it with other blankets and bags and drives to a spot deep in the woods several miles away. He knows about this place because he hunts there. He digs a shallow hole, deposits the body, covers it, smooths the dirt, and places additional branches and leaves over the grave. He leaves and checks into a hotel in a neighboring city.

Three days later, a friend of the wife comes to the house to check on her welfare. She has not been at work and she does not answer her phone, respond to texts, or answer voicemail. When the friend gets no response to knocking, she calls the police. A police unit arrives to make a welfare check. They make entry and discover the house in disarray and copious amounts of dry blood in several locations. The first responder (the police officer making entry) searches the house and surrounding area to be sure no one is there and that medical assistance is not required. He then calls for assistance, which includes the department's crime scene processing unit. The official investigation begins.

Within hours, detectives learn of the arguments and events of the night of the victim's disappearance and they begin to search for the husband. They learn that he was seen at a large box store the day after the victim went missing and obtained video from the store to confirm. The police department declares him a "person of interest" and publishes his photo. A week later, the manager of the motel sees the photo on social media and recognizes the man as one of his guests and calls police. The suspect is arrested within hours of the call. The police find his red pickup truck at the motel, impound it, and deliver it to the crime laboratory for processing.

Over the course of the investigation, police learn that the suspect likes to hunt and fish at a lake on a remote parcel of forest outside the city. Nearly a month after the woman's disappearance, they take a trained cadaver dog to the site, along with volunteers to help search

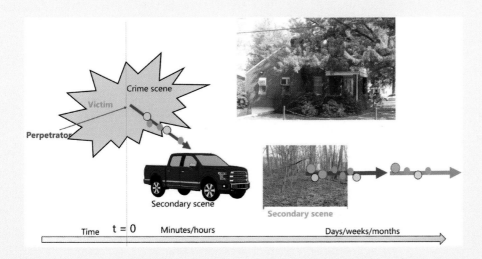
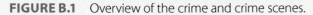

FIGURE B.1   Overview of the crime and crime scenes.

the area. After extensive searches of the area over several days, they locate a remote area that appears to have been recently disturbed and a fire pit area that contains ashes, wood remnants, and partially burnt beer cans. They contact the crime scene unit once again and within a few hours of their arrival, they find the victim in a shallow grave. The body is wrapped in a layer of black plastic covered by a thick blue tarp. Both the plastic and the tarp appear to be relatively new. The body appears to have been burned, and what little tissue remains is significantly decomposed and partially skeletonized. The recovery team thinks the body is that of a female, but the degree of decomposition and damage is such that they cannot tell for certain if the remains are that of the wife. The body is carefully excavated and removed, along with insects and soil from the immediate area of the shallow grave. The body is taken the medical examiner's office, where an autopsy and toxicological tests are performed.

In this fictional case, two competing theories (stories) develop. The first is that of the investigators, who believe that the suspect killed his wife, put her body in the truck, drove to the remote area, and buried the body before attempting to hide out in the hotel where he was arrested. The other story is that of the suspect, who claims that although he did argue with his wife and drive off in a rage, he did not kill her. He claims that he was feeling ill and stayed in the hotel for a few days to rest and to give his wife time to "settle down" after their argument. He is adamant that she was alive and well when he left her at the house, alone.

As we go through this and later chapters, we will examine the types of evidence that might be found in such case, how it would be analyzed, what it could mean in the context of the case, and which "story" the evidence supports—if either. In the rest of this chapter, we discuss the scene at the home in detail. The evidence recovered from the truck is shown in Figure B.2. Figure B.3 shows the burial site and evidence collected there.

Also found in the bed: partial roll of black plastic used for mulch

FIGURE B.2   Found in the truck and truck bed: A hammer, paper receipt, hair on upholstery, blood on the door, and dirt in tire treads.

FIGURE B.3   Burial site.

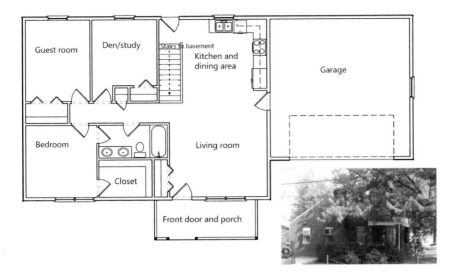

FIGURE 3.2   The macroscropic crime scene.

The objectives of a crime scene investigation are to recognize, preserve, and collect physical evidence at a crime scene with the goal of reconstructing the likely chain of events that generated this evidence. Because we know the story of what happened in the example scenario, we can identify key evidence. The first people on the scene have no such advantage. Often, the biggest challenge for the CSI is to determine what is pertinent evidence and what is extraneous; this is where the skill and experience of CSIs is invaluable. Investigations are always a team effort in which law enforcement, forensic scientists, and CSIs play a role.

Figure 3.2 shows the house, which in the example scenario is the **primary crime scene** and a **macroscopic crime scene**. The primary scene is where the initial criminal act occurred, and a macroscene is on a large scale, such as the house and lot. **Secondary scenes** are created by acts that follow the original crime. In the example scenario, the truck could be considered a secondary scene, as could the burial site. Sometimes smaller parts of a scene, such as a room, are referred to as **microscopic scenes** or microscenes. The room in which the initial assault took place could be referred to as a microscene.

The first responder in the example scenario is the police officer, but in other situations, the first responders may be the fire department or emergency medical personnel. Regardless of who arrives first, their actions will be critical in any subsequent investigation and crime scene processing. The primary responsibility of the first responder is always the safety of themselves and any victim they find, and to render aid and assistance. While doing so, the scene is inevitably disturbed and altered. CPR may be performed, IVs inserted, and the victim stabilized for transport. First responders would be interviewed later to document their actions, so that their impact on the scene can be considered.

In our example, the police officer entering the house would have noticed the signs of struggle and bloodstain evidence. Immediately, he would have moved to ensure his safety as he performed a quick search of the house. Once he was sure that no one else was present and that no medical help was needed, his next steps would be to secure the scene and call for a crime scene response team. He would have documented that he entered the house, where he went, what he touched, and what he observed. A log would be started that indicated who entered the scene and why. Later, this list would be used to collect **elimination samples**. These are samples such as fingerprints and DNA that are collected from people who enter the scene as part of the investigation. For example, if paramedics are first to arrive at a scene, their DNA and fingerprints will be present and elimination samples are critical for sorting out what is relevant and what is irrelevant physical evidence.

3.2 INITIAL CRIME SCENE INVESTIGATION

Once the first responders have completed their initial work and ensured the safety of anyone found at the scene, the next step is to secure it and call for the appropriate assistance.

The police officer in our example would call for help, to include a crime scene response unit. Typically, the first thing the CSI or crime scene investigation team will do is a preliminary scene survey or **walk-through**. The CSI and the first responder will often do the scene survey together. This serves several purposes. First, it helps to establish the scope of the crime scene and how much of the surrounding area must be secured. In our example, the scope of the scene is the house and the yard, but the CSIs will not know this until they do the initial inspection. This survey is also when the first images and videos are usually taken, since this is as close to the original condition that the scene will ever be. The CSI will note if there are special considerations that must be immediately addressed; for example, if it is raining, evidence that is outside will have to be attended to immediately or protected.

As the survey moves toward the house in our example, the team will be looking for entry and exit points to and from the house. In Figure 3.2, these points would be the garage, front and back doors, any basement window exits, and windows. The condition of these will be noted and documented (open/closed, locked, etc.). Once entry is made into the house, the CSIs identify key areas and any microscopic scenes, such as individual rooms. Images and videos will also be collected. Once the survey is completed, a plan will be developed for further processing. In the example scenario, the CSI identified six zones of interest or microscenes because of the initial survey (Figure 3.3): zone 1, the garage and driveway zone; zone 2, the front door and porch; zone 3, the office; zone 4, the master bedroom; zone 5, the kitchen and living zone; and zone 6, the front and side yard. With this survey complete, the critical tasks are documentation of the scene and evidence collection.

3.3 CRIME SCENE DOCUMENTATION, MEASUREMENT, AND IMAGING

Properly documenting a crime scene is the single most important task of the CSI. The purpose of this documentation is to permanently record the condition of the crime scene and its physical evidence in as close to original condition as possible. The documentation freezes the scene in time and should provide all the information needed to reconstruct the story of how that scene was created. Documentation is the most time-consuming activity at the scene and requires attention to detail. The major tasks of documentation are note taking and sketching, photography (digital imaging), and videography. Closely associated with documentation is crime scene measurement and mapping. A photo of a piece of evidence is of no use if its location is not recorded; often, the site where evidence is found provides critical information and context.

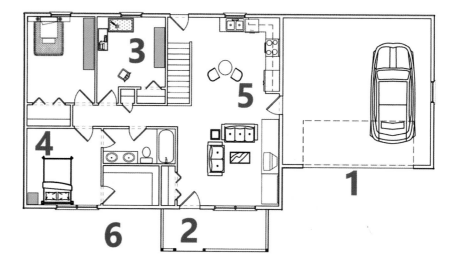

FIGURE 3.3 Zones of interest identified by the crime scene team.

The notes taken at the scene will include a log of digital images that describes where a photo was taken and in what order. The CSI will also note observations and create a rough sketch of the areas investigated. It is critical that documentation and imaging be recorded before anything is moved or any evidence is collected. An example of a common method of identification and proper documentation of evidence is shown in Figure 3.4. A cartridge case has been located and identified with a crime scene marker. Note that the marker has rulers to show the dimension of the evidence and its orientation. The CSI would place the marker, take one or more images, and note each picture taken in the log, along with a description of the evidence and sufficient information to locate it within the larger scene. This additional information could come from a sketch and from larger-scale images.

Beyond the photo log, effective notes provide a written record of all the crime scene activities. The notes are taken in real time as the scene is processed to ensure that all critical details are captured. Notes for a CSI would include how they were notified of the scene, how and when they arrived, who else was at the scene, a description of the scene and conditions, supporting photos, a narrative of actions taken, and evidence identified.

3.3.1 DIGITAL IMAGING AND VIDEOGRAPHY

One of the most striking changes in crime scene documentation in the past 25 years is the advent of digital imaging, which has replaced traditional film cameras in all but a very few instances. Digital image technology provides the CSI with powerful tools for capturing, analyzing, and storing the record of the crime scene and its physical evidence. These digital image tools complement the traditional video and still photography used in crime scene documentation. The advantages of digital images include instant access to the images, easy integration into existing electronic technologies, and no need for the often expensive film processing equipment and darkrooms. Some disadvantages for the use of digital image technology are centered on issues of court admissibility because of easy image manipulation using programs such as Photoshop®. This problem has become less of an issue as software now creates notations and data associated with the image that indicate any changes or alternations made to the original.

Video recording (videography) of the crime scene has become a routine procedure for crime scene documentation. Its acceptance is widespread and is due to a virtual appearance of the scene and increased availability of affordable equipment with user-friendly features like DVD recording, built-in stability, digital zoom lenses, and compact size. Jury acceptability and expectation has also added to the recognized use of video recording of the crime scene investigations.

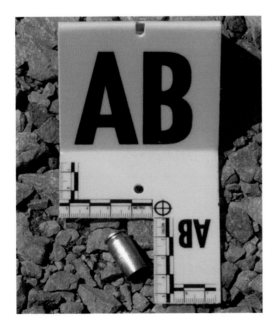

FIGURE 3.4 Cartridge case with evidence marker and scale for photography.

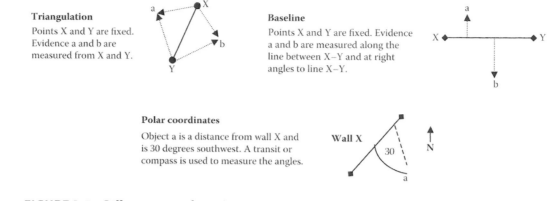

Triangulation

Points X and Y are fixed. Evidence a and b are measured from X and Y.

Baseline

Points X and Y are fixed. Evidence a and b are measured along the line between X–Y and at right angles to line X–Y.

Polar coordinates

Object a is a distance from wall X and is 30 degrees southwest. A transit or compass is used to measure the angles.

FIGURE 3.5   Different types of coordinate systems that can be used for crime scene documentation.

3.3.2 SKETCHING THE CRIME SCENE

The final task to be performed in the documentation of the crime scene is the sketching, and a critical aspect of sketching is obtaining and recording accurate measurements. Sketching the crime scene is the assignment of units of measurement or correct perspective to the overall scene and the relevant physical evidence identified within the scene. Locations of objects and features must be specified relative to a point of reference such as a wall. Angles and distances can be used, as shown in Figure 3.5. There are two basic types of sketches as part of crime scene investigations—preliminary, or rough, and final. An example is shown in Figure 3.6. Note how the final sketch includes measurements that show locations relative to the fixed reference points of the room. A critical reason for doing a sketch is to measure and locate objects in space. Because of the larger scale and distances that may exist, such as between a piece of evidence and a wall or other landmark, sketching can be the best and easiest way to provide measurements that allow for exact placement of the object in the scene.

3.3.3 FORENSIC MAPPING AND LASER SCANNING

Another forensic application to come out of the digital revolution is the use of imaging technology and GPS to compile a **crime scene map** that can, in some cases, be rendered in three dimensions. There are several incarnations of crime scene mapping hardware, but most include a method of electronic distance determination, height and slope measurements, mapping capability, and the ability to locate points in three dimensions (i.e., including elevation information). The data is downloaded to a program that then reconstructs the data and generates a three-dimensional (3D) map of the scene. This equipment is like that used by surveyors, and the practice of recording scenes this way is sometimes referred to as **forensic mapping**.

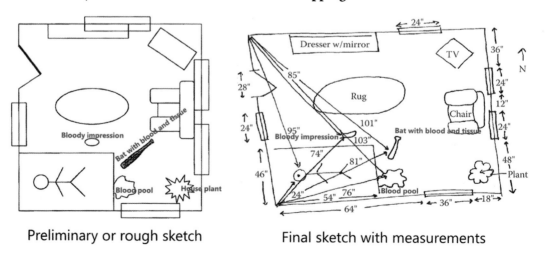

Preliminary or rough sketch   Final sketch with measurements

FIGURE 3.6   An example of a preliminary and final sketch of a murder scene in a single room.

Three-dimensional scanning and imaging has also become available to crime scene investigations and investigators. It can be combined with mapping technologies in a method referred to as 3D forensic mapping or laser scanning. The technology is based on LiDAR (light detection and ranging), and the equipment includes a laser, a scanner, and a GPS device that can record location (Figure 3.7). Once the equipment is placed and located, laser light pulses are sent out in the scanned area and reflect to the system's detector. The time it takes for the light to reflect depends on how far away the object it reflects off is from where the light originated. Because the reflection data is collected from multiple points, it is possible to image an object in three dimensions. When the images are calibrated with scaling devices or objects of known sizes, a remarkably accurate portrait of the scene can be created (example shown in Figure 3.8). As this technology advances and costs continue to fall, this approach to crime scene documentation is rapidly becoming standard practice.

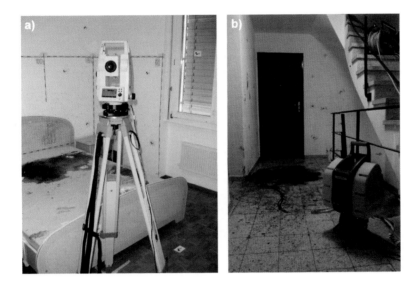

FIGURE 3.7   (a) Laser scanning device. (b) Scanner on the floor. (Reprinted from Buck, U., et al., 3D Bloodstain Pattern Analysis: Ballistic Reconstruction of the Trajectories of Blood Drops and Determination of the Centres of Origin of the Bloodstains, *Forensic Science International* 206, no. 1–3 [Mar 2011]: 22–28. With permission from Elsevier.)

FIGURE 3.8   Example of compiled data obtained from a laser scanner. The diagram corresponds to the scene shown in Figure 3.7(a). (Reprinted from Buck, U., et al., 3D Bloodstain Pattern Analysis: Ballistic Reconstruction of the Trajectories of Blood Drops and Determination of the Centres of Origin of the Bloodstains, *Forensic Science International* 206, no. 1–3 [Mar 2011]: 22–28. With permission from Elsevier.)

3.3.4 CRIME SCENE SEARCHES

The preliminary crime scene search was an initial quasi-search for physical evidence present at the crime scene. That search is for the obvious items of evidence, and it is done for orientation purposes before the documentation begins. Once the scene documentation as described above is completed, then a more efficient and effective search for less obvious or overlooked items of evidence must be done. This intensive search is done after documentation but before the evidence is collected and packaged. If any additional items of evidence are found, then they must be subjected to the same documentation tasks that were done earlier.

Crime scene search patterns are varied and different in style, but they share a common goal of providing organization and systematic structure to ensure that no items of physical evidence are missed or lost. There is no single method for specific types of scenes; the method is selected that best suits the situation or scene. Common crime scene search patterns are shown in Table 3.1.

Most search methods are geometric patterns. The six patterns are link, line or strip, grid, zone, wheel or ray, and spiral methods. Each has its advantages and disadvantages, and some are better suited for outside versus indoor crime scenes. Before any intensive crime scene search is done, care must be taken to instruct the members of the search party. It is tempting for search party members to touch, handle, or move the items of evidence found during the search, but searchers

TABLE 3.1   Crime Scene Search Patterns

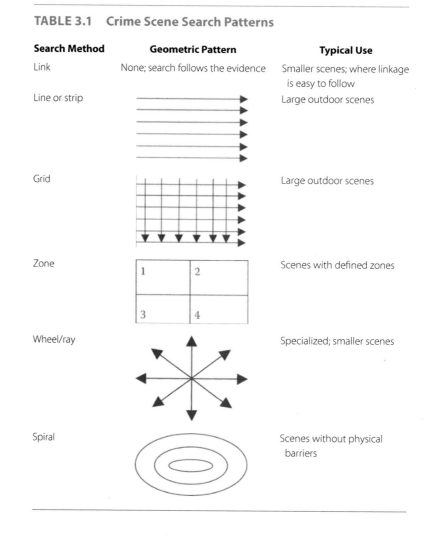

Search Method	Geometric Pattern	Typical Use
Link	None; search follows the evidence	Smaller scenes; where linkage is easy to follow
Line or strip		Large outdoor scenes
Grid		Large outdoor scenes
Zone		Scenes with defined zones
Wheel/ray		Specialized; smaller scenes
Spiral		Scenes without physical barriers

are advised ahead of time to notify, not move. Documentation of the found items must be done before any evidence can be moved or collected; this is the birth of the item as evidence and marks the beginning of **chain of custody** procedures.

When evidence is collected, this activity is always accompanied by initialization of the chain of custody protocols for that object or sample. Each agency has its own version of the chain of custody form that includes identifiers, time and date of collection, number of objects, a basic description, and the name and signature of the person collecting it. In some cases, the chain is written on the outer bag containing the evidence or on a sticker that can be placed on the other container. Once collected, the evidence is in the custody of the person who collected it and signed the form. This evidence must be securely stored and accounted for throughout its life in the system. The CSI will likely deliver the evidence to the laboratory, where the evidence technician or other personnel will take custody, dating and signing below the CSI's signature. Anytime the evidence is removed from secure storage and given to someone else, an entry must be made on the document to indicate transfer of custody and responsibility for safekeeping. There can be no gaps in the chain; if there are, then the evidence becomes suspect since it was not under direct control for some period. If evidence is returned or destroyed, this is also documented as the final disposition of the material.

Figure 3.9 is a flowchart that shows how evidence is managed at the scene. Notice that this is the same evidence as was shown in Figure 3.4, but with a sketch and measurements added to show where it was found at the scene, as shown in Figure 3.3. Distances and angles on the sketch ensure that the casing's position is pinpointed relative to fixed points of reference on the house. The chain of custody is started by the CSI and would be logged in a list of evidence as well. The CSI would be personally responsible for the packaged evidence until it is turned over to the evidence section at the laboratory, as shown in this example. Once transferred, the person who signed for it, here the evidence technician, is responsible and would secure the evidence in a locker, vault, or other location until turned over to the next person in the chain, such as a firearms examiner. Once the case is settled, the evidence may be stored, returned, or disposed of based on the policies and procedures of the agency responsible.

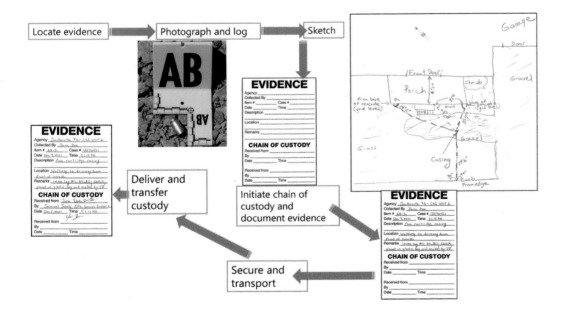

FIGURE 3.9 How evidence is located, documented, and managed at a scene. This is the same cartridge as shown in Figure 3.4.

LASER MAPPING AND CRIME SCENE RECONSTRUCTION

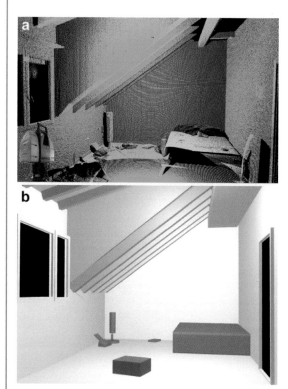

Crime scene reconstruction can be used to compare stories to evidence and evaluate different versions of events. Laser scanning and mapping have made it possible to do digital reconstructions as adjuncts to experiments and tests. In the case reported in this article a young man claimed that he accidentally shot his father as he (the son) fell backward after being pushed. The alternative story was that the shooting was purposeful, and the son was not falling when he shot his father. The scene was scanned, and the data used to generate a 3D representation is shown in the figures. This data was combined with findings from the autopsy and a series of virtual and actual reconstructions to determine which version of events was correct. At autopsy, the victim's body showed that the bullet grazed the forearm before lodging in the pelvis. Models were constructed digitally to represent the possible scenarios, focusing on whether the suspect was falling backwards, which would support his version of an accidental shooting. Images were taken during reconstructions of different scenarios as recreated by investigators and through digital modeling, which allowed investigators to view the scene from different perspectives. The actual bullet trajectory could not have been produced by the positions of the victim and suspect as

Raw data obtained from the initial scans of the scene. The scanning unit can be seen in the lower left corner of the top frame (a). (Reprinted from Buck, U., et al., Accident or Homicide—Virtual Crime Scene Reconstruction Using 3D Methods, Forensic Science International 225, no. 1–3 [Feb 2013]: 75–84. With permission from Elsevier.)

shown. Accordingly, the reconstruction does not support the suspect's version of events since the bullet trajectory could not be produced under these circumstances.

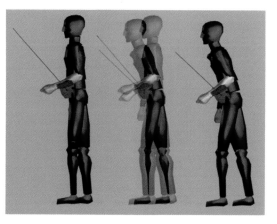

Possible bullet paths based on autopsy data. Variations in the trajectory based on how the victim might have been standing (straight or bent over) are shown. (Reprinted from Buck, U., et al., Accident or Homicide—Virtual Crime Scene Reconstruction Using 3D Methods, Forensic Science International 225, no. 1–3 [Feb 2013]: 75–84. With permission from Elsevier.)

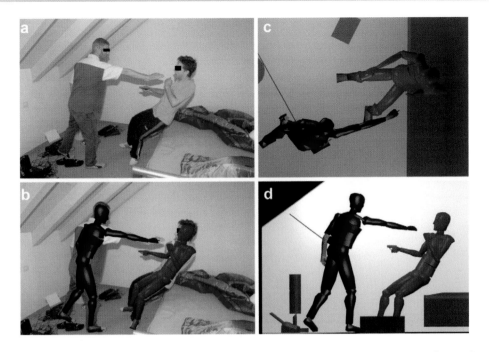

Physical reconstruction combined with digital modeling. The scene was recreated at the site based on the suspect's version of events that (a) and (b) he was pushed and fired accidentally. When recreated digitally, the scene was viewed from above (c) and from the side (d) with the bullet trajectory shown.

Source: Buck, U., et al. Accident or Homicide—Virtual Crime Scene Reconstruction Using 3D Methods. *Forensic Science International* 225, no. 1–3 (Feb 2013): 75–84. Reprinted with permission from Elsevier.

3.3.5 PACKAGING AND PRESERVATION

How evidence is packed differs based on many factors. Different types of physical evidence will require specific or special collection and packaging techniques, and this will be addressed in later chapters as appropriate. A few general rules apply. First, most items of evidence at the crime scene will be packaged into a **primary container** (Figure 3.10) that is then placed inside a **secondary container** (Figure 3.11). The outer containers are then sealed with tamper-resistant tape and initialed (Figure 3.12). If this seal is broken, it will be obvious; note how the initials are on the tape and the back. The outer container is marked with information about the item, identification about the collector, and date/time/location of collection of the item. The sealing tape or evidence tape should completely cover the opening of the outer container. It is marked with the initials of the collector, the date, and the time of collection (Figure 3.12). Notice that some forms of outer containers also have a chain of custody form as part of the package. Each item of evidence should be packaged separately to prevent cross-contamination. To the extent possible, elimination and control samples are also collected at the same time and documented. In the example scenario, elimination samples of fingerprints and DNA would have to be collected from the police officer that first responded to the scene. **Control samples** are samples that are not thought to be related to the event in question, but necessary for proper interpretation of the evidence collected. Arson scenes (Chapter 13) are an example of a type of scene where control samples are critical. Suppose a fire is intentionally set on a carpeted floor. Samples of the burned area would be collected along with control samples of unburned carpet. Both will be analyzed to ensure that any conclusions drawn based on the burned carpet consider results from the unburned carpet control.

Biological evidence such as bloody clothing must be collected and stored in such a way that air drying can occur. Wet stains in an airtight container such as a plastic bag will quickly degrade and decompose and compromise the evidence. Paper bags are typically used. In contrast, evidence

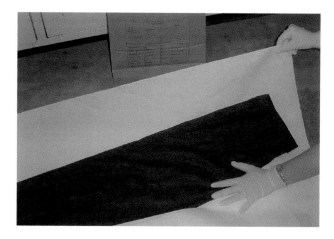

FIGURE 3.10   Packing evidence in paper to protect it and preserve any trace evidence. This paper will be folded and placed in secondary container, such as a paper bag (Figure 3.11).

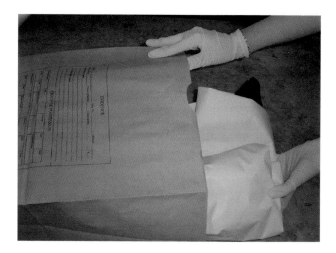

FIGURE 3.11   Locations of objects and features must be specified relative to a point of reference such as a wall. Angles and distances can be used, as shown in Figure 3.5.

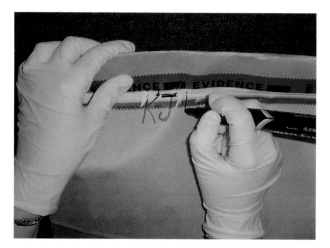

FIGURE 3.12   Secondary packaging secured with evidence tape with the initials of the person who collected it.

such as the cartridge casing (Figure 3.9) could be stored in a plastic bag. We will discuss more about how diverse types of evidence are collected in subsequent chapters.

3.4 EXAMPLE SCENE AND PROCESSING

For the example scene laid out in the insert and Figures 3.2 and 3.3, bloodstains throughout the scene alerted the first responder to the fact that the home was a crime scene. Since no body was found, this suggests to investigators that the victim or victims, dead or injured, have moved or been moved elsewhere. The CSIs have been called and, after doing the walk-through, identified six microscenes. They will have completed the initial photographs and video and done a rough sketch. At this point in the processing of the scene, they have documented and recovered several pieces of evidence, including those shown in Figures 3.13 through 3.15. The chain of custody for

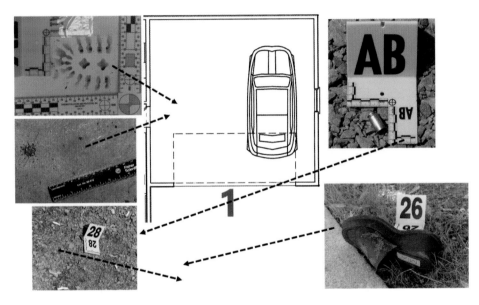

FIGURE 3.13 Evidence identified, documented, and photographed in zone 1 of the scene (Figure 3.3).

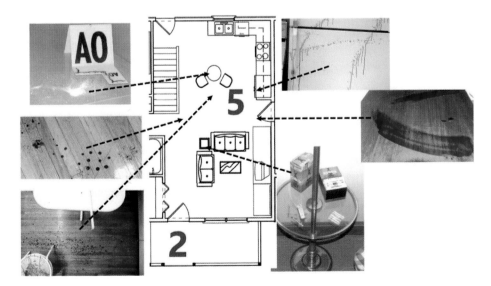

FIGURE 3.14 Evidence identified, documented, and photographed in zones 2 and 5 of the scene (Figure 3.3).

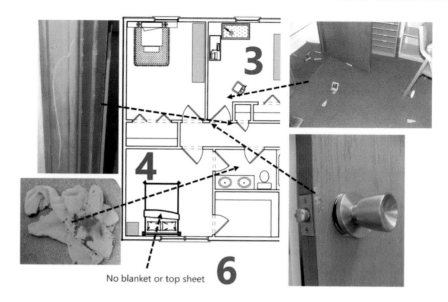

FIGURE 3.15   Evidence identified, documented, and photographed in zones 3, 4, and 6 of the scene (Figure 3.3).

the collected items has been started and CSIs will have taken the evidence to be securely stored until assigned to the appropriate forensic scientists for analysis.

The items found as shown in these figures represent many types of physical evidence that would be analyzed by different analysts in a laboratory. For example, in Figure 3.13, cigarette butts (seen in the lower left frame) might be analyzed for DNA (forensic biology), the shoeprints and likely the shoe by trace evidence personnel, and the cartridge casing by firearms experts. Possible drugs and paraphernalia (Figure 3.14) would be the responsibility of forensic chemists and seized drug analysts. The tool marks on the door shown in Figure 3.15 would be the assigned to tool mark specialists.

In any complex scene such as this, different forensic scientists would work on different pieces of evidence. In some laboratories, a **case manager** would be assigned to oversee and collect the analyses and assist in collating the information that comes from the analysts. There are also likely to be items of evidence that would not be analyzed; deciding what needs to be analyzed and for what type of information can be a challenging job. The role of the CSIs is to document and collect all relevant evidence; collecting irrelevant data and articles can be as harmful as not collecting enough. Thus, a skilled professional CSI is critical in the forensic science process. Future chapters will return to this example crime to follow how the investigation might unfold as evidence is analyzed and facts uncovered.

3.5 REVIEW MATERIAL

3.5.1 KEY TERMS AND CONCEPTS

Case manager
Chain of custody
Control samples
Crime scene map
Crime scene reconstruction
Elimination samples
First responder
Forensic mapping
Macroscopic crime scene/macroscene
Microscopic crime scene/microscene

Primary container
Primary crime scene
Secondary container
Secondary scene
Walk-through

3.5.2 QUESTIONS

1. Why are elimination and control samples critical?
2. In the example crime scene scenario laid out in this chapter, the victim moved or has been moved. How would you classify the vehicle in which that person was moved?
3. You are an analyst assigned to study the shoe shown in Figure 3.12. What type of information would you collect?
4. Why is evidence considered tainted if there is a break in the chain of custody?
5. Suppose you are responsible for organizing a crime scene search of a large open field. You have three assistants. Which method would you use and why?
6. In the example sketch (Figure 3.6), what type(s) of coordinate system(s) (Figure 3.5) is (are) used and where? Compare this to the sketch shown in Figure 3.9 and note any differences in the approach used.
7. Apply Locard's principle and identify how contacts between each of these could result in some form of physical evidence that could be recovered to link one to the other. One box is completed as an example.

	Victim	Suspect	Truck	Burial Site
Victim				
Suspect				
Truck				
Burial Site		Soil traces in shoes; shoe wear impressions at site		

3.5.3 ADVANCED QUESTIONS AND EXERCISES

1. Look at the image in the middle left of Figure 3.14. What do you think created this pattern? What type of logic did you use to come to this conclusion? How would you test your hypothesis?
2. When evidence is delivered to an analyst and is packaged as shown in Figure 3.12, the analyst will cut the bag at a different location than is taped. Why wouldn't they just cut it open through the seal?
3. What steps could the crime scene investigation team assigned to the scenario in the chapter do to minimize the chances that they will develop biases?

BIBLIOGRAPHY AND FURTHER READING

BOOKS

Baxter, E. *Complete Crime Scene Investigation Handbook*. Boca Raton, FL: CRC Press, 2015. ISBN: 978-1498701440.
Fisher, B. A. J., and Fisher, D. R. *Techniques of Crime Scene Investigation*. 8th ed. Boca Raton, FL: CRC Press, 2012. ISBN: 978-1439810057.
Jones, P. *Practical Forensic Digital Imaging*. Boca Raton, FL: CRC Press, 2011. ISBN: 978-1420060126.
Robinson, E. M. *Crime Scene Photography*. 2nd ed. Boston: Elsevier, 2010. ISBN: 978-0123757289.
Russ, J. C. *Forensic Uses of Digital Imaging*. 2nd ed. Boca Raton, FL: CRC Press, 2016. ISBN: 978-1498733076.

ARTICLES

Buck, U., Naether, S., Rass, B., Jackowski, C., and Thali, M. J. Accident or Homicide—Virtual Crime Scene
Reconstruction Using 3D Methods. *Forensic Science International* 225, no. 1–3 (Feb 2013): 75–84.

Crispino, F. Nature and Place of Crime Scene Management within Forensic Sciences. *Science and Justice*
48, no. 1 (Mar 2008): 24–28.

Holowko, E., Januszkiewicz, K., Bolewicki, P., Sitnik, R., and Michonski, J. Application of Multi-Resolution
3D Techniques in Crime Scene Documentation with Bloodstain Pattern Analysis. *Forensic Science
International* 267 (Oct 2016): 218–227.

Komar, D. A., Davy-Jow, S., and Decker, S. J. The Use of a 3-D Laser Scanner to Document Ephemeral
Evidence at Crime Scenes and Postmortem Examinations. *Journal of Forensic Sciences* 57, no. 1 (Jan
2012): 188–191.

Ribaux, O., A., Baylon, Lock, E., Delemont, O., Roux, C., Zingg, C., and Margot, P. Intelligence-Led Crime
Scene Processing. Part II: Intelligence and Crime Scene Examination. *Forensic Science International*
199, no. 1–3 (Jun 2010): 63–71.

Ribaux, O., Baylon, A., Roux, C., Delemont, O., Lock, E., Zingg, C., and Margot, P. Intelligence-Led Crime
Scene Processing. Part I: Forensic Intelligence. *Forensic Science International* 195, no. 1–3 (Feb 2010):
10–16.

WEBSITES

Link	Description
http://www.theiai.org/	The International Association for Identification; certifies CSIs and bloodstain pattern analysts
http://www.necrosearch.org/	Organization that helps locate bodies
https://www.nij.gov/topics/law-enforcement/investigations/crime-scene/guides/death-investigation/pages/evaluate.aspx	Guide to managing crime scenes with a death involved; published by the National Institute of Justice

Bloodstain Patterns

4

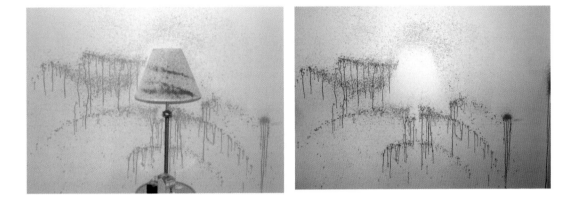

4.1 BLOODSTAIN PATTERN EVIDENCE

When a crime involves violence, there likely will be blood at the scene. The basic sciences behind bloodstain pattern analysis (BPA) are physics and geometry—physics describing how blood in motion behaves, and geometry used to calculate areas of origins for the patterns. The nature of blood as a fluid also plays an important role. There are several challenges that must be addressed at a crime scene, including recognition of bloodstains and bloodstain patterns, meticulous documentation, measuring, and using these measurements to estimate angles of impact and area where the blood that created a pattern originated from in two- and three-dimensional space. Often, follow-up experiments are useful for evaluating hypotheses about how stains could be formed, based on types of weapons and many other factors. Patterns of bloodstain are one of the best types of evidence for recreating past events and sorting through conflicting stories regarding what may or may not have happened to create them. This area of forensic science is referred to as bloodstain pattern analysis.

4.2 HUMAN BLOOD

4.2.1 BIOLOGICAL PROPERTIES

Fluid blood circulates throughout the body via the heart, arteries, veins, and capillaries. It transports oxygen, electrolytes, nourishment, hormones, vitamins, and antibodies to tissues, and transports waste products from tissues to the excretory organs. When fresh blood is centrifuged, it separates into two components—a fluid or **plasma** and cellular components. There are three types of cells in blood—red cells, white cells, and platelets. **Red blood cells** (RBCs or **erythrocytes**) transport oxygen from the lungs via the arterial system and return carbon dioxide through the veins. **White blood cells** (WBCs or **leukocytes**) assist with defense against foreign substances and infection. The nuclei of the WBCs are the sources of DNA in the blood. **Platelets** are major components of the clotting mechanism of blood. In normal individuals, cellular components comprise approximately 45% of the total blood volume, which ranges in healthy adults from

4.5 to 6.0 L (about 1.2 to 1.6 gallons). A person who loses significant amounts of blood can die by bleeding to death or **exsanguination**.

4.2.2 PHYSICAL PROPERTIES

Blood, whether a single drop or large volume, is held together by **surface tension** (Figure 4.1), which is defined as the force that pulls the surface molecules of a liquid toward its interior, decreasing the surface area and causing the liquid to resist penetration. Water has a high surface tension, and blood has slightly less. Surface tension plays a significant role in determining the shape of a blood drop in air, which is roughly spherical. Compared with water, blood also has a higher **viscosity**, meaning it has a higher resistance to flow than water.

A **passive drop** of blood is created when the volume of the drop increases to a point where gravity overcomes surface tension at the source, breaks free, and falls. Blood dripping from a bloody nose creates passive drops, which is different than cases in which blood is propelled away from a source, such as a swinging baseball bat. The source of the drop is also important; the volume of a passive drop of blood falling through air from a fingertip will be larger than a drop that originates from a hypodermic needle and smaller than a drop originating from a surface such as a baseball bat. Examples are shown in Figure 4.2. The volume of a typical or average drop of blood is about 0.05 mL (0.0017 ounces), with an average diameter of about 4.5 mm (while in air). These reported measurements vary as a function of the surface from which the blood has fallen and the rate at which the blood accumulates. The height from which it falls will influence the shape of the pattern created when it impacts a surface, as does the surface itself. In general, free-falling drops of blood will produce bloodstains of increasing diameters with increasing height; this is due to the higher impact speed. However, it is not possible to determine how far a drop has fallen since the volume of the original drop is not known.

4.3 FORMATION OF BLOODSTAINS AND BLOODSTAIN PATTERNS

4.3.1 SURFACE CONSIDERATIONS

The factors that act on a falling passive drop of blood are shown in Figure 4.3. As it flows down the surface of the knife blade and collects as a drop, surface tension drives formation of the spherical shape. Once the drop becomes large enough that gravity causes it to fall, it will retain its shape. The only opposing force is air resistance. No matter how far a drop of blood falls, it will not break into smaller droplets or spatters unless something disrupts the surface tension. One factor in breaking the surface tension of a blood drop is the physical nature of the target surface the drop strikes.

FIGURE 4.1 A paper clip made of steel with copper metal plating floats on water. Notice that the water bends inward but does not give way; this is due to surface tension. (Credit: Armin Kübelbeck.)

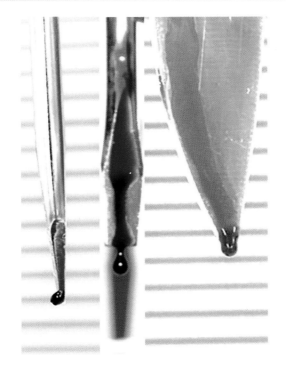

FIGURE 4.2 Variation of drop size between a screwdriver (left, side view and middle) and tip of a knife. .

Generally, a hard, smooth, nonporous surface, such as clean glass or smooth tile, will create little if any spatter (Figure 4.4a), in contrast to a surface with a rough texture, such as wood or concrete, which can create a significant amount of spatter (Figure 4.4b). Rough surfaces have protuberances that rupture the surface tension of the blood drop and produce spatter and irregularly shaped parent stains with spiny or serrated edges.

4.3.2 SIZE, SHAPE, AND DIRECTIONALITY

The geometry of individual bloodstains will generally allow the analyst to estimate their direction of flight prior to impacting an object. This is done by examining the edge characteristics

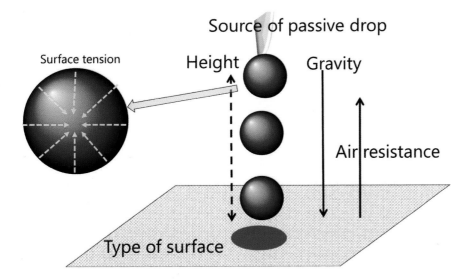

FIGURE 4.3 Factors that influence blood dropping. Surface tension creates the spherical shape; the height determines the speed at impact.

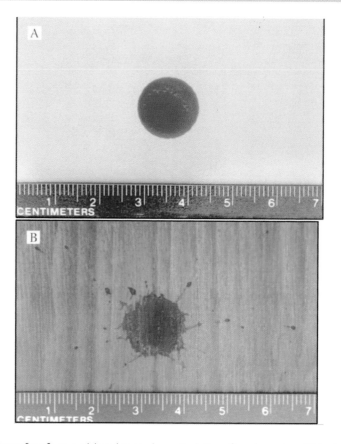

FIGURE 4.4   Effect of surface on bloodstain characteristics after dropping 30 inches. (a) Smooth tile surface. (b) Wood paneling.

of individual stains (Figure 4.5). The narrow end of an elongated bloodstain usually points in the direction of travel. After this **directionality** of several bloodstains has been determined, an **area of convergence** may be established by drawing straight lines through the long axes of the bloodstains (Figure 4.6). The area where these lines converge represents the relative location of the blood source in a two-dimensional perspective. This area of convergence will be an area, *not* an exact point. An analogous operation allows for determining the **area of origin**, which is in three dimensions (Figure 4.7).

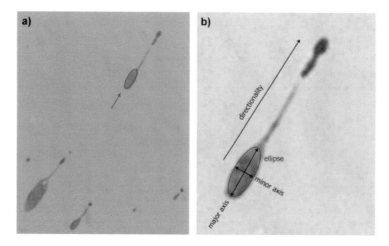

FIGURE 4.5   (a) Directionality of a blood spatter pattern. (b) The major axis is the length (L) and the minor axis is the width (W). (Reproduced from Buck, U., et al., 3D Bloodstain Pattern Analysis: Ballistic Reconstruction of the Trajectories of Blood Drops and Determination of the Centres of Origin of the Bloodstains, *Forensic Science International* 206, no. 1–3 [Mar 2011]: 22–28. With permission from Elsevier.)

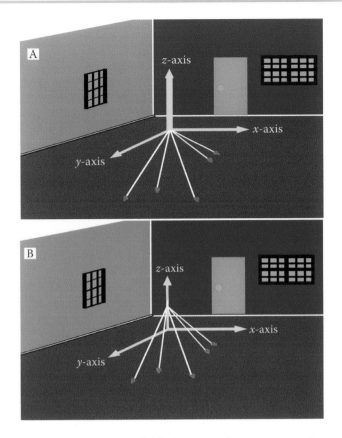

FIGURE 4.6 Graphical representations of (a) the area of convergence in two dimensions and (b) area of origin in three dimensions.

If the **angle of impact** is 90°, the resulting bloodstain generally will be circular in shape, as in Figure 4.4. Blood drops that strike a target at an angle less than 90° will create elliptical bloodstains with a shape that varies based on the angle of impact and the distance the blood drop fell (Figure 4.8). The dimensions of the ellipse can be used to estimate the angle of impact for the original spherical drop. This calculation is accomplished by measuring the width and length of the bloodstain, as shown in Figure 4.5. The width measurement is divided by the length measurement to produce a ratio number less than 1. This ratio is the arcsine of the impact angle, and the impact angle of the bloodstain can be determined by simple calculations. For a circular

FIGURE 4.7 An example of estimation of the area of origin using laser scanning and computer measurement. This is the same scene as shown in Figure 3.9. (Reproduced from Buck, U., et al., 3D Bloodstain Pattern Analysis: Ballistic Reconstruction of the Trajectories of Blood Drops and Determination of the Centres of Origin of the Bloodstains, *Forensic Science International* 206, no. 1–3 [Mar 2011]: 22–28. With permission from Elsevier.)

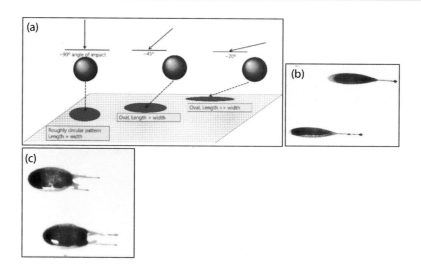

FIGURE 4.8   (a) Effect of angle of impact on shape of a bloodstain. (b) 20° angle, 16-inch drop. (c) 35° angle, 16-inch drop.

bloodstain, the width and length are equal, and thus the ratio is 1.0, which corresponds to an impact angle of 90°.

After establishing the angle of impact for each of the bloodstains, the three-dimensional origin of the bloodstain pattern can be determined. One method, known as **stringing**, involves placing elastic strings at the base of each bloodstain and projecting these strings back to the axis that has been extended 90° up or away from the two-dimensional area of convergence. Stringing is rapidly being replaced with laser scanning and other computer-based methods (Figure 4.7), but the concept is the same. For manual measurements, a protractor is placed on each string and then lifted until it aligns with the previously determined impact angle. The string is then secured to the axis placed at the two-dimensional area of convergence. This is repeated for each of the selected bloodstains; an example is shown in Figure 4.9.

4.3.3 SPATTERED BLOOD

Blood spatter is created when sufficient force is available to overcome the surface tension of the blood. The amount of force applied to a source of blood and the size of the resulting spatter vary considerably with gunshot, beating, and stabbing events (Figure 4.10). These patterns were created on a target about 30 cm from the sponge. The first column shows the overall pattern; the second and third show the magnified box and the box after processing. The size range of spatter produced by any one mechanism may also vary considerably. As

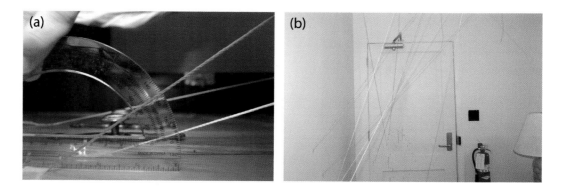

FIGURE 4.9   Stringing as a method of estimating areas of origin and convergence. (a) Use of the protractor. (b) Stringing in progress.

9 mm

Bat (half-strength)

Bat (full-strength)

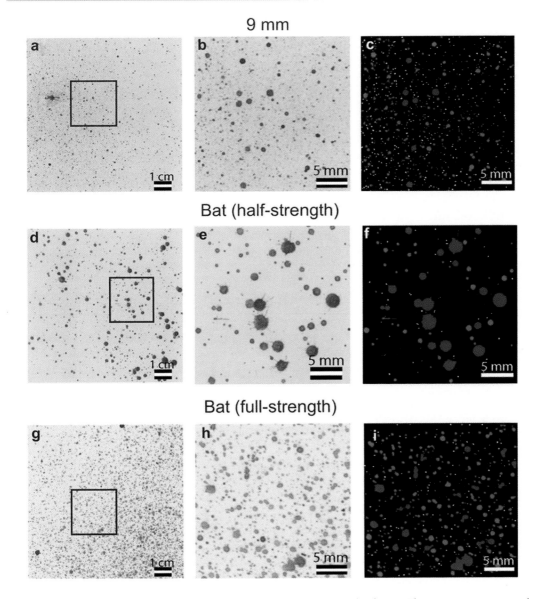

FIGURE 4.10   Variation of pattern and drop size by weapon and velocity. The patterns were made by hitting or shooting a blood-soaked sponge such that the spatter went forward onto a paper target. (From Siu, S., et al., Quantitative Differentiation of Bloodstain Patterns Resulting from Gunshot and Blunt Force Impacts, *Journal of Forensic Sciences* 62, no. 5 [Sep 2017]: 1166–1179. Reproduced with permission from Wiley.)

shown in Figure 4.11, droplets from the three types of weapons were generally smaller and varied less than droplets from the bat and crowbar. In this figure, the small black error bars indicate variation.

When stains are encountered at a scene, it is important that the analyst first identify a group of stains as spatter; an isolated drop does not constitute a pattern. The size, quantity, and distribution of these spatters vary depending on factors such as the quantity of blood involved, the force of the impact, and the texture of the surface where the blood was deposited. Terms such as *low-*, *medium-*, and *high-velocity impact patterns* were used in the past to differentiate stains that origi-nated from different events; high velocity and misting are consistent with a gunshot wound, for example, while medium velocity might describe a beating with a blunt object. However, because of the inherent variability in events, weapons, and conditions, this terminology is considered out-dated now and the Scientific Working Group on Bloodstain Pattern Analysis (SWGSTAIN) has removed these terms from its definitions, instead using only **impact stains**.

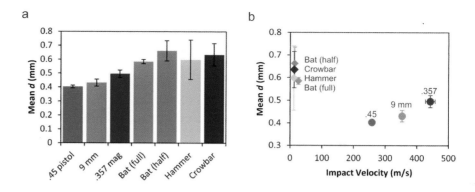

FIGURE 4.11   Drop size based on velocity. Notice that droplets from the gunshots tend to be smaller and vary less than droplets from the bat and crowbar. (Reproduced from Siu, S., et al., Quantitative Differentiation of Bloodstain Patterns Resulting from Gunshot and Blunt Force Impacts, *Journal of Forensic Sciences* 62, no. 5 [Sep 2017]: 1166–1179. With permission from Wiley.)

As seen in Figure 4.10, gunshot wounds do tend to create a relatively unique type of stain pattern on nearby surfaces.

When a bullet enters the body at high velocity, two spatter processes can occur—**back spatter** (Figure 4.12) and **forward spatter** (Figure 4.13). For these studies, bullets were fired through a blood-soaked sponge encased in silicon. At the point of bullet impact, blood and tissue can be ejected in a manner similar to that of a meteor collision to produce backward ejections of material (back spatter). Notice how tiny and how widely distributed the droplets are in Figure 4.12. The dispersion increases with distance. Any surface close to the impact site will be spattered with fine droplets that are sometimes referred to as **misting**. If the bullets or fragments exit the body, blood and tissue will travel with it, as shown in Figure 4.13. Again, the droplets tend to be small, consistent with the data shown previously in Figures 4.10 and 4.11. Given that bullets travel so fast, the diameter of the droplets tends to be smaller than that of other blunt force weapons; however, one stain diameter alone is not sufficient to assess the weapon type.

In certain situations, multiple mechanisms of stain formation may exist, such as any combination of gunshot, beating, stabbing, expired blood, or satellite spatter resulting from blood dripping onto surfaces. There is often more than one plausible explanation for the creation of a given pattern, and this is not a problem if it is noted and considered. The evidence can still be useful in the elimination of other hypotheses or scenarios.

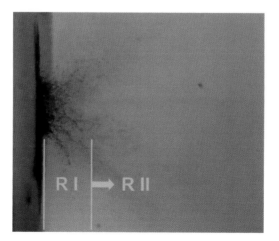

FIGURE 4.12   Back spatter 0.1 ms after impact. The two areas R1 and R2 divide the region into an area of relatively large droplets (R1) and finer droplets (R2). (Reproduced from Comiskey, P. M., et al., High-Speed Video Analysis of Forward and Backward Spattered Blood Droplets, *Forensic Science International* 276 [Jul 2017]: 134–141. With permission from Elsevier.)

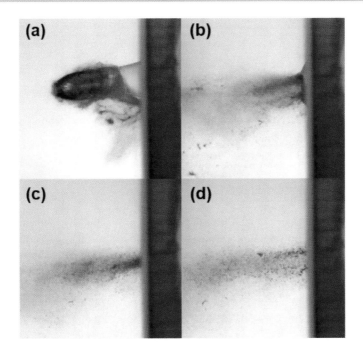

FIGURE 4.13   Forward spatter produced at (a) 0.2 ms, (b) 0.8 ms, (c) 1.1 ms, and (d) 1.5 ms after impact. (Reproduced from Comiskey, P. M., et al., High-Speed Video Analysis of Forward and Backward Spattered Blood Droplets, *Forensic Science International* 276 [Jul 2017]: 134–141. With permission from Elsevier.)

MYTHS OF FORENSIC SCIENCE

You can tell what weapon was used to create a bloodstain pattern.

Not always. While some weapons create distinctive patterns and drop sizes, such as a gunshot at close range, there is overlap in the appearance of patterns. A particularly vicious beating with a bat can produce small spatter that can appear similar to gunshot patterns. Similarly, a castoff pattern that comes from a golf club may look very similar to that from a bat or large wrench. It can be challenging to differentiate some types of spatter from castoff patterns. As always, BPA is part of a larger whole and has to be integrated into the context of the investigation for it to be put to best use.

4.3.4 SATELLITE SPATTERS

Single drops of blood will produce small spatters around the **parent stain** because of striking a rough target surface. Spatter produced in this manner is referred to as **satellite spatter** or satellite stains. When multiple free-falling drops of blood are produced from a stationary source onto a horizontal surface, **drip patterns** will result from blood drops falling into previously deposited wet bloodstains or small pools of blood. These drip patterns will be large and irregular in shape, with small satellite spatters around the periphery of the central parent stain on the horizontal and nearby vertical surfaces. Satellite spatters are the results of smaller droplets of blood that have detached from the main blood volume at impact. These satellite spatters of blood are circular to oval. Several factors influence the appearance of satellite spatter, including the blood drop volume, freshness of blood, surface texture, and the distance of the vertical target from the impact site.

An example containing all these types of stains is shown in Figure 4.14. This pattern was created by a person creating drops of blood from a dropper while walking quickly across the paper surface. The drops are roughly circular, so the angle of impact was close to 90°. All the parent drops have multiple small satellite spots and in many overlap, creating drip patterns. The crown-like pattern around the edge of each drop is referred to as **scalloping.**

4.3.5 CASTOFF BLOODSTAIN PATTERNS

During a beating with a blunt object, blood does not immediately accumulate at the impact site with the first blow. As a result, no blood is present on the weapon to be spattered or cast from the first blow. Spatter and **castoff patterns** are created with subsequent blows to the same general area where a wound has occurred and blood has accumulated. The castoff can be seen on floors, ceilings, and walls; an example is shown in Figure 4.15. Blood will adhere in varying quantities to the object that produces the injuries. If the force generated by swinging the weapon is great enough, blood will be flung from the object to form a castoff bloodstain pattern. For example, suppose a victim is hit with a baseball bat on the top of the head, causing a severe tear and bleeding. If hit again in the same area, the bloodied bat can now create castoff as it is raised or swung again.

The blood that is flung (cast off) will strike objects and surfaces, such as adjacent walls and ceilings in the vicinity, at the same angle from which it is flung or cast. The size, distribution, and quantity of these castoff bloodstains vary. Castoff patterns are often seen in conjunction with impact spatters, and a study of each may help determine the relative position of the victim and the assailant at the time the injuries were inflicted. Castoff bloodstains are not always present at scenes where blunt or sharp force injuries have occurred. The arc of the back or side swing may be minimal, especially in the case of a heavy blunt object.

FIGURE 4.14   Stain pattern created by rapid walking and passive dripping.

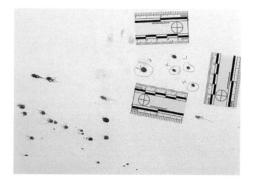

FIGURE 4.15   Castoff pattern created on a wall by a baseball bat.

4.3.6 EXPIRATED BLOODSTAIN PATTERNS

Because of trauma, blood will often accumulate in the lungs, sinuses, and airway passages of the victim. In a living victim, this accumulation of blood will be forcefully expelled from the nose or mouth to free the airways. This type of bloodstain is referred to as an **expirated bloodstain** pattern (Figure 4.16). The size, shape, and distribution of an expirated bloodstain pattern are often like the patterns that are observed with impact spatter associated with beatings and gunshots. Since impact spatter due to a gunshot or beating mechanism can closely resemble expirated bloodstain patterns, the deciding factor may be the case history. An expiratory bloodstain pattern cannot be produced unless the victim has blood on their face or in their mouth or nose, or some type of injury to their chest or neck that involves the airways.

Expirated bloodstains may appear diluted if mixed with sufficient saliva or nasal secretions. If the blood has been recently expelled, there may be visible air bubbles within the stains due to the blood being mixed with air from the airway passages or lungs. When the bubbles rupture and the bloodstains dry, the areas of previous air bubbles will appear as **vacuoles**.

AGING OF BLOODSTAINS

Forensic scientists have long sought a method that would allow them to estimate how much time has elapsed since a bloodstain was deposited. Although this is still not possible, progress is being made in research that looks at biochemical degradation and, as shown here, specialized imaging of stains. In this study, the breakdown of the light reflected off a stain was analyzed in the visible wavelengths of light. There are slight differences between the spectra of new and old blood that can be analyzed to estimate the age of a stain. Using specialized image collection and processing, the researchers could differentiate experimental stains based on age.

Source: Edelman, G., et al., Hyperspectral Imaging for the Age Estimation of Blood Stains at the Crime Scene, *Forensic Science International* 223, no. 1–3 (Nov 2012): 7277. Reproduced with permission. Copyright Elsevier.

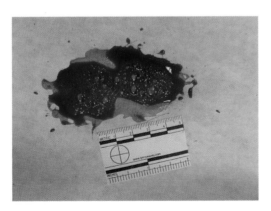

FIGURE 4.16   Expirated blood with bubbles.

4.3.7 ARTERIAL BLOODSTAIN PATTERNS

When an artery is breached, blood is projected from it in varying amounts. The size of arterial bloodstains varies from very large gushing or spurting patterns to very small spray types of patterns (Figure 4.17). The type of arterial pattern observed depends on the severity of the injury to the artery, the size and location of the artery, whether the injury is covered by clothing, and the position of the victim when the injury was inflicted. These patterns are usually very distinctive due to the overall quantity of bloodstains observed. There may be a wavy pattern or other type of repetitive pattern that corresponds with the heart beating.

4.3.8 TRANSFER PATTERNS

When blood on one surface contacts another, a transfer pattern is created. The image at the beginning of Chapter 3 is an example; a bloody gun created a transfer pattern on paper. Often the patterns are not as obvious. It is not unusual to have transfer patterns on clothing that occur as part of a violent crime, and such cases are good examples of how the substrate (fabric) can complicate the identification and interpretation of transfer patterns. Figures 4.18 through 4.20 illustrate how the substrate can make identification and interpretation of transfer patterns particularly challenging with fabric. The apparatus used (Figure 4.18) created spatter and transfer patterns for comparison. The patterns in this study were created using pig blood, which is frequently used in laboratory studies. Two types of stains were created on a fabric—simple spatter onto the surface, and spatter onto another surface (glass or leather) that was then pressed against the fabric to create a transfer pattern. The fabric used was knit fabric widely used in clothing. The low-magnification results are shown in Figure 4.19 and the high-magnification images in Figure 4.20. To the naked eye (Figure 4.19), the difference between the spatter and transfer is not obvious. Figure 4.20 shows stains with additional magnification; in the magnified areas in the right column, the pressing action of the transfer is evident in the bottom two frames. Had this transfer taken place on a smooth substrate, the interpretation would be easier.

FIGURE 4.17   Arterial spurt pattern.

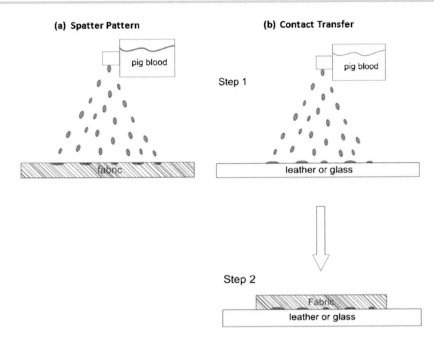

FIGURE 4.18 An experimental apparatus to create and compare tranfer patterns. (Reproduced with permission from Cho, Y., et al., Quantitative Bloodstain Analysis: Differentiation of Contact Transfer Patterns versus Spatter Patterns on Fabric via Microscopic Inspection, *Forensic Science International* 249 [Apr 2015]: 233–240. Copyright Elsevier.)

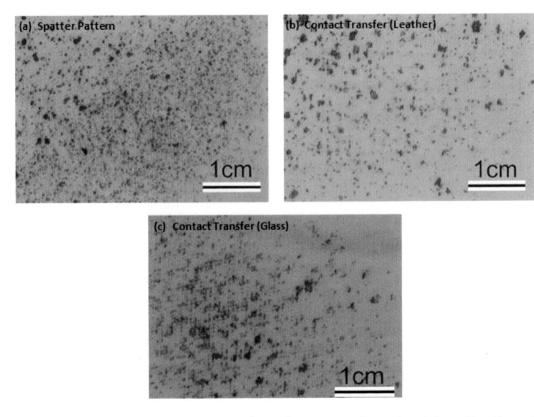

FIGURE 4.19 Low-magnification view of transfer patterns. (Reproduced from Cho, Y., et al., Quantitative Bloodstain Analysis: Differentiation of Contact Transfer Patterns versus Spatter Patterns on Fabric via Microscopic Inspection, *Forensic Science International* 249 [Apr 2015]: 233–240. With permission from Elsevier.)

Spatter

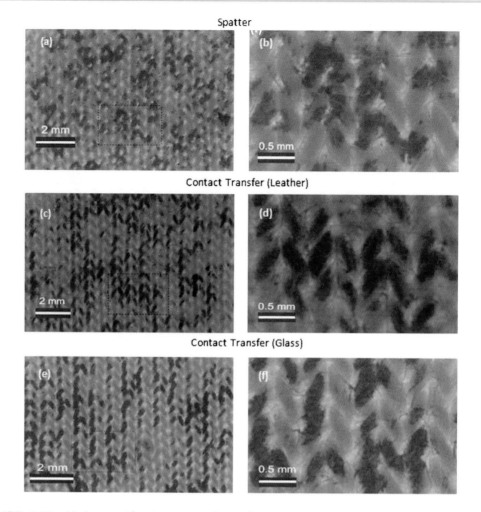

Contact Transfer (Leather)

Contact Transfer (Glass)

FIGURE 4.20 High-magnification view of transfer patterns. (Reproduced from Cho, Y., et al., Quantitative Bloodstain Analysis: Differentiation of Contact Transfer Patterns versus Spatter Patterns on Fabric via Microscopic Inspection, *Forensic Science International* 249 [Apr 2015]: 233–240. With permission from Elsevier.)

A **wipe** pattern is another type of transfer pattern that occurs when a bloody object such as hair or clothing is dragged across another surface. An example is shown in Figure 4.21.

4.4 ALTERED BLOODSTAINS

Bloodstains deposited on surfaces at a scene are subject to various forms of change from their original appearance at the time the blood-shedding event occurred. When blood exits the body, the processes of drying and clotting begin. The drying time of blood is a function of its volume, the nature of the target surface texture, and the environmental conditions. Small spatters and light transfers of blood will dry within a few minutes under normal conditions of temperature, humidity, and air currents. Larger volumes of blood may take considerable time to completely dry. Drying is accelerated by increased temperature, low humidity, and increased airflow. Initially, the outer rim or perimeter of the bloodstain will show evidence of drying, which then proceeds toward the middle of the stained area.

When the center of a dried bloodstain flakes away and leaves a visible outer rim, the result is referred to as a **skeletonized stain** (Figure 4.22). Another type of skeletonized bloodstain occurs when the center of a partially dried bloodstain is altered by contact or a wiping motion that leaves the periphery intact. This can be interpreted as movement or activity by the victim or assailant when or after injuries were inflicted. The alteration of bloodstains by heat, fire, or smoke may

FIGURE 4.21   Example wipe pattern.

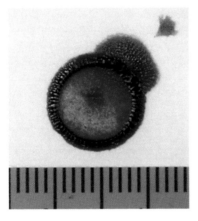

FIGURE 4.22   Skeletonized stain.

also cause problems for the analyst. Bloodstains covered with soot may be entirely missed at the scene of a homicide that preceded a fire. Heat and fire may also cause existing bloodstains to fade, darken, or be destroyed.

Void areas or patterns are absences of bloodstains in otherwise continuous patterns of staining. These patterns are commonly seen where items have been removed from an area previously spattered with blood. This permits the analyst to establish sequencing and identify alterations within a crime scene. At a scene containing a considerable amount of spattered blood, the void areas may be utilized to recognize the general location where the spatter-producing event(s) occurred. The two photos shown at the start of this chapter illustrate how void patterns are formed.

CONTEXTUAL INFORMATION AND BPA

A recent study by researchers in New Zealand evaluated the effect of contextual information on the interpretation of bloodstain patterns by 39 analysts with varying degrees of experience. The authors first evaluated psychological characteristics of the investigators, such as decision-making style and need for closure using standard tests. Next, they provided the analysts with a color photo of a castoff stain created in a laboratory (shown) and asked the analysts to characterize the stain in terms of what they considered more or less favored mechanisms of formation and the analysts could select more than one possibility and weigh the likelihood of each possibility. The mechanisms were castoff, drip stain/trail, expiration pattern, projection (arterial spurt), gunshot, swipe, blunt force impact, saturation, wipe, or transfer pattern. Analysts were then allowed to request contextual information, one item at a time, if they thought it would assist them. The information items are shown in the first table. Note that all these bits of contextual information are fictional since the stain pattern was artificially produced. The analysts were allowed to change their opinions based on each piece of contextual information provided to them. All the analysts asked for at least one piece of contextual information and 49% requested all six pieces of contextual information. However, the way the study was designed, the analysts were only asked to classify the pattern and, as such, no contextual information was needed. The authors were interested in how contextual information would impact the analysts' opinion, if at all. Interestingly, only four of the analysts (10%) did not alter conclusions based on the contextual information. The results are summarized in the second table. If an analyst at first excluded a mechanism and then changed their opinion to include it, that would fall in the first column. For example, if an analyst at first excluded gunshot as a mechanism and then upon reading about the medical findings decided to include it as a possibility, that would count in the first column of the table. Medical information was the most influential while witness statements were the least. The authors also found that in this study, the psychological factors didn't help to determine which analysts were more or less prone to changing their opinion. The authors recommended that BPA analysts should carefully document their analytical process and how contextual information is used to ensure that the information does not evolve into bias.

Contextual information for pattern classification task.

Context item	Details
Police briefing	The deceased is Mr. Robert Harold Sing. A work colleague, Mr. Simon Peters, has gone to Mr. Sing's home after he failed to show for his 7 am work shift, and has found his body in the back yard. Mr. Peters called police, who launched a homicide investigation. A scene examination revealed bloodstaining on the walls and floor in the hallway leading to Mr. Sing's bedroom, and also on the west facing bedroom wall.
Medical findings	An autopsy shows that Mr. Sing died of a combination of blunt force trauma to the head and subdural haemorrhaging. Injuries to the eyes and nose were consistent with an object impacting Mr. Sing's face with considerable force. It is likely that he was alive for several hours after the injuries were sustained.
Other bloodstain patterns	An analysis of other bloodstains in the hallway and bedroom show patterns consistent with drip stains, leading from the bedroom to the hallway and outdoor area where the body was discovered.
Other forensic evidence	A bloodstained fingerprint pattern was analysed and found to match to Mr. Sing's brother in-law Mr. Jacob Walters. Several shoeprints found in the hallway of Mr. Sing's home, and on the path leading to the back yard matched to that of a woman Mr. Sing had been in a relationship with, Ms. Jessica Patel. A partial bloodied palm print was found on the west facing bedroom wall above the blood panel that is under analysis; however, police were not able to match it to any known person. A search of the homes of Mr. Jacob and Ms. Patel did not uncover any further evidence.
DNA statement	The blood on the wall panel being analysed matches to that of Mr. Sing [deceased].
Witness statement	Neighbours heard an argument between a man and a woman around 11 pm on the night before Mr. Sing was found dead. About 2 am that night they awoke to loud bangs and another intense argument between two men before a car was heard speeding away. The neighbours did not recognise the voices.

Number of exclude-after-include and include-after-exclude decisions.

Context item	Exclude-after-include	Include-after-exclude
Medical findings	24	3
DNA statement	8	1
Police briefing	5	2
Other bloodstain patterns	15	0
Other forensic evidence	2	0
Witness statement	0	0
Total changes (1870)	54 (31 analysts)	6 (5 analysts)

Source: Osborne, N. K. P., et al., Bloodstain Pattern Classification: Accuracy, Effect of Contextual Information and the Role of Analyst Characteristics, *Science and Justice* 56, no. 2 (Mar 2016): 123–128. Reproduced with permission. Copyright Elsevier.

4.5 RETURN TO THE SCENE OF THE CRIME

The crime scene presented in Chapter 3 (Figures 3.2 and 3.3 and the insert) revealed several types of bloodstain patterns that allow investigators to begin developing theories as to what could and could not have occurred in the house. This example illustrates how BPA could be undertaken and contribute to the investigation as it evolves. The copious amounts of blood, particularly in the kitchen, coupled to the lack of a body has sent investigators looking for the woman who lives at the address. They have noted evidence that indicates the woman, if indeed the victim, is likely deceased and suspect that the body was transported in the suspect's vehicle. They have also confirmed that the husband drives a red truck, the one seen leaving the scene the night before the first responder arrived on scene. Locard's principle dictates that there will be evidence of the perpetrator left at the scene as well as evidence from the scene and the crime on the perpetrator and in the vehicle.

Bloodstain evidence would be very useful in this case in many ways. The bloody shirt (Figure 3.15) and finding it in a bathroom suggests severe injury, but early in the investigation, it will not be known if the blood is that of the victim or the suspect. The passive blood drop and the bloody shoeprint (Figure 3.13) found in the garage support the hypothesis that the injured person was in the garage and that someone walked through blood to create the shoeprint. Notice how the texture of the concrete effected the appearance of that bloodstain compared with impacts on a smooth surface. The bloodstain patterns in Figure 3.14 tell the story of a violent and likely fatal struggle and probable movement of a victim after injury was inflicted and bleeding occurred. Investigators could apply stringing or lasers to different stain patterns to estimate areas of origin, identify any castoff patterns, and begin to form theories as to what weapons might have been used in the crime.

4.6 REVIEW MATERIAL

4.6.1 KEY TERMS AND CONCEPTS

Angle of impact
Area of convergence
Area of origin
Arterial spurt
Back spatter
Castoff patterns
Directionality
Drip patterns
Erythrocytes

Expirated bloodstain pattern
Exsanguination
Forward spatter
Impact stains
Leukocytes
Misting
Parent drop
Passive drop
Plasma
Platelets
Satellite spatter
Skeletonized stain
Stringing
Surface tension
Vacuoles
Viscosity
Void areas

4.6.2 QUESTIONS

1. What properties of blood determine the shape of a blood drop in flight?
2. What is the most important factor governing the degree of distortion and amount of spatter created when a blood drop strikes a surface?
3. What factors influence the stain diameter produced by a free-falling drop?
4. How are the physical characteristics of spatter utilized to determine their angle of impact?
5. What variables can affect the size, quantity, and distribution of spatters created by an impact mechanism such as beatings and shootings?
6. Explain the mechanism of castoff bloodstain patterns.
7. What bloodstain pattern evidence shown in Figure 3.14 would lead investigators to believe that the victim is deceased rather than injured?
8. Refer to Figure 3.8. What type of stain pattern is this?
9. How old would you say the stains are in Figure 4.14? Why? What type of logic did you use to come to this conclusion—inductive, abductive, or deductive?
10. What is the difference between castoff and spatter patterns? Compare the stains in Figures 4.11 and 4.15; both were produced by a bat but are different stain types.

4.6.3 ADVANCED QUESTIONS AND EXERCISES

1. How could contextual information create the potential for bias in BPA? Use the crime scene scenario from Chapter 3 and illustrate by example.
2. What is the difference between area of origin and point of convergence? Why isn't the term *point of origin* used?
3. Go back to the crime scene scenario in Chapter 3 and look at the bloodstain patterns. Hypothesize about what type each stain is and give your reasoning.
4. Related to the previous question, do you think your interpretation could be biased because you already know the basics of what happened at the scene?

BIBLIOGRAPHY AND FURTHER READING

BOOKS

James, S. H., Kish, P. E., and Sutton, T. P. *Principles of Bloodstain Pattern Analysis—Theory and Practice.* Boca Raton, FL: CRC Press, 2005.

Wonder, A. Y. *Blood Dynamics.* San Diego: Academic Press, 2001.

Wonder, A. Y. *Bloodstain Pattern Evidence–Objective Approaches and Case Applications.* New York: Elsevier, 2007.

ARTICLES

Adam, C. D. Experimental and Theoretical Studies of the Spreading of Bloodstains on Painted Surfaces. *Forensic Science International* 229, no. 1–3 (Jun 2013): 66–74.

Attinger, D., Moore, C., Donaldson, A., Jafari, A., and Stone, H. A. Fluid Dynamics Topics in Bloodstain Pattern Analysis: Comparative Review and Research Opportunities. *Forensic Science International* 231, no. 1–3 (Sep 2013): 375–396.

Basu, N., and Bandyopadhyay, S. K. 2D Source Area Prediction Based on Physical Characteristics of a Regular, Passive Blood Drip Stain. *Forensic Science International* 266 (Sep 2016): 39–53.

Bremmer, R. H., de Bruin, K. G., van Gemert, M. J. C., van Leeuwen, T. G., and Aalders, M. C. G. Forensic Quest for Age Determination of Bloodstains. *Forensic Science International* 216, no. 1–3 (Mar 2012): 1–11.

Camana, F. Determining the Area of Convergence in Bloodstain Pattern Analysis: A Probabilistic Approach. *Forensic Science International* 231, no. 1–3 (Sep 2013): 131–136.

Cho, Y., Springer, F., Tulleners, F. A., and Ristenpart, W. D. Quantitative Bloodstain Analysis: Differentiation of Contact Transfer Patterns versus Spatter Patterns on Fabric via Microscopic Inspection. *Forensic Science International* 249 (Apr 2015): 233–240.

Comiskey, P. M., Yarin, A. L., Kim, S., and Attinger, D. Prediction of Blood Back Spatter from a Gunshot in Bloodstain Pattern Analysis. *Physical Review Fluids* 1, no. 4 (Aug 2016).

Doty, K. C., McLaughlin, G., and Lednev, I. K. A Raman "Spectroscopic Clock" for Bloodstain Age Determination: The First Week after Deposition. *Analytical and Bioanalytical Chemistry* 408, no. 15 (Jun 2016): 3993–4001.

Edelman, G., van Leeuwen, T. G., and Aalders, M. C. G. Hyperspectral Imaging for the Age Estimation of Blood Stains at the Crime Scene. *Forensic Science International* 223, no. 1–3 (Nov 2012): 72–77.

Gao, W. Y., Wang, C., Muzyka, K., Kitte, S. A., Li, J. P., Zhang, W., and Xu, G. B. Artemisinin-Luminol Chemiluminescence for Forensic Bloodstain Detection Using a Smart Phone as a Detector. *Analytical Chemistry* 89, no. 11 (Jun 2017): 6161–6166.

Hakim, N., and Liscio, E. Calculating Point of Origin of Blood Spatter Using Laser Scanning Technology. *Journal of Forensic Sciences* 60, no. 2 (Mar 2015): 409–417.

Holowko, E., Januszkiewicz, K., Bolewicki, P., Sitnik, R., and Michonski, J. Application of Multi-Resolution 3D Techniques in Crime Scene Documentation with Bloodstain Pattern Analysis. *Forensic Science International* 267 (Oct 2016): 218–227.

Illes, M., and Boue, M. Robust Estimation for Area of Origin in Bloodstain Pattern Analysis via Directional Analysis. *Forensic Science International* 226, no. 1–3 (Mar 2013): 223–229.

Joris, P., Develter, W., Jenar, E., Suetens, P., Vandermeulen, D., Van de Voorde, W., and Claes, P. Hemovision: An Automated and Virtual Approach to Bloodstain Pattern Analysis. *Forensic Science International* 251 (Jun 2015): 116–123.

Kabaliuk, N., Jermy, M. C., Morison, K., Stotesbury, T., Taylor, M. C., and Williams, E. Blood Drop Size in Passive Dripping from Weapons. *Forensic Science International* 228, no. 1–3 (May 2013): 75–82.

Kabaliuk, N., Jermy, M. C., Williams, E., Laber, T. L., and Taylor, M. C. Experimental Validation of a Numerical Model for Predicting the Trajectory of Blood Drops in Typical Crime Scene Conditions, Including Droplet Deformation and Breakup, with a Study of the Effect of Indoor Air Currents and Wind on Typical Spatter Drop Trajectories. *Forensic Science International* 245 (Dec 2014): 107–120.

Kettner, M., Schmidt, A., Windgassen, M., Schmidt, P., Wagner, C., and Ramsthaler, F. Impact Height and Wall Distance in Bloodstain Pattern Analysis—What Patterns of Round Bloodstains Can Tell Us. *International Journal of Legal Medicine* 129, no. 1 (Jan 2015): 133–140.

Kim, S., Ma, Y., Agrawal, P., and Attinger, D. How Important Is It to Consider Target Properties and Hematocrit in Bloodstain Pattern Analysis? *Forensic Science International* 266 (Sep 2016): 178–184.

Kunz, S. N., Adamec, J., and Grove, C. Analyzing the Dynamics and Morphology of Cast-Off Pattern at Different Speed Levels Using High-Speed Digital Video Imaging. *Journal of Forensic Sciences* 62, no. 2 (Mar 2017): 428–434.

Kunz, S. N., Grove, C., and Adamec, J. Forensic Analysis of Bloodstain Impact Patterns from a Ballistic Point of View. *Rechtsmedizin* 25, no. 6 (Dec 2015): 548–555.

Laan, N., de Bruin, K. G., Slenter, D., Wilhelm, J., Jermy, M., and Bonn, D. Bloodstain Pattern Analysis: Implementation of a Fluid Dynamic Model for Position Determination of Victims. *Scientific Reports* 5 (Jun 2015).

Larkin, B. A. J., and Banks, C. E. Bloodstain Pattern Analysis: Looking at Impacting Blood from a Different Angle. *Australian Journal of Forensic Sciences* 45, no. 1 (Mar 2013): 85–102.

Larkin, B. A. J., and Banks, C. E. Preliminary Study on the Effect of Heated Surfaces upon Bloodstain Pattern Analysis. *Journal of Forensic Sciences* 58, no. 5 (Sep 2013): 1289–1296.

Lee, R., and Liscio, E. The Accuracy of Laser Scanning Technology on the Determination of Bloodstain Origin. *Canadian Society of Forensic Science Journal* 49, no. 1 (2016): 38–51.

Li, B., Beveridge, P., O'Hare, W. T., and Islam, M. The Age Estimation of Blood Stains up to 30 Days Old Using Visible Wavelength Hyperspectral Image Analysis and Linear Discriminant Analysis. *Science and Justice* 53, no. 3 (Sep 2013): 270–277.

Mapes, A. A., Kloosterman, A. D., de Poot, C. J., and van Marion, V. Objective Data on DNA Success Rates Can Aid the Selection Process of Crime Samples for Analysis by Rapid Mobile DNA Technologies. *Forensic Science International* 264 (Jul 2016): 28–33.

Mapes, A. A., Kloosterman, A. D., van Marion, V., and Depot, C. J. Knowledge on DNA Success Rates to Optimize the DNA Analysis Process: From Crime Scene to Laboratory. *Journal of Forensic Sciences* 61, no. 4 (Jul 2016): 1055–1061.

Miles, H. F., Morgan, R. M., and Millington, J. E. The Influence of Fabric Surface Characteristics on Satellite Bloodstain Morphology. *Science and Justice* 54, no. 4 (Jul 2014): 262–266.

Osborne, N. K. P., Taylor, M. C., Healey, M., and Zajac, R. Bloodstain Pattern Classification: Accuracy, Effect of Contextual Information and the Role of Analyst Characteristics. *Science and Justice* 56, no. 2 (Mar 2016): 123–128.

Osborne, N. K. P., Taylor, M. C., and Zajac, R. Exploring the Role of Contextual Information in Bloodstain Pattern Analysis: A Qualitative Approach. *Forensic Science International* 260 (Mar 2016): 1–8.

Park, J. Y., and Kricka, L. J. Prospects for the Commercialization of Chemiluminescence-Based Point-of-Care and On-Site Testing Devices. *Analytical and Bioanalytical Chemistry* 406, no. 23 (Sep 2014): 5631–5637.

Peschel, O., Kunz, S. N., Rothschild, M. A., and Mutzel, E. Blood Stain Pattern Analysis. *Forensic Science Medicine and Pathology* 7, no. 3 (Sep 2011): 257–270.

Sant, S. P., and Fairgrieve, S. I. Exsanguinated Blood Volume Estimation Using Fractal Analysis of Digital Images. *Journal of Forensic Sciences* 57, no. 3 (May 2012): 610–617.

Sheppard, K., Cassella, J. P., Fieldhouse, S., and King, R. The Adaptation of a 360 Degrees Camera Utilising an Alternate Light Source (ALS) for the Detection of Biological Fluids at Crime Scenes. *Science and Justice* 57, no. 4 (Jul 2017): 239–249.

Stotesbury, T., Taylor, M. C., and Jermy, M. C. Passive Drip Stain Formation Dynamics of Blood onto Hard Surfaces and Comparison with Simple Fluids for Blood Substitute Development and Assessment. *Journal of Forensic Sciences* 62, no. 1 (Jan 2017): 74–82.

Sun, H. M., Dong, Y. F., Zhang, P. L., Meng, Y. Y., Wen, W., Li, N., and Guo, Z. Y. Accurate Age Estimation of Bloodstains Based on Visible Reflectance Spectroscopy and Chemometrics Methods. *IEEE Photonics Journal* 9, no. 1 (Feb 2017).

Taylor, M. C., Laber, T. L., Kish, P. E., Owens, G., and Osborne, N. K. P. The Reliability of Pattern Classification in Bloodstain Pattern Analysis, Part 1: Bloodstain Patterns on Rigid Non-Absorbent Surfaces. *Journal of Forensic Sciences* 61, no. 4 (Jul 2016): 922–927.

Taylor, M. C., Laber, T. L., Kish, P. E., Owens, G., and Osborne, N. K. P. The Reliability of Pattern Classification in Bloodstain Pattern Analysis, Part 2: Bloodstain Patterns on Fabric Surfaces. *Journal of Forensic Sciences* 61, no. 6 (Nov 2016): 1461–1466.

Thanakiatkrai, P., Yaodam, A., and Kitpipit, T. Age Estimation of Bloodstains Using Smartphones and Digital Image Analysis. *Forensic Science International* 233, no. 1–3 (Dec 2013): 288–297.

Vitiello, A., Di Nunzio, C., Garofano, L., Saliva, M., Ricci, P., and Acampora, G. Bloodstain Pattern Analysis as Optimisation Problem. *Forensic Science International* 266 (Sep 2016): E79–E85.

Wang, C., and Zhang, L. T. Simulations of Blood Drop Spreading and Impact for Bloodstain Pattern Analysis. *CMES—Computer Modeling in Engineering & Sciences* 98, no. 1 (Mar 2014): 41–67.

WEBSITES

Link	Description
http://www.iabpa.org/	International Association of Bloodstain Pattern Analysts
https://www.ameslab.gov/mfrc/bloodstain-pattern-analysis-video-collection	Excellent collection of videos on how bloodstain patterns form
http://www.swgstain.org/	Scientific Working Group on Bloodstain Pattern Analysis
http://www.forensicsciencesimplified.org/blood/	Simplified Guide to Bloodstain Pattern Analysis

Medicolegal Investigation of Death

5

In the last few chapters, we covered crime scenes, evidence, and documentation. In forensic science, crime scenes are often associated with one or more homicides. The crime scene investigation in such cases is also a death investigation. The next few chapters will focus on that topic, also referred to as the medicolegal investigation of death. We begin by discussing forensic pathology and the role forensic pathologists play in death investigation, most famously through the performance of an autopsy. Closely associated with the autopsy is forensic toxicology, specifically postmortem toxicology. The information provided by the autopsy is combined with toxicological data to determine the cause and manner of a death. Sometimes, the death occurs long before the body is discovered. In these cases, decomposition complicates the investigation and often the skills of forensic anthropologists are needed to study decomposed and skeletal remains. This section concludes with an exploration of how insects interact with bodies and how insect evidence can contribute to death investigation. This is forensic entomology, and as we will see, all four areas—pathology, toxicology, anthropology, and entomology—overlap to some extent. This chapter begins this part of the book with a discussion of death investigation and forensic pathology.

5.1 INVESTIGATION OF DEATH

The system of **medicolegal investigation of death** is triggered when a **questionable death (equivocal death)** is reported. A questioned death is one in which the circumstances and cause are not clear or obvious and which requires additional investigation. Many factors go into determining if a death is equivocal, such as age of the deceased, health, clinical history, and circumstances of the death. If an older person with a terminal illness dies in a hospice, this would not be a questionable death. However, if a young and healthy person dies in a hospital after an appendectomy, this might well be a questioned death. Investigations are needed to determine if this is the case, and what the implications are in terms of the legal system as well as public health. For example, if that young person died because of an antibiotic-resistant bacterium, many more people in that hospital are at risk. Unattended deaths frequently fall into the category of questioned deaths, as do those in which violence is obvious. The context of the death is critical in determining if it should be considered questionable and what type of investigation follows. A flowchart outlining a typical death investigation flow is shown in Figure 5.1.

As we will see shortly, death investigation systems vary across the United States. In many jurisdictions, a trained **medicolegal death investigator** (MDI) is the first investigative person to respond to a death scene, much as the crime scene investigator is the first forensic person to examine a crime scene. In homicide cases, it is not unusual for both CSIs and MDIs to be present at the scene, and the processes and procedures used by MDIs are like what was discussed in Chapter 3, except that the MDI will focus on the body as well as the scene. For example, the MDI will measure body temperature as well as environmental temperature and check for stiffening and other **postmortem** (after death) changes. MDIs also interview family and other key personnel as part of their work in understanding the context of the death. In some cases, the MDI will need to track down the identity of the deceased and may be responsible for notifying the next of kin. Gathering medical records and other data about the person is undertaken as well. If an autopsy is deemed necessary, the MDI will often coordinate transport of the body to the place where the autopsy will be completed.

As an examine, consider a case in which a man (45 years old) is found dead at his apartment by a friend who sees no sign of violence and thus asks for an ambulance and paramedics. These first responders declare the man dead and because the death was unattended, they call for an MDI. The death investigator is the person responsible for documenting and collecting evidence and working with other relevant personnel to determine if an autopsy is needed. It may turn out that the investigator finds many types of heart medications at the scene and a call to the decedent's doctor may reveal that he had a long history of heart problems. The friend might reveal that his friend had complained of shortness of breath a day before but refused to take medication. The context of this death supports the hypothesis that the man died of a heart problem, but because he was relatively young, an autopsy might be ordered to be sure this suspicion is correct.

DEATH INVESTIGATION AND PUBLIC HEALTH

We think of death investigation as being related to criminal activity, but death investigation arose as much from public health concerns as it did from public safety. A famous example occurred during an outbreak of cholera in London in 1854. Cholera is a bacterial disease associated with unsanitary conditions and sewage mixing with drinking water. However, this was not known in 1854 when an outbreak struck part of London. Dr. John Snow (1813–1858, shown below), a physician who lived near the heart of the outbreak, mapped the cases and talked with people in the neighborhood. He eventually tied the outbreak to a water pump in the center of the red area. Once the pump was disabled, the outbreak subsided. The map he created is shown below. Note the clustering of cases near the pump.

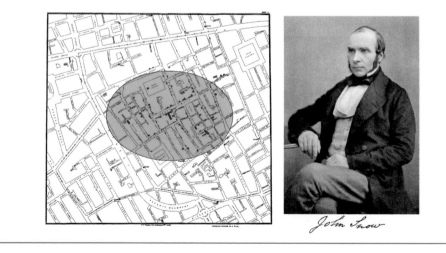

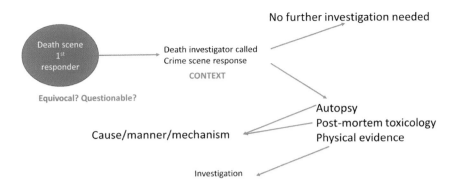

FIGURE 5.1   The flow of death investigation.

5.2 SYSTEMS OF DEATH INVESTIGATION

In many ways, the history of forensic science is the history of death investigation. From early in human history, the cause of a death was important in determining inheritance and taxes as well as any crime or public health issue. For example, death investigation was critical to identifying epidemics of infectious disease and still represents the first line of defense in such cases. Thus, death investigation has a public health role as well as a forensic one.

In English-speaking countries, the **coroner** is a government agent charged with responsibility for death investigations. The office of the coroner has existed in England since before the 10th century. These early coroners focused less on the possibility of murder and more on assessing taxes and the disposition of the property and wealth of a person who had died. The crowner of the king (the source of the word *coroner*) had many duties in medieval times, including the investigation of the causes of deaths. English law carried over to the American colonies, where death investigation was considered a local, governmental, or county function. States and other jurisdictions enacted their own coroner's laws, leading to a variety of systems across the country. Some coroners are appointed, some are hired as employees, and some are elected. In some jurisdictions, coroners are required to have medical training, but many do not have such a requirement.

5.2.1 MEDICAL EXAMINER SYSTEM

As medical science advanced, one of the weaknesses of the coroner system became apparent. While the coroner was charged with a quasi-judicial function and determined the causes of deaths, no specialized training was required. Massachusetts was the first state to license nurses, physicians, and lawyers, and in 1877, the state legislature passed a statute that replaced coroners with **medical examiners** (MEs). Such examiners were required to be licensed physicians. The ME system of death investigation was adopted by cities such as Baltimore, Richmond, Virginia, and New York around the time of World War I. Generally, this change was set in motion by local scandals arising from deaths that were improperly investigated by coroners. The establishment of the Office of the Chief Medical Examiner (OCME) in New York City in 1918 was a significant milestone, and over time, more jurisdictions adopted ME systems. As of 2011, the number of people in the United States served by MEs and those served by coroners were about evenly split.

MYTHS OF FORENSIC SCIENCE

Coroners and medical examiners are the same.

These two terms are often used interchangeably and incorrectly. An ME is a physician with additional professional training in pathology and forensic pathology. MEs are hired (appointed) to their positions in the same way other forensic scientists apply for and get a job in a forensic laboratory. They conduct autopsies and integrate information from the case to make the final determination of the cause, manner, and mechanism of death. They work in an ME system. Depending on the jurisdiction, the coroner may not have any medical training. In the coroner system, the coroner may be appointed or elected and the requirements for medical background vary widely. A funeral director can be a coroner, as can a sheriff, or someone with no background in forensic science, law enforcement, or death investigation. Where the coroner lacks the necessary medical training, the chances of a faulty death investigation and missed crimes increase. Unfortunately, where someone dies can lead to incorrect assignment of the cause and manner of death, which means murders may go undetected, accidents may be missed, and suicides may be incorrectly assigned.

5.2.2 FORENSIC PATHOLOGY AND FORENSIC PATHOLOGISTS

The key player in modern death investigation is the **forensic pathologist**. Forensic pathologists are physicians specializing in pathology, the diagnosis of disease, and who then subspecialize in the borderline area between law and medicine that emphasizes the determination of the cause of death. Pathologists began to appear in hospitals in Europe and the United States in the middle of the 19th century after advances in the use of microscopes to examine tissues from patients led to the employment of physicians who used these new methods. These doctors came to be called pathologists from the Greek *pathos*, meaning "suffering" or "disease," and *logos*, meaning "word" or "writing." Thus, a pathologist studies disease, its causes, and its diagnosis. Early pathologists examined tumors removed from patients to determine whether the tumors were cancerous. They also examined the bodies of deceased persons to determine the causes of death. By the middle of the 20th century, most pathologists specialized. Anatomic pathologists performed autopsies and examined tissues under microscopes. Clinical pathologists managed laboratories where body fluids were tested. Most of these physicians worked in hospitals.

The police and the coroners recognized that pathologists were needed to perform autopsies and determine the causes of deaths of people who died suddenly and unexpectedly. Thus, pathologists began doing autopsy examinations for the police, coroners, and MEs. By the end of World War II, the formal specialty of **forensic pathology** was recognized by the American Board of Pathology (ABP). Today, in most large cities in the United States, the ME is required to be a forensic pathologist. Forensic pathologists also handle autopsies for coroners in rural areas.

5.3 TIME, MANNER, AND CAUSE OF DEATH

The purpose of investigating a death is to determine the cause and manner of death, which are two different pieces of information that must be included in a death certificate. Drawing such conclusions requires several sources of information, including investigative, death scene information, case history, and autopsy results. The goal of death investigation is to determine the cause, manner, mechanism, and to the extent possible, date and time of the death.

5.3.1 CAUSE AND MECHANISM OF DEATH

The **cause of death** is the disease or injury that initiated the lethal chain of events, however brief or prolonged, that led to death. It is the underlying cause, the one event that ultimately caused a person to die, even though several complications and contributing factors may have been involved. The **mechanism of death** is a biochemical or physiologic abnormality produced by the cause of death that is incompatible with life.

JAMES BRADY (1940–2014)

James Brady was the press secretary for President Ronald Reagan. He was with the president on March 30, 1981, when he was shot in an attempted assassination. Brady was also shot in the head and was paralyzed. He advocated for gun control legislation for the rest of his life. He died in 2014 and the ME in Washington ascribed the manner of death as homicide given that he died of complications associated with the shooting, even though 33 years had passed from the injury to the death.

For example, suppose a middle-aged man is shot twice during a robbery and is taken to the hospital. He undergoes surgery to repair the organs damaged by bullets. The man's condition improves, but then he develops pneumonia followed by kidney, liver, and finally heart failure. An autopsy reveals that the man had preexisting severe lung and heart diseases. In this case, the *cause* of death was multiple gunshot wounds because those injuries set in motion the lethal chain of events. The *manner* of death was multiple organ failure, complicated by the fact that the man had preexisting heart and lung diseases, without which the man probably would have survived. The preexisting diseases were not the cause of death because injury takes precedence over disease in determining cause of death. In other words, injury trumps disease.

5.3.2 MANNER OF DEATH

A forensic pathologist also is called upon to determine the **manner of death**. There are hundreds or thousands of potential causes of death, but only four manners of death (or five, if undetermined is included). The four manners of death are *natural, accidental, homicidal,* and *suicidal.* Sometimes the acronym **NASH** is used to express the first four. Natural deaths are caused solely by disease, without the intervention of trauma. Accidental, homicidal, and suicidal deaths all involve trauma. Thus, the manner of death describes the circumstances of the death, not its cause. For the example in the previous section regarding the man who died because of the shooting, the cause of death was organ failure, the contributing factor was his preexisting conditions, and the manner of death was homicide. The difference between suicide and homicide is merely the person who acted. If the deceased took the action, the death is a suicide. If someone other than the deceased took the action, the death is a homicide.

5.3.3 TIME OF DEATH

Although we often think of death as an event, it is best described as a process. Currently, death is defined as cessation of brain activity or brain death. A person's heart may stop beating or they may stop breathing, but this alone does not mean they are dead because the condition could be reversible. Heartbeat can sometimes be restored using electric shock (defibrillation) and breathing can be restarted as well. If they are not restored, then brain death occurs. Even brain death is a process rather than an on/off situation. The brain slowly loses function as oxygen levels drop, but at some point, the damage is so severe that it is irreversible. The death of cells in the body follows, which is the beginning of the decomposition process.

Estimates of time of death are made based on knowledge of how and when predictable postmortem changes generally occur. The time that has elapsed since death is referred to as the

postmortem interval (PMI), and it is not possible to identify the exact time of death from this information alone, but understanding these processes make it possible to estimate it. Many of these and the rate at which they occur are affected by the environment (principally temperature), so forensic investigations must take this into account. The three key changes are **rigor mortis**, **livor mortis**, and **algor mortis**.

Rigor mortis is the stiffening of the muscles that occurs following death. This is a chemical reaction that occurs when the glycogen normally found in muscles is used up following death and is not reformed. Glycogen is used to provide energy for the contraction of muscles, and it is depleted slowly after death. Generally, rigor mortis is seen about 4 hours after death. It can occur sooner if the glycogen has been depleted by the exercise of muscles just before death, such as a victim running from an attacker. Rigor generally disappears during the period from 24 to 36 hours after death as further decomposition of the muscles leads to their loss of ability to remain fixed in rigor.

MYTHS OF FORENSIC SCIENCE

The time of death was 10:03 p.m.!

In all but a very few cases, the best that can be made is an estimate of the time of death, and in general, the more time that passes from the death to discovery of the body, the more difficult the estimate becomes. Circumstances in which a time of death is more certain would be if the death was witnessed by someone capable of discerning the difference between death and unconsciousness, a stopped watch (rare in the digital age), or some other factor. As an example, the picture shows the encased body of a person recovered from the city of Pompeii, which was destroyed when Mt. Vesuvius (near Naples, Italy) erupted in 79 AD. If we knew the time when the eruption occurred, the time of death of this individual could be stated with confidence. Short of such circumstances, the time of death has to be considered an estimate and is often presented as a range.

Livor mortis is the discoloration of the body that occurs from the settling of red blood cells after the blood stops circulating. This can be seen within minutes of death where the blood cells have an increased sedimentation rate from infectious or other disease. Generally, **lividity** develops within an hour or so after death. Lividity becomes fixed, meaning that finger pressure will not blanch the lividity, about 12 hours after death. Lividity slowly disappears with decomposition after 36 hours.

Algor mortis is the cooling of the body that occurs after death, assuming the ambient temperature is lower than the body temperature. The general rule of thumb for a nearly nude body exposed to 18°C–20°C is 1.5°C of temperature drop per hour for the first 8 hours. The normal body temperature is 37°. Thus, if a person has been dead for 4 hours, the temperature will be 31°. Unfortunately, many factors can cause body temperature to deviate from 37° at the time of death; in addition, the environmental conditions (heat, cold, etc.) all play a role in cooling. Thus, the rule of thumb is truly that—a rough estimate and nothing more.

5.4 TOOLS OF DEATH INVESTIGATION

The forensic pathologist, coroner, and others involved in a death investigation should have several sources of information available to help determine the cause, manner, and mechanism of a death.

Usually the final responsibility for making the call rests with the forensic pathologist. Although the autopsy is the best-known tool of death investigation, and one we will discuss in detail shortly, it is by no means the only tool used. Obtaining past medical history and understanding the issues raised by that history are important parts of the process of death investigation. Indeed, a medical history is generally the starting point of any investigation. For certifying cause of death, forensic pathologists do not recognize a statute of limitations for fatal injuries (see the sidebar above regarding James Brady). Careful study of medical records is required to properly determine the causes and manners of death of persons with histories of trauma.

Knowing what witnesses recall of the activities of the deceased prior to death or injury is important to a forensic pathologist. First, this information helps determine jurisdiction in cases where injury is not obvious. Also, since forensic pathology deals with recreating the circumstances of death, knowing what witnesses say happened is extremely valuable in developing questions to be answered. Understanding the contents of witness statements allows a forensic pathologist to know what questions will be asked. In the example of the middle-aged man we discussed above, the fact that his friend reported that the victim was experiencing symptoms before he died would be critical information for the death investigator.

Information from the scene is also important in providing context and information to the death investigator, and in determining whether a crime has been involved. If a person is found dead surrounded by drug paraphernalia, this is essential information. Similarly, if someone dies in their room and a scene investigation reveals an empty bottle of pills, this can suggest routes of investigation. Finally, because of the efficiency of the rescue squads in the United States and Europe, most persons who die of injuries have been treated for the injuries even if the decedent showed no signs of life. Treatment includes the insertion of needles, the creation of a small or large incision, and even bone fractures such as broken ribs from CPR. Although it is generally possible to discern between injuries produced after death and those produced before death, such distinction can be difficult when vigorous resuscitation takes place.

DR. CHARLES NORRIS (1868–1935)

Dr. Norris was the first appointed ME for the City of New York and a pioneer of forensic pathology. In addition to his medical skills, Norris was a New York native and grasped the problems faced by the enormous city. His tenure spanned a horrific explosion on Wall Street and the implosion of the stock market in 1927, resulting in many suicides. Gang warfare was rampant, and Prohibition caused problems for law enforcement as well as medical detectives. He is remembered particularly for his organizational and managerial skills; in effect, he pioneered the structure of ME offices.

5.5 AUTOPSY

The dissection of the human body to determine the cause of death has been practiced since the Early Middle Ages. **Autopsy** means to "look at oneself," so that term hardly seems appropriate. A more technically correct term for the dissection is **necropsy**, or "looking at the dead." In the United States, the term *autopsy* generally refers to human dissection, while *necropsy* refers to the dissection of animals.

5.5.1 AUTOPSY PROCESS

Autopsy examinations generally entail the removal, through incisions, of the internal organs of the chest, abdomen, and head. Typically, an incision is made beginning at each shoulder, extending to the midline of the body in the lower chest, and extended to the top of the pubic bone (Figure 5.2).

Examination of the brain entails an incision from behind one ear to behind the other ear, reflection of the scalp by peeling it upward and backward, and then sawing of the skull in a circular cut, followed by removal of the skull cap (Figure 5.3). The brain is sometimes dissected immediately, or it may be put in a solution of formaldehyde for a week to "fix" the tissue for better dissection and examination. Fixation is a chemical process that causes proteins to harden; it preserves the tissue and prevents further decomposition.

After removal, organs are weighed and then dissected to determine disease or injury. Figure 5.4 shows a heart removed from a body in preparation for dissection. Additional dissections may be done but are not generally considered part of a routine autopsy. For example, dissection of the spine and removal of the spinal cord may be required in a case involving possible spinal injury. Occasionally, especially in suspected child abuse cases, a posterior neck dissection is done, to show injury to the muscles, ligaments, and spinal cord.

5.5.2 DOCUMENTATION AND SPECIMENS

Proper documentation of an autopsy and its findings is as important to death investigation as it is to crime scene investigation. In addition, the pathologist must collect the samples needed by other forensic professionals, such as toxicologists and DNA analysts. Toxicological sampling is the most extensive. In most forensic autopsies, specimens are removed for toxicology testing, the subject of Chapter 6. The usual method of obtaining urine is shown in Figure 5.5. A syringe and needle are used to remove urine for testing.

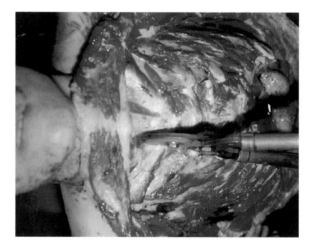

FIGURE 5.2 Typical initial incision during an autopsy.

FIGURE 5.3 Cutting the skull.

FIGURE 5.4   Removal of the heart.

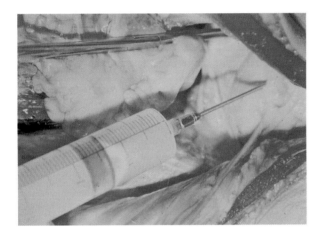

FIGURE 5.5   Removal of urine from the bladder using a syringe.

Blood is usually taken from the aorta and sometimes from large veins. Some drugs redistribute in the postmortem (after death) period and venous blood is considered more reliable than heart or aorta blood for many drugs. We discuss postmortem redistribution in more detail in Chapter 6. Bile is taken from the gallbladder. Blood and urine are routinely used to determine the presence of common drugs of abuse. Alcohol is generally measured in the blood. Small portions of the internal organs are put into a solution of formaldehyde to "fix" them and preserve them for further study using microscopy. Samples for DNA testing and reference are also collected.

5.6 INVESTIGATION OF TRAUMATIC DEATH

Death caused by trauma (injury) can be natural, accidental, suicidal, or homicidal, but whatever the underlying manner, traumatic death is frequently the subject of a death investigation. Types of trauma include mechanical, thermal, and electrical. A mechanistic classification termed asphyxial death overlaps the other causes. Asphyxial death is caused by interference with the oxygenation of the brain. This asphyxia can occur from mechanical causes (strangulation), drug or poison (cyanide poisoning), and electrical causes (low-voltage electrocution).

5.6.1 MECHANICAL TRAUMA

5.6.1.1 SHARP FORCE INJURY

Mechanical trauma occurs when applied physical force exceeds the tensile strength of the tissue to which the force is applied. Sharp force refers to injuries received from sharp implements, such as knives, swords, and axes. The amount of force required for a sharpened instrument to exceed the tensile strength of tissue is significantly less than the force required with a blunt object.

Blunt objects produce **lacerations** and sharp objects produce **incised wounds**. Examining a wound allows one to know whether a sharp or blunt object caused the wound. Figure 5.6a shows an incised wound with clean edges, and Figure 5.6b shows a laceration with torn and uneven edges. Figure 5.6c shows representative stab wounds and incised wounds of the neck. A stab wound is produced by a sharp object whose longest dimension is depth as opposed to other surface dimensions. It is difficult to determine the size of a sharp object from examination of the characteristics of the wound. Death from blunt and sharp trauma arises from multiple mechanisms, but sharp trauma most commonly causes death by exsanguination (bleeding to death).

5.6.1.2 FIREARMS

Firearm injuries are the most common suicidal and homicidal wounds seen in the United States because of the availability of firearms and the lethality of these weapons. Figure 5.7 is an x-ray of a person who received a high-speed projectile injury to the right upper chest. The dark area is missing tissue. Around the missing tissue are white fragments of lead, often called a **lead snowstorm**, that are diagnostic of a high-speed projectile. High-speed projectiles are only seen in wounds inflicted by high-powered hunting rifles and military rifles.

Often, a key piece of needed information to understand the circumstances of a shooting is the distance from the shooter to the victim. When a firearm is discharged, the bullet is ejected from the barrel along with hot expanding gases, unburned propellant, and other residues. How far each component travels is the basis for determining the distance of the barrel from the deceased at the time the weapon was discharged. The gases, including the heavy metals and some smoke from unburned but gaseous carbon, are projected only a few inches. The effects of the gas produce what can be discerned as **contact** or near-contact wounds. These wounds show blackening of the skin and laceration because the gas blown into the wound tears the skin apart. Figure 5.8a shows a typical contact gunshot wound.

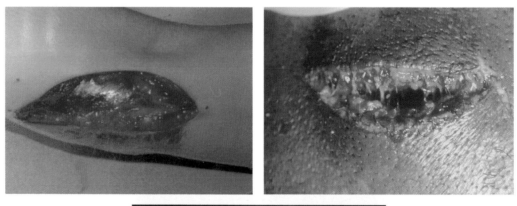

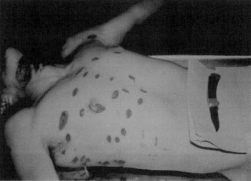

FIGURE 5.6   Top left: An incised wound. Top right: Laceration, Bottom: Stab wounds to the torso and an incised neck wound.

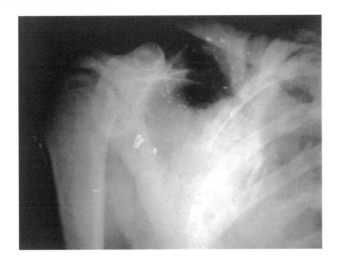

FIGURE 5.7   Massive destruction caused by a high-speed bullet; this pattern is referred to as a lead snowstorm.

Figure 5.8b shows a contact gunshot wound of the head. Because of the tearing characteristics of the scalp and the reflection of gases by the skull, large lacerations are characteristic of head contact wounds. As the distance from the barrel to the skin increases, the effect of the gas diminishes and only the unburned powder and bullet can penetrate the skin. Unburned powder that penetrates the skin produces **stippling** or **tattooing** around the defect produced by the bullet (Figure 5.8c).

5.6.1.3 OTHER BLUNT FORCE INJURY

Most common blunt force injuries are from collisions, usually motor vehicle collisions. Generally, except for gunshot wounds, homicidal blunt force trauma in an adult requires a lethal head injury.

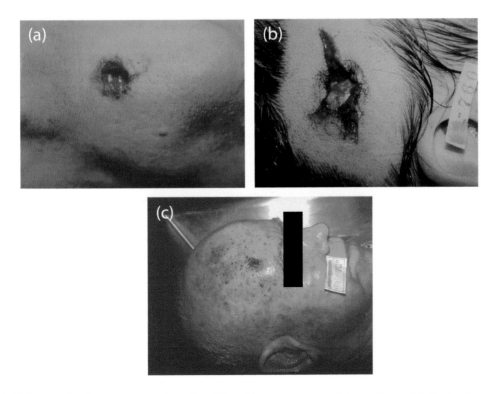

FIGURE 5.8   (a) Contact wound to the chin with soot and small lacerations. (b) Contact wound. (c) Example of stippling.

Injuries to other areas of the body rarely produce death. In children, lethal battery is most commonly due to head injury, but chest and abdominal traumas with laceration of internal organs, such as the spleen, liver, and heart, are also seen.

DR. BARNARD SPILSBURY (1877–1947)

The United States had Dr. Charles Norris and England had Dr. Bernard Spilsbury. His career was the subject of several books, including *The Scalpel of Scotland Yard: The Life of Sir Bernard Spilsbury*. His public life was one of fame, his private later life one of tragedy. Two of his sons, also doctors, died during the war period. He died by committing suicide using carbon monoxide.

Spilsbury worked many famous cases in his tenure as a forensic pathologist. He became famous based on the 1910 case involving another physician. The accused was a fellow physician named Crippen whose wife had been reported missing in early January.

The badly decomposed body was found in July, buried in their cellar. His work was hampered because the killer had removed the head, bones, limbs, and sexual organs from the body. All Spilsbury had to work with was a decomposed torso. Analysis of the tissue revealed the presence of a compound called hyoscine (scopolamine). This drug is a naturally occurring alkaloid that is readily absorbed through the skin, but it is not as toxic as other poisons common in the day. Spilsbury's testimony centered on the challenging task of identification, which he was able to do using a scar found on the abdomen. Crippen was convicted. In 1923, after several more high-profile cases and convictions, he became Sir Bernard Spilsbury. He continued his public service role throughout his lifetime.

5.6.2 ASPHYXIA

Asphyxia occurs when the oxygen supply to the brain is cut off. A familiar example is drowning, which kills several thousand people a year. Drowning is death by asphyxiation from immersion in water or other liquid. Some deaths from immersion are not asphyxial and result from hypothermia. Exposure of a person to water temperatures below 20°C (68°F) will result in death from hypothermia after exposure of many hours. Exposure to water temperatures near 0°C (32°F) will produce death in a matter of a few minutes.

A drowning victim typically attempts to keep his or her head above water to breathe. When this becomes difficult, he or she struggles to maintain the airway, and this increases the need for oxygen. Inhalation of water adds to the excitement. Water that enters the back of the throat is reflexively swallowed. This transmits the negative pressure associated with trying to inhale water to the middle ear via the Eustachian tubes that open during swallowing. The swallowed water enters the stomach, and this can be detected at autopsy, as shown in Figure 5.9. Further efforts to breathe cause water to enter the upper air passages, triggering coughing and additional reflex inhalation.

As the water enters the smaller air passages, the lining muscles go into spasm, thus protecting the alveoli or small air sacs from the entry of anything but air. The spasms create the equivalent of a severe acute asthma attack that traps air in the lungs. Loss of consciousness generally occurs within 1–2 minutes of the onset of the struggle, although consciousness may be prolonged if some air can be obtained. Loss of consciousness may be followed by involuntary inhalation

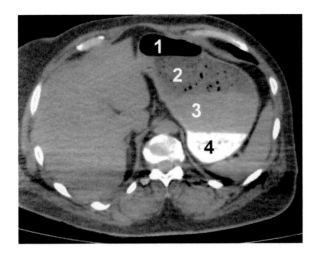

FIGURE 5.9   Postmortem CT scan of a drowning victim. This is a cross section of the abdomen that shows areas of (1) air, (2) froth, (3) fluid, and (4) debris. (Reproduced from Plaetsen, S. V., et al., Post-Mortem Evaluation of Drowning with Whole Body CT, *Forensic Science International* 249 [Apr 2015]: 35–41. With permission from Elsevier.)

attempts and vomiting. Heart cessation occurs a few minutes later. While the heart continues to beat, the pressure that the heart produces in the circulation of the lungs increases greatly, and the right side of the heart dilates from the increased pressure and perhaps from the increase in blood volume from absorbed water from the lungs.

The autopsy findings from drowning will vary depending on whether the drowning event followed the full sequence of events described above. If a person is unconscious when he enters the water, many effects of the excitation phase are not seen because an unconscious person cannot become excited.

The excitation results in transmittal of the negative pressure from the upper airway to the middle ears. The negative pressure along with other asphyxial changes in blood clotting factors results in hemorrhage into the mastoid air sinuses. In addition, water and materials in the water will be found in sinuses and in the stomach (Figure 5.9).

Asphyxia has many other possible causes, including **manual strangulation** (with the hands) and strangulation by **ligature** used to tie around the neck. Manual strangulation constricts the airway by compressing the neck. In some instances, this results in fracture of the **hyoid bone**, which is a fragile bone in the throat. Ligature strangulation, whether from hanging or garroting, characteristically does not involve fracture of the hyoid. Generally, the only findings are asphyxia and the presence of a furrow in the neck.

5.7 VIRTUAL AUTOPSY

As we have seen in other areas of forensic science, computers and digital information are increasingly being integrated into forensic pathology and death investigation. X-ray imaging has long been routine in autopsy procedures (e.g., Figure 5.7) but with the development of other scanning techniques, such as computer-aided tomography (**CT** or cat scans) and magnetic resonance imaging (**MRI**), forensic pathologists now have many more tools available. This type of imaging allows for three-dimensional (3D) reconstructions of the body and easier visualization of wound tracks and traumatic injury. A **virtual autopsy** is one in which noninvasive imaging is used to obtain the same types of information as the types of dissection we have been discussing throughout this chapter.

Currently, the imaging techniques are often used in conjunction with traditional autopsy methods, but there are other situations in which a purely virtual autopsy is ideal. For example, there are many cultures and religions for which a traditional dissection autopsy is forbidden. In others, such as trauma due to accidents, imaging could be sufficient.

In Figure 5.7, an x-ray film showed the damage associated with a gunshot wound to the shoulder. A similar type of image is shown in Figure 5.10. In this case, a homicide, the final position of the bullet plus fragments can be seen in the x-ray, along with a skull fracture pattern. Figure 5.11, from the same study, shows how data from CT and MRI images can be combined to show a bullet trajectory.

Virtual autopsy methods are also valuable in cases of blunt force trauma, such as a motorcycle accident (Figure 5.12). Here, 3D images are recreated from the postmortem imaging that clearly show the extensive damage to the victim's skull, including a fracture at the base of the skull that runs completely around it. As these methods and instruments become less expensive and more widely available to forensic pathologists, the use of virtual autopsy will inevitably expand.

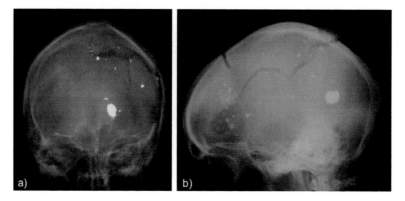

FIGURE 5.10   A 3D reconstruction showing a bullet trajectory. (Reproduced from Peschel, O., et al., Postmortem 3-D Reconstruction of Skull Gunshot Injuries, *Forensic Science International* 233, no. 1–3 [Dec 2013]: 45–50. With permission from Elsevier.)

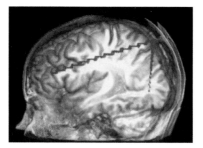

FIGURE 5.11   A 3D reconstruction showing the bullet trajectory. (Reproduced from Peschel, O., et al., Postmortem 3-D Reconstruction of Skull Gunshot Injuries, *Forensic Science International* 233, no. 1–3 [Dec 2013]: 45–50. With permission from Elsevier.)

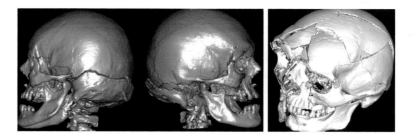

FIGURE 5.12   A 3D rendering of the skull of a motorcycle accident victim. (Reproduced from Moskala, A., et al., The Importance of Post-Mortem Computed Tomography (PMCT) in Confrontation with Conventional Forensic Autopsy of Victims of Motorcycle Accidents, *Legal Medicine* 18 [Jan 2016]: 25–30. With permission from Elsevier.)

5.8 RETURN TO THE SCENE OF THE CRIME

Refer to the crime scene presented in Chapter 3 (Figures 3.2 and 3.3 and the insert). There are several key questions that can be answered through doing an autopsy. The first would be to establish an identification. Depending on the degree of decomposition, fingerprints might be one method used, although in this example scenario, where decomposition is advanced, discernible patterns on the fingers may be unrecoverable. In addition, if the victim has not been fingerprinted in the past, there may not be reliable reference prints for comparison. A blood/tissue sample for DNA will be important for many reasons, such as comparing to blood found at the scene. Another tool that can be used for identification is dental records. Teeth are among the toughest materials in a human body and are not subject to any significant decomposition over a time frame that would be relevant.

After identification, the goal of the autopsy and other testing conducted at the ME's office would be to determine the cause, manner, and mechanism of death. In the scenario we are using, there is some question as to whether a shot was fired at the house. The autopsy will tell if the woman was shot or not and if she was shot, if that was the cause of death. Another critical piece of information would be an estimate of the time of death, which is challenging when decomposition is advanced. Fortunately, there are other tools available, which will be discussed in chapters to come. For now, assume the following information is revealed by the autopsy:

> The ME conducts a thorough autopsy and determines through DNA analysis that the deceased is the wife of the couple in question. The forensic pathologist finds evidence of blunt force trauma around the head. The report also notes abrasions and bruising on the face and arms that were likely inflicted near the time of death (perimortem). The is no evidence of a gunshot wound and no livor or rigor mortis. Decomposition was advanced, and it was not possible to determine stomach contents. No signs of diseases such as heart disease were found. Samples were collected and sent to the toxicology lab for further analysis.

5.9 REVIEW MATERIAL: KEY CONCEPTS AND QUESTIONS

5.9.1 KEY TERMS AND CONCEPTS

Algor mortis
Asphyxia
Autopsy
Cause of death
Contact wound
Coroner
Equivocal death
Forensic pathologist
Hyoid bone
Incised wound
Laceration
Lead snowstorm
Ligature
Livor mortis/lividity
Manual strangulation
Mechanical trauma
Mechanism of death
Medicolegal death investigator
Medicolegal investigation of death
Necropsy
Perimortem

Postmortem
Postmortem interval
Questioned death
Rigor mortis
Stippling
Tattooing
Virtual autopsy

5.9.2 REVIEW QUESTIONS

1. What is the difference between cause of death and manner of death?
2. What is the key difference between a medical examiner and a coroner?
3. If a weapon such as a brick is used to inflict trauma, would you predict that the wounds that result are incised or lacerations? Why?
4. What would an MDI do at a crime scene that is different from what a CSI would do? Where would their jobs overlap?
5. Why is medical history important in death investigation?
6. Compare the different features of a contact gunshot wound with a wound resulting from a shot fired several feet away from the victim.

5.9.3 ADVANCED QUESTIONS AND EXERCISES

1. Scenario: A couple is out on a boating excursion on a family boat. They anchor in a small cove. In the morning, the husband's body is found floating some distance away from the boat. The water was relatively warm, and the husband was not a good swimmer. Two theories are developed by investigators: (1) The man was intoxicated, passed out, fell overboard, and drowned. (2) The man was intoxicated but conscious and the wife forced him into the water after a brief struggle. What evidence could be gathered from the autopsy that could help sort out which theory, if either, is the most plausible? Justify your answer based on this and earlier chapters.
2. Use the same scenario except assume that the water temperature is ~40°F.
3. In the example scenario laid out in Chapter 3, do the autopsy findings tend to support or refute the two competing versions of the event?

BIBLIOGRAPHY AND FURTHER READING

BOOKS

Catanese, C. Color Atlas of Forensic Pathology. 2nd ed. Boca Raton, FL: CRC Press, 2016. ISBN: 978-1466585904.
Maloney, M. S. Death Scene Investigation. Boca Raton, FL: CRC Press, 2012. ISBN: 978-1439845905.
Wayne, J. M., Schandl, C. A., and Presnell, S. E. Forensic Pathology Review: Questions and Answers. Boca Raton, FL: CRC Press, 2017. ISBN: 978-1498756389.

ARTICLES

Advenier, A. S., Guillard, N., Alvarez, J. C., Martrille, L., and de la Grandmaison, G. L. Undetermined Manner of Death: An Autopsy Series. Journal of Forensic Sciences 61 (Jan 2016): S154–S158.
Amadasi, A., Mazzarelli, D., Merli, D., Brandone, A., and Cattaneo, C. Characteristics and Frequency of Chipping Effects in Near-Contact Gunshot Wounds. Journal of Forensic Sciences 62, no. 3 (May 2017): 786–790.
Blessing, M. M., and Lin, P. T. Identification of Bodies by Unique Serial Numbers on Implanted Medical Devices. Journal of Forensic Sciences 63, no. 3 (May 2018): 740–744.

Cecchetto, G., Bajanowski, T., Cecchi, R., Favretto, D., Grabherr, S., Ishikawa, T., Kondo, T., et al. Back to the Future—Part 1. The Medico-Legal Autopsy from Ancient Civilization to the Post-Genomic Era. *International Journal of Legal Medicine* 131, no. 4 (Jul 2017): 1069–1083.

Ferrara, S. D., Cecchetto, G., Cecchi, R., Favretto, D., Grabherr, S., Ishikawa, T., Kondo, T., et al. Back to the Future—Part 2. Post-Mortem Assessment and Evolutionary Role of the Bio-Medicolegal Sciences. *International Journal of Legal Medicine* 131, no. 4 (Jul 2017): 1085–1101.

Fleischer, L., Sehner, S., Gehl, A., Riemer, M., Raupach, T., and Anders, S. Measurement of Postmortem Pupil Size: A New Method with Excellent Reliability and Its Application to Pupil Changes in the Early Postmortem Period. *Journal of Forensic Sciences* 62, no. 3 (May 2017): 791–795.

George, A. A., and Molina, D. K. The Frequency of Truly Unknown/Undetermined Deaths: A Review of 452 Cases over a 5-Year Period. *American Journal of Forensic Medicine and Pathology* 36, no. 4 (Dec 2015): 298–300.

Hlavaty, L., and Sung, L. Applying the Principles of Homicide by Heart Attack. *American Journal of Forensic Medicine and Pathology* 37, no. 2 (Jun 2016): 112–117.

Karakasi, M. V., Nastoulis, E., Kapetanakis, S., Vasilikos, E., Kyropoulos, G., and Pavlidis, P. Hesitation Wounds and Sharp Force Injuries in Forensic Pathology and Psychiatry: Multidisciplinary Review of the Literature and Study of Two Cases. *Journal of Forensic Sciences* 61, no. 6 (Nov 2016): 1515–1523.

Kirchhoff, S. M., Scaparra, E. F., Grimm, J., Scherr, M., Graw, M., Reiser, M. F., and Peschel, O. Postmortem Computed Tomography (PMCT) and Autopsy in Deadly Gunshot Wounds—A Comparative Study. *International Journal of Legal Medicine* 130, no. 3 (May 2016): 819–826.

Makino, Y., Yokota, H., Nakatani, E., Yajima, D., Inokuchi, G., Motomura, A., Chiba, F., et al. Differences between Postmortem CT and Autopsy in Death Investigation of Cervical Spine Injuries. *Forensic Science International* 281 (Dec 2017): 44–51.

Morgan, L. O., Johnson, M., Cornelison, J. B., Isaac, C. V., deJong, J. L., and Prahlow, J. A. Autopsy Fingerprint Technique Using Fingerprint Powder. *Journal of Forensic Sciences* 63, no. 1 (Jan 2018): 262–265.

Oliver, W. R. Effect of History and Context on Forensic Pathologist Interpretation of Photographs of Patterned Injury of the Skin. *Journal of Forensic Sciences* 62, no. 6 (Nov 2017): 1500–1505.

Prahlow, S. P., Arendt, A., Cameron, T., and Prahlow, J. A. Accidental Trauma Mimicking Homicidal Violence. *Journal of Forensic Sciences* 61, no. 5 (Sep 2016): 1250–1256.

Rowbotham, S. K., and Blau, S. Skeletal Fractures Resulting from Fatal Falls: A Review of the Literature. *Forensic Science International* 266 (Sep 2016): 582.e1–582.e15.

van Daalen, M. A., de Kat, D. S., lo Grotebevelsborg, B. F., de Leeuwe, R., Warnaar, J., Oostra, R. J., and Duijst-Heesters, W. L. M. An Aquatic Decomposition Scoring Method to Potentially Predict the Postmortem Submersion Interval of Bodies Recovered from the North Sea. *Journal of Forensic Sciences* 62, no. 2 (Mar 2017): 369–373.

Yang, M. Z., Li, H. J., Yang, T. T., Ding, Z. J., Wu, S. F., Qiu, X. G., and Liu, Q. A Study on the Estimation of Postmortem Interval Based on Environmental Temperature and Concentrations of Substance in Vitreous Humor. *Journal of Forensic Sciences* 63, no. 3 (May 2018): 745–751.

WEBSITES

Link	Description
http://www.thename.org/	National Association of Medical Examiners
http://britannia.com/history/coroner1.html	History of coroners in Britain
https://www.nlm.nih.gov/visibleproofs/galleries/cases/examiner.html	History of death investigation and the difference between coroners and MEs
http://www.abmdi.org/	American Board of Medicolegal Death Investigators

Postmortem Toxicology

6

6.1 TOXICOLOGY AND DEATH INVESTIGATION

There are two forensic disciplines that work with drugs and poisons—toxicology and seized drug analysis. Both use many of the same analytical tools, but their task differs significantly. The primary job of forensic toxicologists is the analysis of biological samples such as blood and tissue for the presence of drugs, poisons, and their metabolites. Seized drug analysts work with physical evidence such as pills and powders and are charged with identifying illegal drugs. In the context of death investigation, the application of toxicology is called **postmortem toxicology**, in which the focus is on determining what a deceased person ingested, when they ingested it, how much they took, and whether this contributed to or caused the death. In many jurisdictions, a toxicology laboratory is in the same building as the medical examiner (ME) or coroner since the samples used in postmortem toxicology are collected during the autopsy. There are other branches of toxicology that are used in forensic science and other areas, such as workplace drug testing and the intoxication of drivers. We will explore all of those and seized drug analysis in Chapter 11. For now, we focus on toxicology as part of the medicolegal death investigation system (Figure 6.1).

Toxicologists work in concert with ME offices and coroners to assist in determining manner and mechanism of death. In any questioned death, samples are collected at autopsy for toxicological analysis in which toxicologists search for the presence of drugs or poisons. When present, these compounds may have directly caused the death, such as in an overdose. However, there are many cases in which compounds are found in the body that may have contributed to the death but not caused it. For example, in a fatal automobile crash it is not unusual to find elevated levels of alcohol in the driver's system. Intoxication by alcohol would be a contributing factor to the death, but the cause of the death would be trauma due to the accident. Finally, there are cases in which drugs are found in the system but do not contribute to the death, such as when a person has taken a daily prescription or a pain reliever like aspirin. The forensic toxicologist analyzes samples taken at autopsy and reports findings to the ME or coroner, who consider all data generated during a death investigation when deciding on the manner and cause of death.

MATHIEU ORFILA (1787–1853), SPANISH-FRENCH FORENSIC TOXICOLOGIST

Matthieu-Joseph-Bonaventure Orfila was born in Spain but moved to France as a medical student, where he worked and became professor of forensic chemistry and dean of the medical faculty at the University of Paris. He began publishing articles describing his work early in his career; his first paper on poisons appeared in 1814 when he was 26 years old. Orfila spent a good deal of time studying poisons, particularly arsenic. As a toxicologist, he concentrated on methods of analyzing poisons in blood and other body fluids and tissues. He became involved in the Lafarge case in 1840. Initial results of the analysis of Charles Lafarge's remains were negative for arsenic, but Orfila was eventually able to detect arsenic in the exhumed remains using the Marsh test. The dead man's widow, Marie Lafarge, was eventually convicted of poisoning her husband after a long and highly publicized trial. Orfila's testimony in the case was one of the earliest examples of sound scientific testimony by a recognized scientific expert in a court of law.

6.2 XENOBIOTICS

The word **xenobiotic** refers to substances that are foreign to the body. Aspirin is a xenobiotic because it does not occur naturally in our system. This is true of most drugs and poisons encountered in forensic science. A **drug** can be defined as a compound that causes a physiological effect. When you take an aspirin, your body responds in specific ways, including in processes that reduce inflammation (swelling) and fevers. In contrast, sugar is not a drug by this definition, nor is flour or cornstarch. A **poison** here is defined as a substance that when **ingested** (taken into the body) results in a toxic or damaging physiological effect or effects. Such substances are referred to as **toxins**. These definitions overlap because any substance taken in a large enough dose is toxic; that is, the dose makes the poison.

6.2.1 DOSES AND TIMING

The dose of any compound is an important consideration in determining the degree of toxicity, but other factors, such as the **mode of ingestion** (how the substance enters the body), how

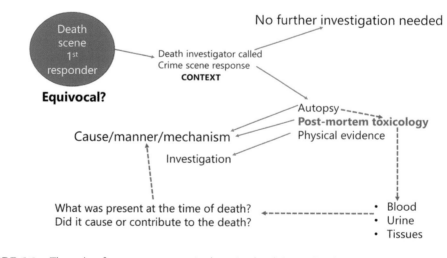

FIGURE 6.1   The role of postmortem toxicology in death investigation.

it is metabolized, the toxicity of the metabolites, and how long they stay in the body, also must be considered. For example, the common over-the-counter (OTC) drug Tylenol® is effective for relieving pain and inflammation with a recommended dose of two tablets, each of which contains 350 mg of the drug, for a total dose of 700 mg per two tablets taken orally. The dose is repeated in 4–6 hours as needed. Two tablets, or 700 mg, is the recommended **therapeutic dose**, which provides the desired degree of pain relief for most people as long as sufficient drug remains in the bloodstream. To maintain a therapeutic concentration in the bloodstream, the dose must be repeated every few hours, ideally on a schedule that keeps the level in the effective range. This pattern of periodic dosing is shown in Figure 6.2. Once the tablets are swallowed, it takes ~20–30 minutes for the drug to enter the bloodstream and arrive at the sites where it can induce the desired effect (pain relief). The amount of drug needed to relieve pain varies; the average value is called the ED_{50}, which corresponds to the effective dose for 50% of a population, and the range of effective doses are bracketed by the lowest (ED_{low}) and highest (ED_{high}) doses required across a large population.

MYTHS OF FORENSIC SCIENCE

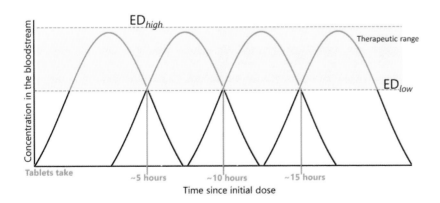

The most toxic substance in the world is …

Plutonium, botulism, nerve gas, arsenic, mercury, … Do an Internet search and you will find many different rankings and poisons listed as the most dangerous. As we have seen, any substance can be poisonous if enough is ingested. Therefore, to identify the most toxic substances, the qualifier *by weight* or *by dose size* should be included for clarification. Carfentanil (Figure 6.6) certainly qualifies as very toxic since a few micrograms can be fatal. For context, a typical large paperclip weighs about a gram (g). Divide that into 1000 pieces and each will weigh about a milligram (mg). Take one of those small fragments and cut it into 1000 more fragments and each will weigh about a microgram (µg). Polonium, a radioactive metal, can be fatal at a dose of nanograms (ng), which is a billionth of a gram, or the paperclip cut into a billion pieces. Botulism toxin can kill with similarly tiny doses. The toxicity of arsenic, perhaps the most famous poison, depends on how it is ingested and its chemical form.

FIGURE 6.2 Timing of doses to keep the blood concentration in the therapeutic range.

6.2.2 LETHAL DOSES

Death caused by drug overdoses has increased dramatically in the last 10 years and is now one of the most common types of deaths that trigger the death investigation system. In 2014, drug overdose deaths outnumbered firearms deaths in the United States, and in some places, the number of overdose deaths is severely straining the death investigation system. The trend for one drug (heroin) is shown in Figure 6.3. There are two types of dose patterns that can lead to an overdose—a single large dose (the most common in fatal cases) and repeated doses taken so closely together that the concentration builds up and exceeds a fatal dose.

Figure 6.4 shows a single fatal dose, using the same type of presentation as was used in Figure 6.2. On the left is a single dose in which the concentration of the toxin in the blood has exceeded the therapeutic levels (blue lines) and has moved above the concentration that is fatal to a small portion of the population (susceptible individuals). The middle frame dose would be fatal to half the population, and the dose shown on the right would be universally fatal. The dose that would kill half of all people who took it is called the **LD_{50}**, the lethal dose for 50% of a population. This does not mean that if someone took the LD_{50} of a drug that they would die; nor does it mean there is a 50% chance they would die. If 100 people took that dose, 50 of them would die. In any population, there are susceptible individuals and those that can tolerate higher doses.

The other way in which fatal concentrations are reached in the blood is shown in Figure 6.5. In this case, a person takes multiple doses, as was shown in Figure 6.2, without allowing time for the blood concentration to drop to a low enough level. The doses build on each other until the fatal dose is reached. In the figure, that point would be reached about 2 hours after the initial dose was taken.

The amount of a drug that constitutes a fatal dose varies widely and depends on factors that include body size and gender. LD_{50} is usually reported in units of milligrams of drug per kilogram of body weight (mg/kg). Often, the LD_{50} is not known for humans but is known for laboratory animals, such as rats or mice. It also depends on how the drug is ingested, a topic that is discussed in the next sections. Table 6.1 shows the approximate LD_{50} for several compounds and the equivalent lethal dose for an adult (150 pounds), a child (80 pounds), and an infant (12 pounds). The smaller the LD_{50}, the greater the toxicity of that substance.

WATER POISONING

Even a substance vital to life and considered safe can be poisonous if too much is ingested. Excessive water intake, particularly by athletes such as marathon runners, can cause a condition called hyponatremia. In such cases, excess water dilutes the blood to the point that the concentration of sodium (Na) drops below safe levels. As a result, cells swell, which if not treated can lead to coma and death. The lethal dose of water (ingested all at once) is about 6 L, or about 1½ gallons. Incidentally, *natrium* is the Latin name for sodium and is why the element is symbolized as Na.

The table shows the significant role of body weight in toxicity. Small aspirin tablets, often referred to as "baby aspirin," are found in many homes. A typical adult could take 14 tablets and end up with a stomach ache, while an infant ingesting that same dose could die.

One of the most serious problems facing law enforcement and forensic science is the startling increase in drug overdose deaths (Figure 6.3). The current scourge is heroin and closely related drugs. Heroin is derived from morphine, which is extracted from the seed pods of the opium poppy plant. Collectively, morphine, heroin, and other drugs obtained from opium are referred to as **opiates**. Another familiar opiate is codeine, which is particularly effective for suppressing

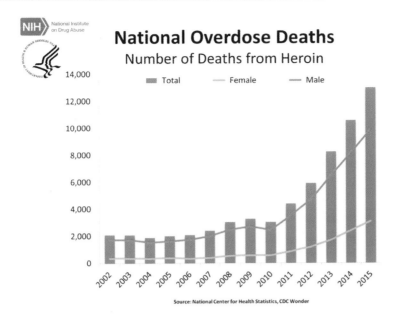

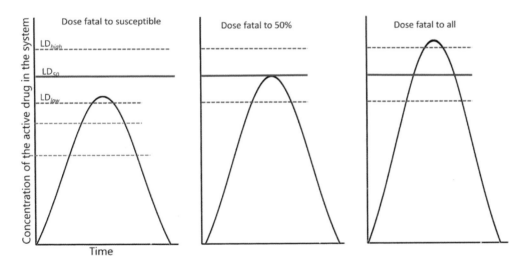

FIGURE 6.3   Heroin death increase from 2002 to 2015, the latest year for which data is available.

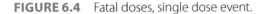

FIGURE 6.4   Fatal doses, single dose event.

coughs. Heroin abuse is not a new problem, dating back to the early 1900s. Heroin is ~3–10 times as potent as morphine, and thus more toxic.

Synthetic opiates (**opioids**) were developed shortly after World War II and include oxycodone, hydrocodone, and hydromorphone, all of which have led to overdose deaths. Currently, **fentanyl** and **carfentanil** are of significant concern because of their extreme toxicity. An example is illustrated in Figure 6.6. Fentanyl is ~10× as potent as heroin, while carfentanil (designed for use on large animals like elephants) is ~10× as potent as fentanyl. Unfortunately, drugs sold on the street as heroin may contain these more potent opioids, which often results in death to the unsuspecting user. The problem has become so severe that police officers, laboratory workers, and dogs used to sniff drugs can be exposed to lethal doses without realizing it. The dose indeed does make the poison.

6.2.3 ADME

The process by which drugs and toxins move through your body is referred to as **ADME** (absorption, distribution, metabolism, and excretion). ADME begins once a substance has been ingested,

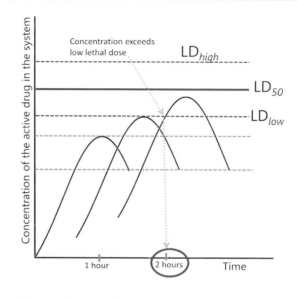

FIGURE 6.5   Repeated doses taken too close to each other.

TABLE 6.1   Comparison of Lethal Doses

Substance	Approximate LD$_{50}$ (mg/kg)	Adult Dose	Child Dose	Infant Dose
Sugar	30,000	4.5 pounds	2.4 pounds	0.4 pounds
Table salt	12,300	1.8 pounds	1 pound	0.1 pound
Water	3000	1.4 gallons	1 gallon	0.1 gallon
Tylenol	400	136 tablets (200 mg each)	73 tablets	11 tablets
Caffeine	300	70 cups of coffee	36 cups	5 cups
Aspirin	200	170 baby aspirin tablets	91 tablets	14 tablets
Heroin	0.2	13.6 mg	7.3	1.1
Fentanyl	0.02	1.4 mg	0.7	0.1
Carfentanil	0.002	0.1 mg	0.007	0.01
Botulism toxin	1.0×10^{-6}	7×10^{-5} mg	4×10^{-5}	5×10^{-6}

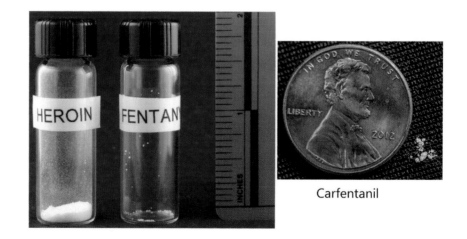

Carfentanil

FIGURE 6.6   Lethal doses of heroin, fentanyl, and carfentanil. (Images by the New Hampshire State Police and the Drug Enforcement Agency.)

although the specifics of that pathway vary depending on many factors. A generic ADME curve is shown in Figure 6.7. The origin of the plot is the time when the dose is ingested; the y-axis is the concentration of that substance in the bloodstream, and the maximum is referred to as C_p (peak plasma concentration). Prior to that point, the ingested substance is absorbed into the blood and, once in the blood, distributed through the system. How a drug is ingested is the key factor in determining how the AD portion of this curve will appear, and we discuss that in the next section. Once in the bloodstream, the drug is subject to metabolism, which changes the molecule and results in a decrease of drug concentration in the blood. Eventually, the drug is eliminated by excretion, usually in the urine.

There are many ways for drugs and poisons to enter the body besides swallowing, and how a substance is ingested can have a significant impact on its toxicity. This is referred to as the mode of ingestion or the **mode of administration**. The most common modes are shown in Figure 6.8. Overdoses can result from any mode of ingestion, but the ones of most forensic interest are swallowing, inhaling smoking, and snorting (substances absorbed through the nasal passage tissue). The differences between these modes of ingestion are significant.

When you swallow a substance such as an aspirin tablet, the drug enters the stomach and the digestive tract. However, if the drug is never absorbed into the bloodstream, it will never have the desired therapeutic effect. How and where a drug is absorbed in the digestive system depends on many factors, such as size, charge on the molecules, and their relative solubility in water and fats, such as found in the membrane that separates the intestines from the bloodstream. Once the substance has crossed the membrane, it enters a vein that delivers it to the liver, which is the primary site of metabolism. Here, the substance undergoes **first-pass metabolism**, the extent of which varies. The net effect of first-pass metabolism is to reduce the concentration of the drug in the bloodstream, and any drug that is ingested orally is subject to this process.

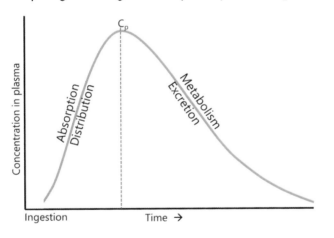
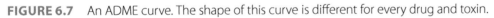

FIGURE 6.7 An ADME curve. The shape of this curve is different for every drug and toxin.

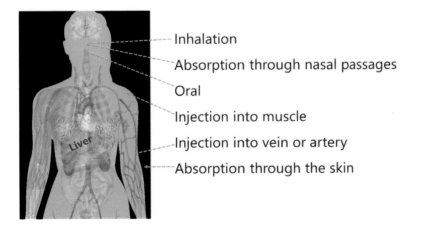

FIGURE 6.8 Common modes of ingestion. The liver is where most metabolism takes place.

The same is not true with other modes of ingestion. When a substance in ingested by smoking or injection into a vein, it avoids first-pass metabolism and the full amount injected is rapidly dispersed in the bloodstream. As a result, when a person smokes or injects a drug into a vein, the effect is felt within seconds. As demonstrated in Figure 6.9, the difference in modes of ingestion can have a fatal outcome.

A recent example involves overdose deaths due to the drug Oxycontin®. These tablets are designed to release the opioid (oxycodone) slowly over time after the pill is taken. The goal is to provide long-term and steady pain relief for people with conditions like cancer. The dose in the tablet is adjusted to account for losses due to first-pass metabolism. Abusers will crush the tablets, extract the oxycodone, and smoke or inject it all at once, defeating the time-release features. This mode of ingestion results in an immediate spike in blood concentrations that can easily exceed fatal doses. A fatal dose of oxycodone is in the range of 500 mg; an abuser who crushes six or seven tablets and injects or smokes the powder would ingest a potentially fatal dose:

$$\frac{500 \text{ mg}}{\frac{80 \text{ mg}}{\text{tablet}}} = 6.25 \text{ tablets} \tag{6.1}$$

6.2.4 MOVING THROUGH THE BODY

One of the reasons that several types of samples are collected at autopsy is that toxins move into different tissues. As we saw in Figure 6.7, the distribution phase is the time during which the ingested toxin is spread to various parts of the body by the blood. Since the liver is the primary location of metabolism, high concentrations of the drugs and metabolites may be found in that organ, depending on how much time has elapsed since the dose. Toxins and metabolites can also get into the urine. How much is found in each type of sample depends on how much of the toxin was ingested, how it was ingested, and how long the person lived after the ingestion. If death is rapid, gastric contents can contain undigested pills or tablets, for example; in contrast, if the death occurs hours after ingestion, little or none of the drug may be in the blood. The process is illustrated in a series of figures.

First, we can track the drug itself through different tissues of the body, as shown in Figure 6.10. In this example, a drug is ingested. It first enters the bloodstream and is distributed throughout the body. Some of the drug is filtered out in the kidneys, where the concentration of the drug increases as it decreases in the blood. In addition, some of the drug may be deposited in tissues such as the brain, hair, or liver. Note that we are discussing only the drug itself, and not any metabolites. Many things can happen to a drug molecule in the bloodstream—it can move into

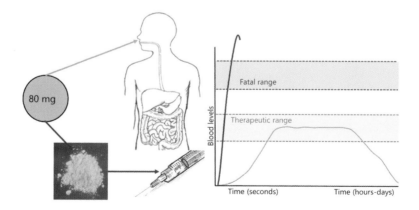

FIGURE 6.9   Effect of different modes of ingestion. When taken as directed, the pill provides hours of steady pain relief; when ingested all at once by smoking, the user immediately ingests the full dose with no loss to metabolism.

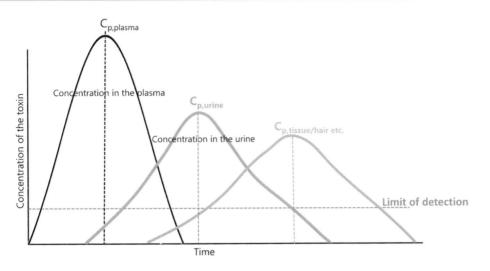

FIGURE 6.10 Movement of a toxin through different tissues. The net effect is a decrease in the blood concentration. The green line represents a concentration below which the drug could not be detected by a forensic analysis.

other tissues unchanged as shown here, or it may be metabolized to a different compound. The net effect is the same—the concentration in the blood goes down.

Next, we can track the drug and its metabolites in the bloodstream. The liver is the primary site of metabolism in the body. When a drug enters it, either directly after being absorbed in the digestive track or circulating after being ingested in another way, metabolism can occur. The parent drug concentration decreases while the concentration of the metabolite or metabolites increases. This is shown in Figure 6.11. The degree to which a toxin is metabolized varies, as does the number of metabolites and how long these compounds stay in the system. Thus, as the time since the dose increases, the concentration of drugs and metabolites in blood, urine, and other tissues changes. While complicated, knowing how a given toxin behaves can provide the forensic toxicologist with vital information on what was ingested, how much, and when. In effect, the "crime scene" for the toxicologist is the dose event, and their goal is to look at the evidence (concentration of toxins and metabolites in postmortem samples) and offer an opinion as to what happened and when.

Understanding this pattern is critical information to the forensic toxicologist. Suppose that a person dies and an autopsy is ordered because the cause and manner of death are not

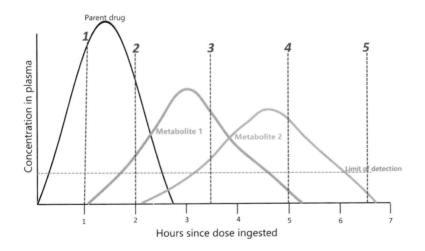

FIGURE 6.11 Blood concentrations of the ingested substance and metabolites compared with time since dose. Once a substances falls below the concentration indicated by the green line, it cannot be detected by forensic toxicological analysis.

clear. A blood sample would be taken as part of the autopsy and subject to toxicological analysis. Suppose that analysis revealed a high concentration of a drug, but no metabolites. This is evidence that the person died shortly after ingestion, before any metabolism took place. This would be consistent with an overdose death, but as always, this is only part of the story that would be revealed by the full death investigation. If no drug or metabolites were found, two possibilities arise: either the decedent never took that drug or the death occurred hours after the ingestion, so long after that none of the toxin or its metabolites could be detected (at about 6½ hours after ingestion in Figure 6.11). Similarly, at 2 hours past ingestion, the drug and one metabolite would be detected; at 3 hours, two metabolites, and at 4 hours, one metabolite.

A final example of this pattern is shown in Figure 6.12, this time showing the blood concentrations of heroin ingested by smoking. While not exact, the general shapes and times reasonably approximate a real dose. Because the drug is smoked, the AD phase of the curve is irrelevant; as soon as the vapor enters the lungs, it enters the blood and is distributed immediately to the rest of the body, including the brain. As it circulates through the liver, it is rapidly converted to 6-monoacetylmorphine (6MAM), which is itself rapidly converted to morphine. Another metabolite, morphine glucuronide (3MG), appears later and is eventually filtered into the urine, to be excreted. Thus, finding heroin in a postmortem blood sample is evidence that death occurred minutes after ingestion, while finding 3MG indicates that more time passed from dose to death.

ARSENIC POISONING AND THE MARSH TEST

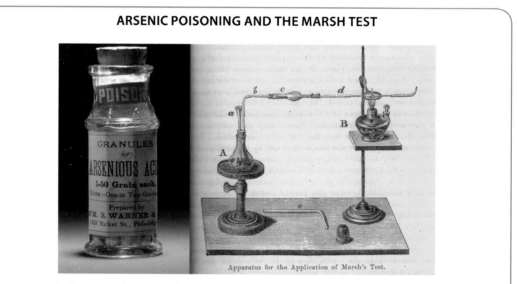

Left: Arsenic bottle. Right: Marsh test apparatus.

For literally thousands of years, arsenic was commonly used as a poison. Because it builds up in the system, clever poisoners could kill their victims gradually over weeks by giving small daily doses. As the arsenic built up, the victim would show symptoms such as nausea, vomiting, and diarrhea, which are common to many diseases, such as cholera and dysentery. Until relatively recently (the mid-20th century), these illnesses were a common cause of death, and as such, these generic symptoms were often mistakenly associated with a natural cause. In 1836, a Scottish chemist, James Marsh, published one of the first reliable tests for arsenic in tissues. The Marsh test relied on simple chemistry and no instrumentation and is primitive by modern standards, but in 1836 it represented a huge breakthrough. The test is illustrated in the figure. Tissue or samples such as stomach contents are added to a container, along with metallic zinc and an acid such as hydrochloric acid. Once heated, the arsenic, if present, is converted to a gas called arsine (AsH_3). The gaseous arsenic travels through the glass tube to another heated zone, where the arsine breaks down into metallic arsenic, which plates out and forms what was commonly called an "arsenic mirror." This was tangible physical evidence that could be shown to a jury. This test and its variants were used countless times in the decades to come to convict murderers and to eventually put an end to widespread use of metallic poisons.

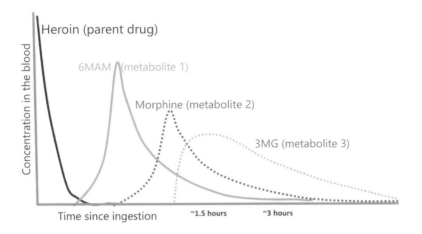

FIGURE 6.12   Fate of heroin ingested by smoking in terms of concentration in the bloodstream.

6.3 TYPES OF POSTMORTEM SAMPLES

Given how a drug or poison moves through the body after ingestion, many types of samples are collected from the body at autopsy. Some, such as blood and urine, are routinely taken, while others may be taken upon recommendation of the pathologist or toxicologist. The most common types of samples collected for postmortem toxicological analysis are described below. Not all samples may be available depending on the state of the body. For example, if a body is badly decomposed, it can be challenging to gather some types of samples.

6.3.1 BLOOD

Because the concentration of toxin present in blood often correlates more closely with lethal outcome than concentrations in other specimens, blood is the most important specimen in postmortem toxicology. At least two blood specimens are customarily collected in postmortem studies. One is taken from the heart (called **central blood**), and the second is taken from a peripheral site away from the heart, such as the femoral vein in the leg. This sample is referred to as **peripheral blood**. Having two different samples is important because the concentration of toxins and metabolites can change after death, a process called **postmortem distribution** (PMR). While the heart is beating and blood is circulating, the concentration of drugs in the bloodstream should be reasonably consistent no matter where the blood is sampled. However, after death, circulation and metabolism cease and substances in the blood can migrate based on such factors as solubility in fat compared with solubility in blood and similar **aqueous** (water-based) systems. Toxicologists can compare the concentration of substances in the central blood with that in the peripheral blood; if these concentrations are significantly different, PMR may have occurred. This can be important because the critical data needed from the toxicological analysis is the concentrations of toxins at the time death, not at the time the samples were taken.

6.3.2 URINE

Where possible, urine is collected. Although the correlation between drug concentration in urine and the effect on the user is not easily summarized, the data can be important in estimating the initial dose and what was taken. Urine should also be collected in postmortem investigation because certain toxins appear in urine in much larger quantities than amounts found in blood, and false-negative findings could be obtained if blood alone were tested. Figure 5.5 shows how urine is collected during the autopsy.

6.3.3 GASTRIC CONTENTS

Testing of **gastric contents** (material found in the stomach) may be beneficial in the case of the sudden death of a person who has copious quantities of a lethal agent in his stomach. For highly toxic substances, very low concentrations may be present in the blood, in contrast with substantial amounts in the stomach. If the manner of death is suicide, large amounts of drugs in the stomach may help to establish this conclusion.

6.3.4 VITREOUS HUMOR

This the gelatinous fluid found in the eyeball, and it can be a valuable sample for postmortem toxicological investigation. This specimen should be collected in postmortem investigation. Since the eye is an isolated bodily area, the vitreous humor is resistant to putrefaction. It is therefore sometimes possible to obtain a reliable measure of a biochemical or drug in vitreous humor at a time when the same biochemical has decomposed in the blood compartment. Vitreous humor may be the only fluid remaining in a decomposed cadaver, and assay of the vitreous may show chemical abnormalities that suggest the cause of death. Postmortem increases of certain substances such as potassium change over time in the vitreous humor, which can provide another tool for estimating the time of death.

6.3.5 BILE AND LIVER

The liver is the organ most heavily involved in drug metabolism. It is likely to contain significant quantities of most drugs and may, on occasion, permit the identification of an agent that caused death even when that substance cannot be found in the blood. Bile drains from the liver and is very rich in certain types of drugs, such as opiates.

6.4 METHODS OF ANALYSIS

Analysis of samples in forensic toxicology follows a common pattern in forensic science. Analysis often begins with a screening test that gives the toxicologist an idea of whether a drug or poison is likely present (or not present) and guides confirmatory testing. This is important in that it ensures that further laboratory analysis focuses on those samples where a drug or poison is likely present.

6.4.1 IMMUNOASSAY

The most widely used screening tests in forensic toxicology are **immunoassays**, which are based on the reaction between an antigen and an antibody. Immunological reactions occur between antigens and antibodies, which are large protein molecules. Your blood type is a description of the types of antigens found on the surface of your red blood cells. If you have type A, this means that there are A antigens on your red blood cells and B antibodies in your bloodstream. This is why it is so important to know blood type before a person is given a transfusion; giving B blood to a person with type A blood would result in catastrophic antigen–antibody reactions that can be fatal.

Antigen–antibody reactions are harnessed in many ways to test for the presence of drugs in postmortem samples. Drugs are small molecules compared with proteins, so left unmodified, a drug will not invoke an immunological reaction. Thus, the first step in an immunoassay scheme is to link the drug to a large protein molecule called an **immunoglobulin** to create a complex large enough to participate in an antigen–antibody reaction. This complex is called an immunogen because it will invoke the production of an antibody. This complex is then injected into laboratory animals to produce antibodies specific to that protein–drug complex.

Figure 6.13 depicts the components of one type of immunoassay referred to as enzyme-linked immunosorbent assay (ELISA). The assay is carried out in a well that is part of a much larger plate, allowing for the analysis of many samples simultaneously. The antibody adheres to the base of each well. In this type of assay, a sample such as urine collected at autopsy is added to the well, along with the target drug with a special label attached. The labeled drug and the drug itself, if present, compete for the binding sites. After a short incubation period, the plate is rinsed thoroughly. If there is no drug present in the sample, all the binding sites will be occupied with the labeled drug. In contrast, if the sample has a high concentration of the drug present, this unlabeled drug will occupy most sites. In other words, the more concentrated the drug, the more it will displace the labeled drug from the binding sites. After rinsing, only the drug form (labeled or unlabeled) will remain in the well.

The last step in the assay is to add a reagent that interacts with the labeled drug as shown in Figure 6.14. In this example, the reagent reacts with the label to generate a yellow color. The more label that remains in the well, the more intense the yellow. The intensity of the yellow is analyzed electronically and can be used to estimate the concentration of the drug in the original sample.

Immunoassay is an ideal screening method for forensic toxicologists because it is relatively simple and fast, requires minimal sample preparation, and is easily automated to manage large numbers of samples. A limitation of the method is **cross-reactivity**, and, in many cases, the inability to definitively identify the compound that is generating a positive result. Cross-reactivity refers to the ability of more than one drug to bind the antibody receptor, producing a positive result. This is partially due to how the immunogen is created; recall that the drug of interest is combined with a comparatively big protein to create a molecule large enough to generate an immune response. For this reason, many immunoassays target drug families rather than specific

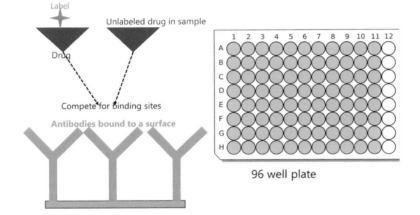

FIGURE 6.13   Components of an immunoassay.

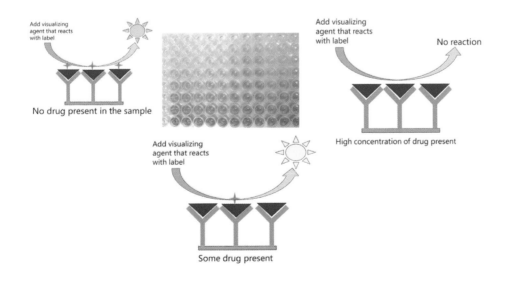

FIGURE 6.14   Outcomes of the assay.

compounds. A positive reaction using an opiate immunoassay could mean the presence of many drugs that are structurally similar or related, such as heroin, morphine, and others. The forensic toxicologist must consider these issues when interpreting the results of any immunoassay. Since the technique is used for screening, a positive result is used to direct further analysis, not to definitively identify a specific compound. A more significant problem is when an immunoassay fails to detect a compound that is present. This has become a concern as more and more new synthetic drugs enter the clandestine markets and has led to the development of other screening techniques to supplement immunoassay.

6.4.2 CONFIRMATORY ANALYSIS

After screening tests are completed, chemical instrumentation is used to determine what compounds are present and at what concentrations. The sample is prepared in the laboratory using many different techniques, depending on the sample type. Blood and tissue are complex samples that require many steps to remove as many extraneous materials as possible (fats, proteins, etc.), while urine is less difficult. The sample preparation step yields a small volume of solvent containing the target analytes. Even the most exacting sample preparation step cannot remove all matrix components or isolate a single compound, so the instrumental analysis typically requires two modules—the first designed to separate the analytes of interest from the matrix and concentrate them, and the second to provide a definitive identification of the compounds. The separation stage is based on chromatography, and the second is based on mass spectrometry (MS) (Figure 6.15).

The word *chromatography* means "color writing" and arises from how the technique was developed. Originally, it was used to separate different colored substances in plants. In chromatography, different compounds are separated based on how they interact with different phases. A phase is a solid, liquid, or gas. An example of chromatographic separation of different colored substances (Figure 6.16) illustrates how the process works. The first phase used is a glass plate coated with a fine silica powder (the solid phase). The second phase (here a liquid) is a solvent mixture (e.g., water and alcohol). A small sample of each ink is dotted on a line ~2 cm above the bottom (the pencil line in Figure 6.16c). The plate is put into the solvent (in a beaker) so that ink spots are above the solvent. The liquid phase is drawn up the plate by capillary action, which is the same reason that dipping a paper towel in water causes the water to be drawn into the towel.

After a few minutes, the solvent has moved up the plate as shown in Figure 16.6b. The components of the inks that have a greater affinity for the silica powder will move the least (purple dye in this example), while the compounds with greater affinity for the liquid phase will move farther (orange dyes here). Because each compound has different affinities for the two phases, they can

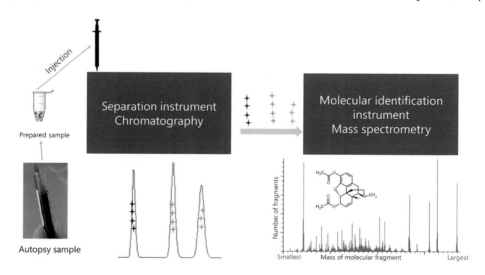

FIGURE 6.15 Schematic of a confirmatory analysis using sample preparation, chromatography, and MS.

(a) (b) (c)

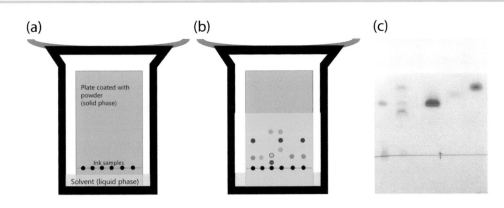

FIGURE 6.16 (a) Start of the analysis. (b) Separation based on different affinities. (c) Example of ink colors separated.

be separated. Figure 16.6c shows an actual separation of components in different inks by this chromatographic method.

In Figure 6.15, the chromatographic method used can be based on gas chromatography (GC) or high-performance liquid chromatography (HPLC). In both methods, the solid phase is inside a small tube inside the instrument and is called the column. In GC, the mobile phase is a gas, usually helium, while in HPLC different solvents can be used. When a prepared sample is injected into the instrument, the compounds separate based on the same mechanism as did the ink components, and as a result, they enter the detector (the mass spectrometer) individually (Figure 6.15).

With the inks, detection of the individual components is easy because you can see and identify colors. In other words, the detector for this type of chromatography is your eyes. In forensic toxicology, the specific chemical compounds must be identified with instrumentation. Identification of toxins and metabolites is confirmed using an instrumental technique called **mass spectrometry**. The basis of identification of a compound by MS is a study of how the molecule fragments under specific conditions. We won't discuss all the types of MS available to forensic toxicology, but only one example, as shown in Figure 6.17, which shows a mass spectrum of caffeine. The molecular weight of caffeine is 194 mass units. This molecule can be fragmented in many ways, depending on where it breaks. In the figure, one example break is shown; when a molecule is fragmented this way, one of those fragments will have a weight of 109 mass units. When a large number of caffeine molecules enter the mass spectrometer, some remain intact while some fragment, as shown in the figure; others fragment differently. As a result, a mass spectrum is a display of the mass of the fragments and how many of each type were detected. For caffeine, a large number never fragment, resulting in a large peak at mass 194; almost as many break down to

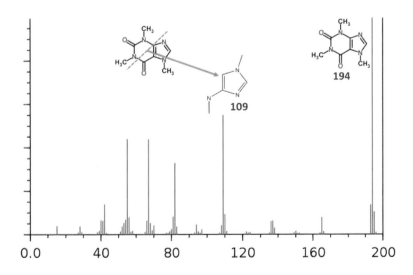

FIGURE 6.17 Mass spectrum of caffeine. The x-axis is mass and the y-axis intensity.

form the 109 fragments. If the conditions used to break up the molecules are consistent, then the mass spectrum of caffeine will always look like the one in the figure. Thus, if a postmortem blood sample was prepared and analyzed by gas chromatography linked to mass spectrometry (GC/MS), a mass spectrum of caffeine will be found and easily identified by comparing it to the known fragmentation pattern of caffeine.

Now we can return to Figure 6.15 and work all the way through it. A blood sample from an autopsy is prepared and injected into the GC instrument. The sample contains three drugs or poisons shown by the colored stars. The compounds separate in the same way as the inks on the glass plate (Figure 6.16), except that the mobile phase is a gas (helium) and is coated on the walls of a thin tube (the column). The compound symbolized with the black star emerges first from the column and enters the mass spectrometer in a concentrated package, where it is fragmented. The resulting mass spectrum is used to identify the compound. A few seconds later, the second compound enters the MS and the process is repeated, and repeated again for the third substance. In Figure 6.15, the example spectrum is that of heroin, which is a larger molecule than caffeine, and as a result, the mass spectrum has many more types of fragments that can be produced. However, if the instrumental conditions are constant, the spectrum of heroin will always be the same, even if analyzed at another laboratory.

SUICIDE BY CYANIDE POISONING

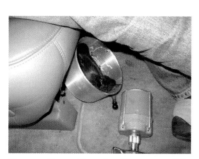

Image from the death scene. The heater is on the floor.

Acid burn damage of the epiglottis, a piece of cartilage that is located near the base of the tongue.

Cyanide is a toxic substance and poison that can be ingested by swallowing or by inhalation of the gaseous form, hydrogen cyanide (HCN).

A report from 2011 described an unusual case in which a man committed suicide by ingesting both forms. The man was not a chemist but apparently was able to get all the information he needed to commit suicide using both chemical forms of the poison.

The victim was found in his locked car, as shown in the figure, along with a cooking stove, acids, and powders. One of the powders was potassium ferrocyanide, which is a nontoxic compound that also contains cyanide. Printouts from an Internet search related to the compounds were also found. He was going through financial difficulties at the time of his death. During the autopsy, powder was found on his mouth and hand, and his throat and respiratory track showed evidence of acid burning. The acid was formed through a two-step reaction from the ferrocyanide, heating, and addition of acid. Also formed is a deep blue compound called Prussian Blue, a common pigment that contains iron that came from the ferrocyanide. Evidence of that compound can be seen in the scene photo. The act showed extensive planning, and the autopsy findings indicated that the man swallowed powdered potassium cyanide and inhaled the HCN formed by the chemical reactions he found through an Internet search.

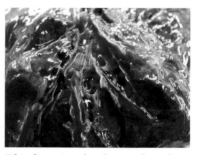

Bleeding in the lungs found at autopsy.

Source: Musshoff, F., et al., An Uncommon Case of a Suicide with Inhalation of Hydrogen Cyanide, *Forensic Science International* 204, no. 1–3 (Jan 2011): E4–E7. Figures reproduced with permission, copyright Elsevier.

6.5 INTERPRETING TOXICOLOGICAL FINDINGS

The goal of the forensic toxicologist in postmortem investigation is to collaborate with the forensic pathologist in determining the cause and manner of death. In simple terms, we infer that a death is due to a specific toxin when appropriate quantities of that toxin are found, when other findings (e.g., congestion in the lungs) are consistent with that conclusion, and no other apparent cause of death is discovered. In addition to explaining the cause of death, the presence of a toxin may help to determine whether the death was an accident, a suicide, or a homicide.

This is a complex task, and it requires that the toxicologist understand ADME, how a drug moves through compartments, and PMR. For example, it is almost impossible to state that a specific death was not poison related. Negative claims infer that all possible toxins were examined. This is not possible in the real world. Even the most comprehensive analytical scheme invariably omits many exotic (but potential) toxins. This is one reason that the context of a death is critical and why a system of medicolegal death investigation is vital. For example, it is very rare for a person to die (natural, suicide, or homicide) because of arsenic poisoning. As a result, a typical postmortem analysis in a modern toxicology lab would not include a test for arsenic. However, if the death investigator reports that a white powder was found at the scene and the autopsy shows signs of heavy metal poisoning, then an arsenic analysis could be conducted.

As an example of how toxicology is used in death investigation, consider this real case example. A 32-year-old man went fishing with friends. He collapsed and complained of dizziness and nausea. He lapsed into a coma and medical assistance was summoned. His respirations became very weak, he became cyanotic (blue color skin), and eventually suffered a cardiac arrest from which he could not be resuscitated. An autopsy revealed no signs of trauma or evidence of any disease condition. The case was referred to a laboratory that found fentanyl in the blood of the decedent at a concentration of 15 ng/mL. In the absence of any other findings and in view of the concentration of fentanyl found, fentanyl was ruled to be the cause of death. Concentrations greater than 12 ng/mL are regarded as sufficient to cause death. Investigation of the decedent's work history revealed that he was a known drug abuser who worked in a funeral home. He had recently handled the body of a cancer patient. Two fentanyl patches on her corpse at the time of her funeral were missing and presumed stolen by the funeral parlor worker. One known pattern of drug abuse is to extract the drug from the patches and then inject it to achieve an opiate high. The patches contain significant quantities of drugs, and unintentional overdose is a very real possibility. The manner of death was ruled an accident.

6.6 RETURN TO THE SCENE OF THE CRIME

Refer to the crime scene presented in Chapter 3 (Figures 3.2, 3.3, and 3.13 through 3.15 and the insert). Focus on Figure 3.14 and you will see evidence of possible drug use, which may or may not be relevant to the case. The body was recovered but with significant decomposition and apparent burn damage. Regardless, postmortem toxicological analysis can be attempted on remaining tissues. This could be important in the case narrative and in determining the cause of death. The nature of the offense changes depending on circumstances; for example, the husband might claim that his wife died at the house while both were engaged in drug use and that his subsequent actions were driven by panic and poor judgment rather than homicidal intent. The next two chapters will show how other approaches can be useful in cases where the body is partially skeletonized, burned, or decomposed.

6.7 REVIEW MATERIAL: KEY CONCEPTS AND QUESTIONS

6.7.1 KEY CONCEPTS

Aqueous
Carfentanil
Central blood
Cross-reactivity

Drug
Effective dose
Fentanyl
First-pass metabolism
Gastric contents
Immunoassay
Immunoglobulin
Ingestion
Lethal dose
Mode of ingestion/mode of administration
Peripheral blood
Poison
Postmortem toxicology
Therapeutic dose
Toxin
Vitreous humor
Xenobiotic

6.7.2 REVIEW QUESTIONS

1. Is caffeine a drug? What about bubble gum?
2. What is the difference between opiates and opioids?
3. In the example shown in Figure 6.11, is there ever a time after the dose was taken that the parent drug and two metabolites would be detectable?
4. Refer to Figure 6.12. If analysis of a postmortem blood sample showed the presence of traces of heroin, 6MAM, and morphine, estimate the time from the dose to the sample being collected.
5. The immunoassay methods described in this chapter are described as "competitive." What does this mean?

6.7.3 ADVANCED QUESTIONS AND EXERCISES

1. Extremely toxic substances such as carfentanil are rarely used in homicidal poisonings. Why?
2. In the crime scene scenario, scrutinize the crime scene evidence. Do you see anything that could support the husband if he claimed that his wife was engaged in drug use and died at the scene?
3. On the same subject, what evidence from the scene is inconsistent with that version of events?

BIBLIOGRAPHY AND FURTHER READING

BOOKS

Klaassen, C., Ed. *Casarett and Doull's Toxicology: The Basic Science of Poisons.* 8th ed. New York: McGraw-Hill, 2013. ISBN: 978-0071769235.

Levine, B. *Principles of Forensic Toxicology.* 4th ed. Washington, DC: American Association for Clinical Chemistry Press, 2013. ISBN: 978-1594251580.

Moffat, A. C., Osselton, M. D., Widdop, B., and Watts, J., Eds. *Clarke's Analysis of Drugs and Poisons.* 4th ed. London: Pharmaceutical Press, 2011. ISBN: 978-0853697114.

Negruez, A., and Cooper, G., Eds. *Clarke's Analytical Forensic Toxicology.* 2nd ed. London: Pharmaceutical Press, 2013. ISBN: 978-0857110541.

ARTICLES

Breathnach, C. S. Biographical Sketch: Orfila. *Irish Medical Journal* 80, no. 3 (Mar 1987): 99.

Drummer, O. H. Post-Mortem Toxicology. *Forensic Science International* 165, no. 2–3 (Jan 2007): 199–203.

Eckert, W. G. Historical Aspects of Poisoning and Toxicology. *American Journal of Forensic Medicine and Pathology* 2, no. 3 (Sep 1981): 261–264.

Gerostamoulos, D., Beyer, J., Staikos, V., Tayler, P., Woodford, N., and Drummer, O. H. The Effect of the Postmortem Interval on the Redistribution of Drugs: A Comparison of Mortuary Admission and Autopsy Blood Specimens. *Forensic Science Medicine and Pathology* 8, no. 4 (Dec 2012): 373–379.

Kennedy, M. Dr. Shipman, Murder and Forensic Toxicology. *Australian Journal of Forensic Sciences* 41, no. 1 (2009): 3–10.

Kennedy, M. C. Post-Mortem Drug Concentrations. *Internal Medicine Journal* 40, no. 3 (Mar 2010): 183–187.

Niyogi, S. K. Historic Development of Forensic Toxicology in America up to 1978. *American Journal of Forensic Medicine and Pathology* 1, no. 3 (Sep 1980): 249–264.

Pappas, A. A., Massoll, N. A., and Cannon, D. J. Toxicology: Past, Present, and Future. *Annals of Clinical and Laboratory Science* 29, no. 4 (Oct–Dec 1999): 253–262.

WEBSITES

Link	Description
https://toxtutor.nlm.nih.gov/index.html	A simple toxicology tutorial from the National Institutes of Health (NIH)
http://www.soft-tox.org/	Society of Forensic Toxicologists

Forensic Anthropology

7

7.1 ANTHROPOLOGY AND DEATH INVESTIGATION

So far, we have discussed death investigations and autopsy based on the assumption that the body was discovered close to the time of death. This is not always the case; sometimes the body has decomposed, leaving little more than bones and bone fragments. Our example crime scene scenario illustrates how this can occur. Skeletal remains are also encountered in cases of severe trauma and fire, such as occurred with the terrorist attacks of September 11, 2001. When skeletonized or partially skeletonized remains are found, medical examiners or coroners frequently call on forensic anthropologists to help determine who it might be and what happened to them. **Forensic anthropology** can be defined most broadly as the application of the theory and methods of anthropology to forensic problems. With skeletal remains, an autopsy such as described in Chapter 5 cannot be completed and postmortem toxicology, with rare exceptions, is also not possible. Thus, the forensic anthropologist becomes a key player in the medicolegal investigation of the death.

With skeletal remains, it is not always obvious how long ago the death occurred. Thus, with this type of human remains, the anthropologist's first tasks are often to determine if the remains are human and, if they are, whether they are of forensic importance. As an example, a skeleton may be found on an old battlefield and determined to have come from a Civil War soldier. These remains are of archaeological and historical interest, but not of forensic interest. Sometimes making such a differentiation is challenging.

The soft tissue of the human body decomposes relatively quickly depending on the environment around that body. In general, the longest-lasting parts of the body are bones and teeth. As a result, knowledge of human **osteology** (study of bones) and knowledge of **odontology** (study of teeth) are key skills that the forensic anthropologist uses in studying the remains.

The adult human skeleton has approximately 206 bones. In children and infants, the number is much higher, because some of the bone **ossification centers** (places where bone growth begins) have not yet fused together. Newborn infants can have as many as 405 bones and bone growth centers. Not all adults have the same number of bones. There is a minor amount of variation in the number of ribs. While most adults have 12 pairs, some have 11 or 13 on one or both sides; this is unrelated to whether the person is male or female. Some persons have an extra vertebra in their lumbar spine. None of these differences cause the person problems while alive, but they can be useful in identifying skeletal remains.

Knowledge about the human physical form and function must then be combined with scientific input regarding postmortem changes (**taphonomy**) to interpret the condition of human remains. In an outdoor scene, for example, such changes might include normal decomposition, alteration and scattering by scavengers such as coyotes, movement and modification by flowing water, freezing, or mummification. Postmortem alterations must be differentiated from the **antemortem** condition of the body (the time prior to death) to estimate time since death, reconstruct the place of death, interpret data regarding cause of death, and sometimes properly identify the remains.

Forensic anthropologists are frequently called upon to participate in or even direct body recoveries in outdoor settings. Often, the same techniques that are used by archaeologists are used in such cases. These practices are sometimes referred to as **forensic archaeology**. Such participation in body recovery is preferred whenever possible because it provides an opportunity to describe and interpret the taphonomic context firsthand. If the forensic anthropologist is not present for the recovery, the examination of the remains should nevertheless include a description and evaluation of the taphonomic condition of the remains, considering any contextual or environmental data gathered by the investigators.

7.2 RECOVERING REMAINS

Anthropologists frequently work with the medical examiner or with law enforcement to recover human remains. Often remains must be recovered from outdoor scenes. They may lie on the surface, they may be scattered on the surface, or they may be buried. When a body has been purposely buried with the intent to disguise the location, the burial is referred to as a **clandestine grave**. In any of these cases, archaeological techniques are used to improve the recovery success and provide careful documentation of how the remains are positioned. Archaeological techniques are used to map a scatter pattern or to document the relationship between body parts and other evidence at the scene. The fundamentals of crime scene processing, as we discussed in Chapter 3, apply in these circumstances as well, although the application to outdoor scenes and buried remains introduces additional procedures.

Processing a scene involving buried remains, for example, requires considerable effort and expertise, particularly if the remains are decomposed or skeletal. If the grave has been disturbed by scavengers, it may be reasonable to set two scene perimeters: one for the immediate grave area and another to encompass the potential scatter area that must be searched. The area surrounding the grave or along the access to it is also examined for footprints or other evidence.

Initially, a grid will be superimposed on the area to be intensively examined to preserve information about the spatial distribution of remains and artifacts within the scene boundaries. The grid may be marked off on the ground or created virtually using methods similar to those for larger crime scenes, and the area must be photographed and documented before any work begins. The area should be examined for insects that may be associated with the body; these are collected and preserved. Any living plants directly associated with the body and indicative of postmortem interval must be collected as well. Metal detectors and specialized radar devices (**ground-penetrating radar [GPR]**) can also be used to find burials and items associated with them.

The remains are removed gradually and carefully documented as the process proceeds. A screening area is selected in a location that has already been thoroughly searched and is somewhat convenient to the grave. Here, material from the grave and surrounding area will be systematically sifted through a screen to reveal human remains, artifacts, fibers, and associated insects (flies, beetles, and larvae).

The next step is to clear a staging area for the excavation, usually consisting of at least several yards in every direction surrounding the grave. Leaf or other vegetation cover is removed and screened, section by section. Bushes and saplings can be cut and removed, unless they are associated with the burial. Sod covering over the grave, if present, can be carefully removed and examined. A second pass with the metal detectors is done to locate the sources of previous hits and new areas of interest. Metal artifacts are documented and removed. Sediment samples are taken from the perimeter area and the grave matrix.

An attempt is frequently made to learn the position of the body prior to excavation. Some graves may need to be excavated from the side to preserve vertical patterns, as this may be critical information. An example would be excavation of mass graves, as are often found in war zones. Excavation of the grave is done with small instruments, such as trowels and brushes, taking care to preserve the original perimeter of the grave and any hairs, fibers, or artifacts associated with the body. If the body is fresh or decomposing, it will be necessary to prevent damage to the deteriorating soft-tissue surfaces. If possible, the body or skeleton should be completely exposed prior to removal. After the body is removed, the base of the grave is excavated and screened, going down several more inches in case small bones, teeth, artifacts, or projectiles have become embedded in the sediment (Figure 7.1).

7.3 TAPHONOMIC ASSESSMENT OF THE BODY AND WHERE IT IS FOUND

7.3.1 TAPHONOMIC CONTEXT OF THE REMAINS

The immediate environment and surroundings where the body is found is called the **taphonomic context**. As with a crime scene, a complete evaluation of the context is a critical part of the scene evaluation. These details help with interpretation of the condition of the remains, estimation of the postmortem interval, and determining if the death occurred at the site where the remains were found or elsewhere. The environmental conditions, such as temperature, moisture, and humidity, will influence the rate of decomposition and the estimate of the postmortem interval. The context is critical in determining if damage patterns seen in bones occurred near or at the time of death (perimortem) or sometime after the death.

7.3.2 TAPHONOMIC CONDITION OF THE REMAINS

It is critical to document the condition of the body, as doing so can provide clues about how long it has been there and what has happened to it since death. Although the anthropologist will also observe the condition of the remains in the morgue or laboratory, seeing the

FIGURE 7.1 Processing of materials at an archaeological excavation. Many of the same techniques are utilized by forensic anthropologists and archaeologists.

remains in the place where they are found is extremely helpful for interpreting postmortem processes. The anthropologist will assess the stage of decomposition, the amount of scattering, and evidence of scavenger chewing, drying and bleaching by the weather and sun, and other changes. Noting any change in the remains as the body decomposes or is affected by the environment is important for a thorough assessment. Examples of remains that have been altered postmortem can be seen in Figures 7.3 and 7.4, which show remains recovered from an Alaskan plane crash site. Figure 7.2 shows where these bones are found in the body. The plane went down in 1954; these samples were recovered in 2012–2013. The ends of the bones in Figure 7.3 show damage by crushing, which probably occurred as the bones were bound in ice and moved with the flow of the glacier. Figure 7.4 illustrates evidence of traumatic injury associated with the crash (perimortem injury) and damage associated with the glacier environment. The fracture likely occurred during the crash, while the other damage was due to movement in the ice.

7.4 BIOLOGICAL PROFILE

A crucial step for identifying unknown remains is a **biological profile**. Developing such a profile entails studying the remains, while noting characteristics of shape and size that may allow an estimation of age, sex, and population or ancestry. Stature (height) is estimated by measuring total body (or skeletal) length or by extrapolating from long bone lengths. Unique antemortem characteristics of the individual that would have been known to family or acquaintances, such as a healed bone fracture or an unusual dental configuration, are also included in the profile.

The goal of developing a biological profile is to describe the individual in such a way that law enforcement or acquaintances can narrow the range of possible identities. Developing it requires the use of statistical descriptions of various populations that have been previously studied. Difficulty can be encountered if the victim is unusual, has mixed ancestry, or belongs to a population that has not been well studied. Regardless, the biological profile is one of the most important pieces of information that the forensic anthropologists can provide in a death investigation.

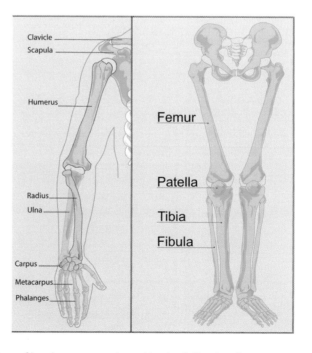

FIGURE 7.2   Location of key bones mentioned in the following figures.

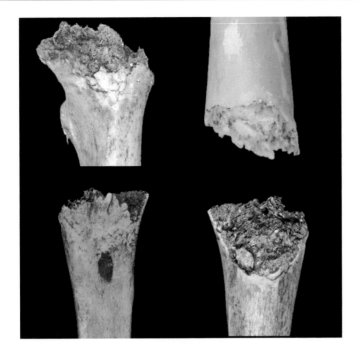

FIGURE 7.3 A femur (left) and a tibia (right). (Reproduced from Pilloud, M. A., et al., Taphonomy of Human Remains in a Glacial Environment, *Forensic Science International* 261 [Apr 2016]. With permission from Elsevier.)

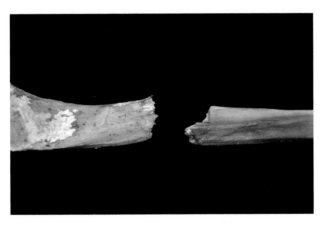

FIGURE 7.4 Fractured and damaged ulna. (Reproduced from Pilloud, M. A., et al., Taphonomy of Human Remains in a Glacial Environment, *Forensic Science International* 261 [Apr 2016]. With permission from Elsevier.)

7.4.1 ESTIMATING AGE

7.4.1.1 GROWTH AND DEVELOPMENT

A forensic anthropologist must be familiar with dental and skeletal development at each stage of life, beginning with the fetus. The skeleton is formed by the development and growth of ossification centers, which gradually replace cartilage. These three centers will ultimately grow together when the individual reaches full size. The timing of the formation, growth, and ultimate fusion of these and other ossification centers is patterned, depending on age, sex, bone element involved, nutritional and hormonal status, and individual variation. By the time a fetus is fully developed into a newborn, approximately 405 ossification centers are present. When an individual reaches adulthood (generally mid-20s), that number decreases to about 206 fully formed ones.

Patterns of bone development differ somewhat in males and females, with females developing a little earlier than males on average. Bone development sequences and timing also differ across populations. Even within well-nourished, homogeneous populations, bone development may differ significantly from person to person. Thus, age estimates are preferably expressed as ranges rather than single values.

Humans, like other mammals, have two sets of teeth: the **deciduous dentition**, so named because it is shed in childhood (baby teeth), and the permanent dentition. A human child has 20 deciduous teeth; each quadrant has 2 incisors, 1 canine, and 2 deciduous molars—a dental formula of 2.1.2. Most human adults have 32 permanent teeth; each quadrant has 2 incisors, 1 canine, 2 premolars, and 3 molars—a dental formula of 2.1.2.3.

Tooth development begins in fetal life with formation of the deciduous tooth crowns within sockets or crypts. Beginning usually around the sixth month after birth, the deciduous teeth begin to erupt. Meanwhile, the permanent teeth begin to form underneath the deciduous dentition. By about the sixth year, the deciduous front teeth begin to be lost and the permanent teeth erupt to take their place. The last teeth to erupt are the third molars, often called wisdom teeth, which begin to emerge in the 18th year in about 70% of the population.

The patterns of tooth development, like those of bone, differ slightly by sex (females develop a bit earlier) and by population. The ranges of variation tend to overlap a great deal. Some dental traits are common in some populations and rare in others. Individuals of Asian or Native American ancestry, for example, commonly have a trait called shoveling, whereby the anterior teeth are slightly thicker (ridged) around the margins of each tooth on the tongue side. However, this trait is not uniform or universal in these populations, and members of other groups occasionally have it.

There are few skeletal collections that include children with documented ages at death, and our knowledge of population differences in bone and tooth development is somewhat limited. Standards for dental development tend to be more useful in estimating age in children, particularly young children, than are those for bone development; hence, forensic anthropologists will generally emphasize dental standards in prepubescent remains. However, it is important to evaluate the complete set of indicators, dental as well as osteological, rather than depending on only one or two.

7.4.1.2 AGE-RELATED CHANGES IN THE ADULT

The bony skeleton is not fixed at adulthood; it changes continually until death, balancing the building and replacing of bony tissue at the cellular level. In general, bone density reaches a peak in the 20s and stays high in the 30s but begins to decline in the 40s. Females experience a bone density drop around menopause; this decline levels off in most women after the age of 55 or 60 but still continues. Adult males over the age of 40 experience a gradual decline into old age. A minority of individuals suffer serious bone density loss (**osteoporosis**), which is more common in females and more prevalent in some populations.

Bone density depends on factors other than age and hormonal status. Weight-bearing exercise and good nutrition (particularly calcium, magnesium, and vitamin D intake) can minimize bone loss in many individuals. Bone density can be observed macroscopically (visually), radiographically, microscopically, or via medical imaging methods. Macroscopic assessment relies on evidence of the thinning of the outer dense bone layer (**cortical bone**), reduced concentration of **trabecular bone** or spongy bone, **remodeling** (changing bone shape), and evidence of fractures. Radiographic methods depend on similar observations and are done using x-ray images. It is also common to see some deterioration in joint integrity connected with use or wear and made worse by inflammation. This is related to the reduction in bone density after the age of 40. The resulting condition, **osteoarthritis**, tends to affect the spine and joints that are overused due to occupation or frequent patterned activity. It is infrequently seen before age 40 and varies among individuals. An example of how evidence of osteoarthritis in bone can be integrated into an estimate of age is shown in Figure 7.5.

One of the most reliable indicators of adult age is the pubic symphysis. The **pubic symphysis** is the area on the pelvis where the right and left pelvis halves join in the front of the body across two surfaces or faces (Figure 7.6). With age, these surfaces change from billowed to more flattened and rimmed. The changes have been divided into age- and sex-associated stages. Several

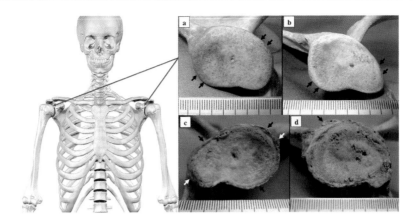

FIGURE 7.5   Changes in the appearance of a joint surface in the shoulder as a result of aging: (a–d) from a younger person through to an older person. (Right figure reproduced with permission from Brennaman, A. L., et al., A Bayesian Approach to Age-at-Death Estimation from Osteoarthritis of the Shoulder in Modern North Americans, *Journal of Forensic Sciences* 62, no. 3 [May 2017]: 573–584. Copyright Wiley. Portion at left from http://lifesciencedb.jp/bp3d/info/userGuide/faq/credit.html.)

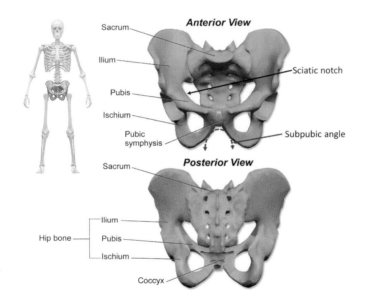

FIGURE 7.6   Location of the pubic symphysis. (Adapted from Blausen.com, Medical Gallery of Blausen Medical 2014, *WikiJournal of Medicine* 1, no. 2 [2014]. https://en.wikiversity.org/wiki/WikiJournal_of_Medicine/Medical_gallery_of_Blausen_Medical_2014.)

published standards exist. The older the individual, the broader will be the estimated age range, however. Other standards have been published for skeletal features related to the fourth rib and the cranium.

As a person ages, wear and tooth loss affect dentition. These processes are, however, dependent on dietary practices, dental hygiene, and genetic background. Although population standards have been developed, application to individual dentitions is problematic.

7.4.2   DETERMINING SEX

Along with many other animal species, male and female humans differ by size, with males being larger on average. *Sexual dimorphism* is the term for the differences in size and shape between the sexes. Size is an indicator of sex, but so are muscularity, overall size, and the presence or absence of certain traits. DNA methods can produce very accurate determinations of biological sex.

The attribution of biological sex to skeletal remains using size, shape, and the presence or absence of skeletal markers is inexact. All methods are limited by the fact that there is some morphological overlap between males and females. It is well known, for example, that males are, on average, taller than females when the entire population is described; however, the ranges of size overlap greatly, and individual males and females may deviate from those tendencies. At puberty, when circulating sex hormones increase greatly, skeletal morphology between males and females differs more. Prior to puberty, although differences exist, they are so small that forensic anthropologists do not attempt to assign sex in those cases. After puberty, differences increase, and the accuracy rate is well over 90% in sex attribution within well-studied populations. The skull and pelvis are the most sexually dimorphic skeletal areas, although it is better to examine the entire skeleton for indicators. The shape of the pelvis is critical for giving birth and enabling upright posture and locomotion. The bowl-shaped human pelvis balances and sustains the weight of the upper body to permit upright posture and efficient movement. The female pelvis has additional breadth and increased diameter of the pelvic inlet and outlet, enabling the infants of our relatively large-brained human species to pass through the birth canal. Associated female traits include a broad, shallow **sciatic notch** and the U-shaped **subpubic angle** (Figure 7.6).

The typical male skull tends to be larger with greater muscularity. It also tends to be more robust at areas of muscle attachment and biomechanical stress (brow ridges, chin) and more right-angled at the lower jaw, and it exhibits larger joint surfaces where the lower jaw connects to the braincase or cranium, and where the head connects to the top of the spine.

The skeleton below the skull is, on average, larger in males. The male tends to exhibit larger weight-bearing joint surfaces (e.g., the size of the hip ball and socket), more accentuated areas of muscle attachment, larger diameters of long bones, and greater stature. These traits vary between populations, though, and are related to nutritional status and behavior (e.g., weight-bearing occupations, strength training). No morphological indicator of sex is perfect. The best approach is to assess the entire skeletal pattern in the context of what may be known about the person or the population.

MYTHS OF FORENSIC SCIENCE

It is easy to tell how tall someone was and their gender using their bones.

Not always. Methodology and data sets have improved over time, but it can still be challenging, especially when the number of available bones is limited. The typical size of individuals varies across regions of the world and depends on age. In addition, interpretation of the anthropologist plays a role. As an example, consider the enduring mystery of Amelia Earhart's disappearance. A recent article revisited the analysis of bones found on an island in the Pacific thought by some to be those of Earhart. These bones were initially studied in 1941 and identified as likely male. The new article, based on current data and methodology, contests this finding and concludes that the bones were from a small female, which Earhart was. The author concludes that the bulk of the evidence indicates that the famous flyer died on the island after she made an emergency landing.

Source: Jantz, R. L., Amelia Earhart and the Nikumaroro Bones: A 1941 Analysis versus Modern Quantitative Techniques, *Forensic Anthropology* 1, no. 2 (2018): 83–98.

7.4.3 STATURE AND DISARTICULATION

The estimation of living height or stature is challenging. Adult stature changes (shrinks) from morning until night and over the course of the lifetime. The accurate measurement of living stature in an individual is a range and not a single number. Stature can be calculated for decomposed or skeletonized remains using several reliable methods. First, if the body is still essentially complete, the length can be measured, keeping in mind that the loss of muscle support will loosen joints and lengthen the body somewhat. If a body is skeletal and the joints are no longer held together with soft tissue (**disarticulated**) but some long bones are present, stature can be estimated using formulas that have been developed. Examples of disarticulation cases are shown in Figures 7.7 and 7.8.

When remains are incomplete, estimates of stature are done by extrapolating from the lengths of individual long bones, or combinations of long bones. This is done by using formulas developed for reference populations of known stature, such as military casualties or modern forensic case databases. The ability to estimate stature from long bone lengths depends on the presence of patterned and proportional relationships between the sizes of body parts, a concept called allometry. Taller people tend to have predictably longer bones. Allometric relationships between bone elements are systematic but not exact. They differ from population to population and from individual to individual. Individuals in any population may or may not conform to the central tendencies. Stature formulas usually specify a presumed ancestral population, such as "African American." They also require the application of measurements of skeletal elements. Forensic anthropologists are trained in how to perform these specialized measurements of the bones, called **osteometry**. Standards exist for how and where to measure each bone element. It is critical when applying stature formulas that the measurement is done the same way as was done for the reference population. The resulting estimate is reported as a range.

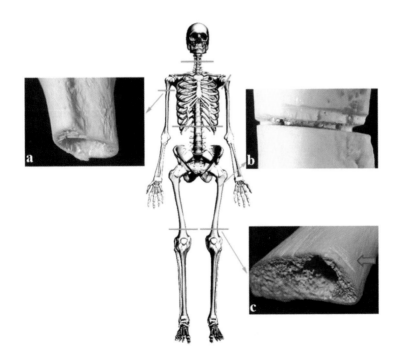

FIGURE 7.7 Disarticulation example showing cut marks clearly evident in the bone. (Reproduced from Porta, D., et al., Dismemberment and Disarticulation: A Forensic Anthropological Approach, *Journal of Forensic and Legal Medicine* 38 [Feb 2016]: 50–57. With permission from Elsevier.)

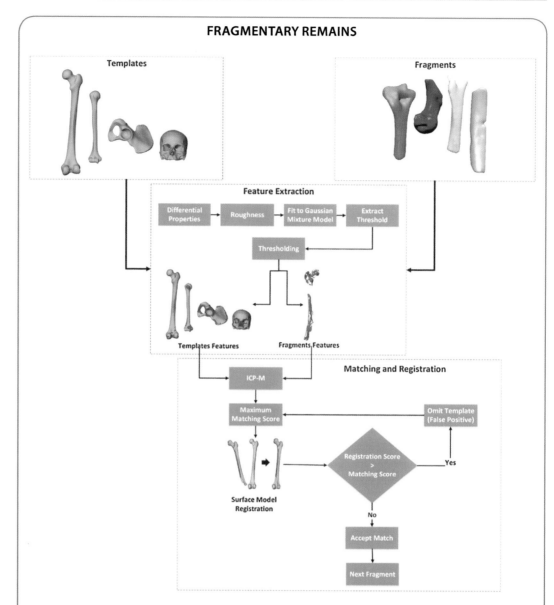

(Reproduced from Mahfouz, M. R., et al., Computerized Reconstruction of Fragmentary Skeletal Remains, *Forensic Science International* 275 (Jun 2017): 212–223. With permission from Elsevier.)

When remains are found, they are often found as fragments that may be spread over a wide area and mixed with fragments from many individuals (comingled remains). One recently reported method has been developed to assist in dealing with fragmentary remains. The remains are first scanned using a computer-aided tomography (CT) and three-dimensional (3D) scanner, and then the results are analyzed by a computer program (Fragmento), which compares the rendering of the fragment to templates of bones. Once the template is identified and the bone fitted to the shape, the program is used to reconstruct the entire bone even in the absence of all the pieces.

The full description of the program is from a 2017 article: Mahfouz, M. R., et al., Computerized Reconstruction of Fragmentary Skeletal Remains, *Forensic Science International* 275 (Jun 2017): 212–223. Copyright Elsevier.

7.5 IDENTIFICATION

Identifications are often based on visual identification or on circumstantial evidence, such as clothing, location, or pathological condition. In cases of suspicious death, most medical examiners and coroners require positive identification—that is, identification beyond a reasonable

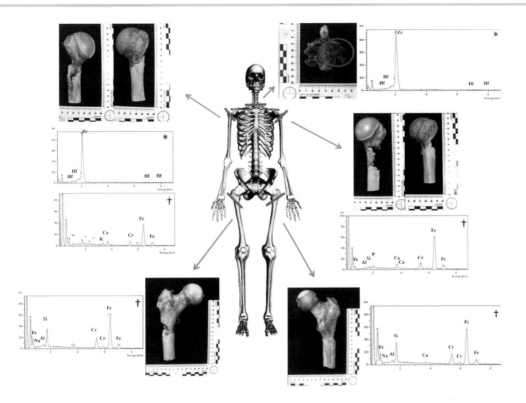

FIGURE 7.8 Disarticulation with chemical analysis in which the body was cut into six segments using two different knives. The chemical analysis is useful in determining which knives were used at a site. Notice that there are two spectral patterns corresponding to the two knives. The blade of one broke off (red circle) in the bone. (Reproduced from Porta, D., et al., Dismemberment and Disarticulation: A Forensic Anthropological Approach, *Journal of Forensic and Legal Medicine* 38 [Feb 2016]: 50–57. With permission from Elsevier.)

doubt. In practice, such identification requires a match using one or more of the following legally accepted techniques: DNA analysis, fingerprints, dental records, x-rays, or a uniquely identifiable medical apparatus, such as an artificial joint.

Occasionally such an identification process begins when the anthropologist or pathologist documents a unique anatomic feature (such as a congenital defect), evidence of a medical condition or surgical procedure, or bony changes associated with certain activities or occupations, but the presence of such characteristics, while indicating a probable identification, is not enough for a positive identification in many U.S. jurisdictions.

A description of one or more individual characteristics may help narrow the number of possible identities for an unknown set of remains. These may include evidence of medical conditions, congenital defects, handedness, and other markers of occupational stress. The body may have unique or unusual features that family and friends were aware of. An anthropologist may also observe features that may never have surfaced, such as an extra lumbar vertebra.

Antemortem medical conditions or disease likewise may or may not have been known; for example, evidence of previous surgeries may include a surgically implanted apparatus, possibly with a unique identification number. On the other hand, evidence of bone inflammation or infection may or may not have produced clinical symptoms that are described in a medical record. Child or other domestic abuse may be revealed by the combined presence of healed and partially healed patterned injuries. Right- or left-hand dominance can often be ascertained, however, at least in individuals who exhibited that dominance behaviorally, and unusual or patterns of use or wear may suggest certain repetitious actions that can be helpful in narrowing identification possibilities.

7.6 TRAUMA

The legal authority to determine cause and manner of death rests with the medical examiner or coroner; however, the anthropologist often contributes critical evidence for these determinations, particularly in the interpretation of skeletal trauma and judgments regarding the timing of the trauma. Human remains frequently exhibit signs of antemortem trauma (healed or healing prior to death) and perimortem trauma (occurring at or around the time of death). By noting the presence of active or previous bone remodeling (e.g., formation of a bony scar or callus at the site of a fracture), the anthropologist can assign a traumatic injury to the antemortem period.

Most remains examined by forensic anthropologists have an extended postmortem period and are decomposed or skeletonized. Such remains may undergo postmortem modification by a wide range of agents, such as carnivores or transport by flowing water, which may damage bone. This taphonomic damage is not related to the cause of death and must be differentiated from perimortem trauma by noting the patterns of bone breakage in relation to moisture and fat loss, differential staining on fracture margins, and signature modifications of scavengers, plants, or geological processes.

Bone damage with no signs of healing, which apparently occurred when the bone was still fresh and for which a taphonomic cause can be ruled out, is described as perimortem trauma. It is not possible to be precise regarding whether bone damage occurred just before or just after death. Unlike soft tissue, bone does not exhibit a detectable vital reaction without several days of healing time.

Bone damage is conventionally divided into blunt force or sharp force trauma categories, depending on the presence or absence of cut surfaces. An example of blunt force trauma of the skull is shown in Figure 7.9. Blunt force damage produces impact marks or fractures and can fragment bone. The force can be delivered at great speed (e.g., gunshot) or slower speed (bludgeoning by hand with a heavy object); these can often be differentiated by the amount of warping of the fragments, termed **plastic deformation**, which is more likely with slower loading. In both blunt and sharp force injuries, the pattern of impact scars, fractures, or cut marks can sometimes indicate object shape (e.g., head of a hammer), trauma type (e.g., frontal motor vehicle impact), or weapon class (e.g., single-bladed knife).

A gunshot wound is a special form of blunt force trauma. Firearm projectiles frequently create signature patterns in bone, particularly in the skull. An anthropologist will generally reconstruct a skull shattered by gunshot to evaluate the injuries. With many gunshot wounds, it is possible to differentiate the entrance from the exit by inward beveling on the former and outward beveling on the latter. Due to the relatively slower speed of the radiating fractures compared with the projectile speed, combined with the fact that fracture lines are halted upon encountering previous fracture lines, it is frequently possible to determine the sequence of multiple gunshot wounds.

In each case, an anthropologist looks at the pattern of antemortem and perimortem injury throughout the skeleton to render an interpretation about cause, timing, trajectory, and weapon characteristics. Taphonomic damage must be excluded. With many suspected instances of trauma, it is necessary to examine bone damage microscopically to rule out postmortem modification. It is possible for different agents of bone modification to produce similar types of damage, a concept known as **equifinality**. These mimics can generally be differentiated, however, by analyzing the pattern throughout the body rather than focusing on a single site.

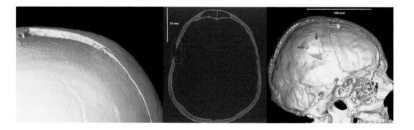

FIGURE 7.9   An example of a blunt force injury as seen in the skull. (Reproduced from Fleming-Farrell, D., et al., Virtual Assessment of Perimortem and Postmortem Blunt Force Cranial Trauma, *Forensic Science International* 229, no. 1–3 [Jun 2013]. With permission from Elsevier.)

MYTHS OF FORENSIC SCIENCE

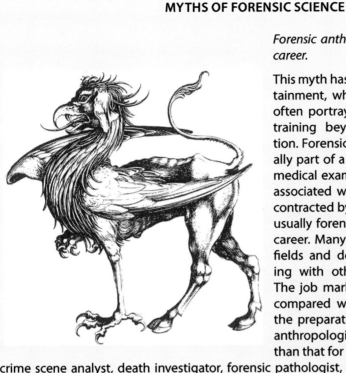

Forensic anthropology is a typical forensic career.

This myth has roots in the media and entertainment, where forensic anthropology is often portrayed as requiring little special training beyond crime scene investigation. Forensic anthropologists are not usually part of a typical forensic laboratory or medical examiner's office; rather, most are associated with universities. They may be contracted by state and local agencies, but usually forensic work is just a part of their career. Many hold PhD degrees in related fields and developed expertise by working with other forensic anthropologists. The job market for the field is very small compared with other forensic fields, and the preparation for working as a forensic anthropologist is significantly different than that for other forensic careers, such as crime scene analyst, death investigator, forensic pathologist, DNA analyst, or forensic toxicologist. It is always important to research different forensic careers if you become interested in pursuing one.

7.7 RETURN TO THE SCENE OF THE CRIME

Review the crime scene scenario presented in Chapter 3. Forensic anthropology could be useful in this case given the state of decomposition of the remains and the damage done by fire. One of the first things that could be established is the sex of the victim, as well as other features of the biological profile. If a DNA type cannot be established from the remains, this information would be critical in identifying the victim as the wife.

7.8 REVIEW MATERIAL

7.8.1 KEY TERMS AND CONCEPTS

Allometry
Antemortem
Biological profile
Clandestine grave
Cortical bone
Deciduous dentition
Disarticulated
Equifinality
Forensic anthropology
Forensic archaeology
Ground-penetrating radar
Odontology
Ossification centers

Osteoarthritis
Osteology
Osteometry
Osteoporosis
Perimortem
Plastic deformation
Pubic symphysis
Remodeling
Sciatic notch
Sexual dimorphism
Stature
Subpubic angle
Taphonomic context
Taphonomy
Trabecular bone

7.8.2 QUESTIONS

1. How would a forensic anthropologist tell if a fracture occurred perimortem or long before the person died?
2. Why is the taphonomic context so critical in forensic anthropology? How does it differ from the context as applied to a crime scene?
3. What role does the forensic anthropologist play in a medicolegal death investigation? Differentiate the role of the forensic anthropologist from those of the forensic pathologist and the medical examiner or coroner.

7.8.3 ADVANCED QUESTIONS AND EXERCISES

1. Assume that a body has been found in a shallow grave in a remote location. Decomposition is nearly complete, but some soft tissue remains. Describe how the different elements of the death investigation system (pathology, anthropology, taphonomy, and entomology) could assist in the investigation.

BIBLIOGRAPHY AND FURTHER READING

BOOKS

Langley, N. R., and Tersigni-Tarrt, M. T. A., Eds. *Forensic Anthropology: A Comprehensive Introduction.* 2nd ed. Boca Raton, FL: CRC Press, 2017. ISBN: 978-1498736121.

ARTICLES

Blau, S. How Traumatic: A Review of the Role of the Forensic Anthropologist in the Examination and Interpretation of Skeletal Trauma. *Australian Journal of Forensic Sciences* 49, no. 3 (Jun 2017): 261–280.

Crowder, C., Rainwater, C. W., and Fridie, J. S. Microscopic Analysis of Sharp Force Trauma in Bone and Cartilage: A Validation Study. *Journal of Forensic Sciences* 58, no. 5 (Sep 2013): 1119–1126.

Fleming-Farrell, D., Michailidis, K., Karantanas, A., Roberts, N., and Kranioti, E. F. Virtual Assessment of Perimortem and Postmortem Blunt Force Cranial Trauma. *Forensic Science International* 229, no. 1–3 (Jun 2013).

Mundorff, A. Z. Integrating Forensic Anthropology into Disaster Victim Identification. *Forensic Science Medicine and Pathology* 8, no. 2 (Jun 2012): 131–139.

WEBSITES

Link	Description
http://theabfa.org/	American Board of Forensic Anthropology
http://www.sfu.museum/forensics/eng/pg_media-media_pg/anthropologie-anthropology/	Virtual Museum of Canada site on forensic anthropology

Forensic Entomology

8.1 ENTOMOLOGY AND DEATH INVESTIGATION

In the first few hours after death, a forensic pathologist can use the condition of the body (e.g., rigor mortis) to estimate the elapsed time since death. However, about 24–48 hours after death, most of these biological changes are over and are of little value in estimating the postmortem interval (PMI). With decomposed remains such as bones, time of death cannot be estimated. Fortunately, **entomology** (the study of insects) can be a useful tool in estimating the PMI in some cases. Even in the very early PMI entomology is still valuable, as flies lay eggs very shortly after a death, so although medical factors are usually used in those first few hours, insect evidence can also be valuable. This is particularly true if only a body part is found, when medical parameters are of no value.

Forensic entomology is most well known for its value in time of death estimations, but studying the insects associated with a corpse can also reveal many other factors about the death, such as whether the body had been moved after death or has been disturbed, the presence or position of wound sites, whether the victim used drugs or was poisoned, and the length of time of neglect or abuse in living victims.

8.2 ESTIMATING THE PMI

Forensic entomologists strive to estimate the minimum elapsed time since death, which is not quite the same thing as the PMI, as we have been discussing up to now. The PMI is best expressed as a range and indicates the least and most time that could have passed since a death occurred. However, with forensic entomology, the lower end of that range is what is estimated. This arises from the methods that are used in this discipline. There are two methods to estimate elapsed time since death using entomological evidence. The first method is based on the predictable development of larval *Diptera*, or flies, primarily blow flies (Figure 8.1), over time. This form of analysis takes advantage of the known passage of time from when the first egg is laid on the remains until

FIGURE 8.1   A blow fly.

the first adult flies emerge from the **puparial** cases and leave the body, making it most valuable in estimating minimum time since death from a few hours to several weeks. The second method is based on the predictable **successional colonization** of the body by a sequence of **carrion** insects. This method can be used from a few weeks after death until only dry bones remain. The approach used will depend on the age of the remains and the types of insects collected.

8.2.1 LARVAL DEVELOPMENT OR MAGGOT AGING

Certain species of insect depend on carrion or dead animal material to survive. Such insects are therefore very adept at locating a dead animal. Although, for the most part, such carrion insects survive on dead wildlife, humans are also animals, and from an insect's point of view a dead human is no different from a dead bear or a dead crow. Hence, insects colonize humans in the same way they colonize a dead animal.

The first insects that are usually attracted to remains are "true flies," or *Diptera* in the blow fly family. As adults, blow flies feed primarily on sugary food sources (e.g., nectar) and protein (e.g., feces, carrion), but as maggots or larvae they need to feed exclusively on a dead animal. Therefore, female blow flies are well evolved to locate carrion. Because their maggots survive best in fresh bodies, the females can locate a dead body immediately after death. Although unnoticeable to a human, a body gives off a variety of chemical cues, such as odors, immediately after death. These cues allow a fly to locate a corpse very quickly; therefore, blow flies will arrive at a corpse very soon after death, if the season and temperatures are appropriate. Blow flies are not generally active during winter. The time of death also influences when a female will lay eggs; if death occurs at night in otherwise favorable conditions, then no eggs will be laid until the following morning. This illustrates why it is so important for a forensic entomologist to always give a minimum estimate of elapsed time since death, because if death occurred at night, then there might have been a considerable delay before the insects arrived. For example, if a fresh victim is found midafternoon and the insect analysis indicates that the eggs were laid on the body at around 8:00 a.m., that means the person has been dead since at least 8:00 a.m., but it is also possible that they could have died during the previous night.

Blow flies are attracted to a wound first, as blood is an excellent and easily obtained protein source for the maggots. Blow flies are so good at locating a wound that they can even locate a needle mark when it is no longer visible to the naked eye. In homicide victims, there are often wounds, so these become the first site of colonization. If there are no wounds, the insects will colonize the natural orifices, as adult human skin is usually too tough for the very small maggots to break. Orifices are lined with a mucosal layer that is very moist and soft and much easier to break than regular skin. This means that if no wounds are present, then the orifices will usually be colonized first.

The blow flies develop from egg through first, second, and third **instars**, or stages, of maggots, and they then enter a puparial stage before becoming adults. Insects are cold blooded, so their development is temperature dependent. As temperature increases, they develop more rapidly, and as it decreases they develop more slowly. This relationship, at temperature optimums, is relatively linear, making it predictable. As the development rates are predictable, an analysis of the oldest insect stage on the body, together with a knowledge of the meteorological conditions and the microclimatic conditions at the scene, can be used to estimate how long insects have been feeding on the body, and hence how long the victim has been dead. Figure 8.2 shows the life cycle of the flies.

The fly eggs hatch into delicate first-instar or first-stage maggots, which feed on liquid protein. After a brief period, primarily dependent on temperature and species, the first-instar larvae will molt to the second instar, shedding the first-instar larval cuticle and mouthparts. The second-instar maggot feeds for a period, then molts to the third instar, again shedding the cuticle and mouthparts of the previous stage. The third-instar maggot is a voracious feeder and frequently aggregates in large masses (Figure 8.3) that can generate a tremendous amount of heat. These masses can quickly remove a large amount of tissue.

After a period of intense feeding, the third instar enters a nonfeeding or wandering stage and leaves the body for other locations, such as the surrounding soil, carpet, or even hair or clothing of

The blow fly life cycle has six parts: the egg, three larval stages, the pupa, and adult.

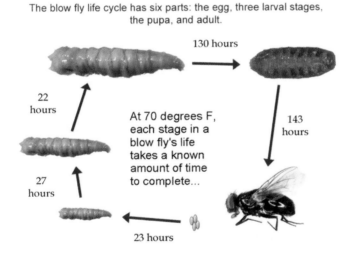

130 hours

22 hours

At 70 degrees F, each stage in a blow fly's life takes a known amount of time to complete...

143 hours

27 hours

23 hours

FIGURE 8.2   Life cycle of blow flies. The times listed will vary depending on environmental conditions and other factors.

FIGURE 8.3   Maggots on a possum carcass.

the corpse. The maggots may burrow down several centimeters into the soil and may crawl several meters away from the remains until they find a suitable pupation site. These maggots then pupate, just as a caterpillar forms a chrysalis. When the adult fly emerges from the puparial case, it spends a few hours drying its body and expanding its wings, then it flies away, leaving behind the hard, dark puparial case as evidence that this cycle has been completed. Newly emerged flies are of great forensic significance, as they indicate that an entire blow fly life cycle has been completed on the body.

The fly's entire life cycle is predictable. It is heavily influenced by temperature, the major variable, in addition to species, nutrition, humidity, and so forth, and so forensic entomology will only *estimate* a minimum elapsed time since death. The science of forensic entomology is based on estimating the length of tenure of insects on a body, rather than the actual time of death. Death precedes insect colonization (except in rare instances), so the insects will indicate a time elapsed since death that is less than the actual time of death.

8.2.2 SUCCESSIVE COLONIZATION

A dead body, whether animal or human, is a rich but limited nutritional resource that supports a rapidly changing ecosystem. Blow flies arrive soon after death. As these insects feed, they change the carcass, and as decomposition progresses the body becomes less attractive to these early colonizers and more attractive to other insects. As the body decomposes, it goes through a sequence of rapid biological, chemical, and physical changes. These changes in decomposition attract a dynamically changing sequence of colonizing insects that continue to feed until there is no nutritional value left in the remains. The process of succession as it proceeds, along with the process of decomposition, is shown in Figure 8.4.

Succession here refers to the evolving group of insects that colonize a body. As an example, once blow flies lay eggs, these can become a food source for another insect, such as a beetle. The beetles feed and reproduce and may themselves become food for another species. While the process is complex and dependent on factors such as location and weather, knowing the order of succession is useful for interpreting what is found and estimating the minimum time since colonization. Note that the figure shows three stages immediately after the death—exposure (the body is exposed to the environment); detection, when the first colonizing insects (usually blow flies) arrive; and acceptance, which occurs after the first arrivals determine that the site is appropriate to their needs. Succession is not exclusive; many species may coexist on the remains at a given time. An example of actual data collected from a pig carcass is shown in Figure 8.5; wherever

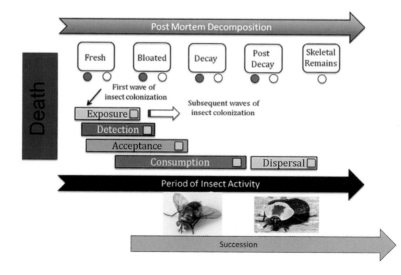

FIGURE 8.4 Decomposition and process of insect succession. (Adapted from Thompson, C. R., et al., Bacterial Interactions with Necrophagous Flies, *Annals of the Entomological Society of America* 106, no. 6 [Nov 2013]: 799–809. Copyright Oxford Academic Press. Image of beetle from Katja Schulz [Flickr].)

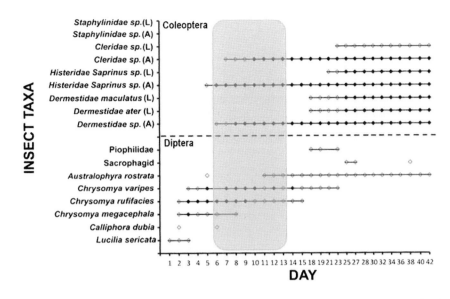

FIGURE 8.5   Insect species present on a pig carcass. (Reproduced from McIntosh, C. S., et al., A Comparison of Carcass Decomposition and Associated Insect Succession onto Burnt and Unburnt Pig Carcasses, *International Journal of Legal Medicine* 131, no. 3 [May 2017]: 835–845. With permission from Elsevier.)

lines overlap, multiple species were found. The term *Diptera* refers to flies, while *Coleoptera* refers to beetles. Note how flies arrive first and that a few days pass before the beetles arrive; this is an example of succession.

When an entomologist studies the large variety of insects found on the remains, the first step is to identify the species that are present at that time. If experimental data are available for the same geographic area and scenario, then the entomologist can say that the insects indicate that the victim has been dead for a certain period; for example, the entomologist might say that a victim has been dead 3–5 months. Then the entomologist will look at the evidence of insects that are no longer on the body, having left when the body passed the stage of decomposition that made it a suitable source of nutrition for that species of insect. This indicates that it is past the time frame in which these species normally are found on a body in this scenario, and the entomologist can estimate that the victim has been dead for at least that long. It might indicate, for example, that the victim has been dead for more than 4 months.

8.3 OTHER APPLICATIONS OF FORENSIC ENTOMOLOGY

8.3.1 INDICATING WHETHER THE BODY HAS BEEN MOVED

A killing may occur on the spur of the moment, leaving the killer with a body that must be hidden in some way. It may take time for the killer to consider how to do so and to make plans. If the body has remained at the original site for any length of time, and the conditions are appropriate, insects will colonize the body. At first, the only sign of colonization will be blow fly eggs in the wound, the corners of the eye, and in the nostrils. When the killer returns, the insects will probably not be noticed. If they are observed, it is unlikely that the killer will consider them as evidence before moving the body to a new location or the dump site, an example of a secondary crime scene, as was discussed in Chapter 3.

When the body is found at the dump site, it will have attracted insects local to the area, but it will also have insects from the death site. These insects often are not consistent with the dump site habitat. They may be species of insects that are most commonly found in an urban area or in a forested area. If the dump site is very different from that, such as a very rural area or an open pasture, the forensic entomologist will be able to comment that the insects indicate that the body has been at the discovery site for a certain time, but also that previously it had been in a very urban environment, for example.

8.3.2 LOCATING THE POSITION OF WOUNDS

Blow flies are the first insects to be attracted to remains and usually lay their eggs close to a wound, so that the first-instar maggots have access to liquid protein for nutrition. Once a body is in a state of advanced decomposition, it is often difficult to work out whether wounds were present. If the weapon does not strike the hard tissue, such as bone and cartilage, it may easily be missed, even though a wound in soft tissue alone can be fatal. A slit throat or a stab wound in the soft tissue of the gut could easily be missed. Insects, however, can locate a wound, no matter how small, because laying eggs close to a wound is a survival strategy that increases the chance of success of the female flies' offspring. Therefore, female flies are genetically programmed to be extremely efficient at locating a wound.

When a body is highly decomposed, we can look at the pattern of maggot colonization on the body. If the maggots began feeding at the natural orifices, then it is likely that there were no open wounds. On the other hand, if the oldest maggots clearly began feeding in, for instance, the stomach area, and only much younger and smaller maggots are found in the orifices, then it is very probable that there was a wound in the stomach. It is not up to an entomologist to state that an area is a wound site, but rather the forensic pathologist, who conducts the autopsy and is qualified to identify a wound and the weapon that caused it. It is up to the entomologist, however, to point out an irregular or atypical insect colonization pattern that might indicate a wound.

8.3.3 DRUGS AND INSECTS: ENTOMOTOXICOLOGY

Carrion insects feed on the tissues of the dead body. This means that they also ingest all the toxins contained in the body at the time of death. This can include poisons and drugs. When a person takes a drug or poison, unless it is immediately fatal, the body begins to metabolize the drug as was discussed briefly in Chapter 6. This process involves breaking the original drug, known as the parent drug, into its component chemicals, known as metabolites. These metabolites are then broken down further and eventually excreted. If a person dies after taking a drug, either due to the actions of the drug itself or for other reasons, such as homicide, then depending on the length of time elapsed between drug consumption and death, various levels of the parent drug and its metabolites will be present. Toxicologists usually analyze body samples to determine what toxins were present at the time of death and then interpret this to assist the investigators in determining how and why a person died, and what behavioral effects might have been expected at a certain time. However, postmortem toxicology, as we discussed in Chapter 6, is only feasible relatively soon after death.

Once the body begins to decompose, the tissues used for drug analyses, such as blood, liver, and kidney tissues, begin to degrade and eventually decompose entirely, making them useless for drug analyses. However, the insects feeding on the body, ingesting the tissue together with all the chemicals in it, are fresh in the sense that they are still alive. Therefore, insects, primarily blow fly maggots but also other insects, can make excellent alternative toxicological specimens that can be analyzed in the same way that the toxicologist would analyze a piece of the human tissue.

There have been many studies and case histories that have shown that insects can accumulate ingested chemicals into their body tissue, allowing them to be used as toxicological specimens throughout much of the life cycle. As such, toxicologists can use maggots, pupae, empty puparial cases, and beetles to identify what the parent drug was and the identity of the various metabolites. However, unlike postmortem autopsy samples, the relative amount of drug and metabolites found in insects feeding on a body cannot be used to reliably estimate intoxication or cause of death of the victim. There are many reasons that prevent this.

Drugs and metabolites can affect insects just as they affect humans, although the mechanisms and processes are not well understood. When insects consume tissues containing drugs and metabolites, this can impact their development, and the effect varies with drug and insect stage. Thus, a particular drug or metabolite concentration in an insect cannot be directly linked to what was present in the victim ante- or perimortem.

There are two roles of DNA in the context of forensic entomology. First, DNA can be used in the usual manner to identify the insect. Insects have many morphological or physical characteristics that make them identifiable using carefully developed insect keys that allow an entomologist to go through each minute feature to identify the animal. Such identifications have been done for hundreds of years and are just as useful today as they were in the past, so in most cases DNA is not required to make a simple adult insect identification. However, DNA is useful in identifying an insect when specimens are broken or badly preserved, such as in the reopening of a cold case or when eggs or very young maggots are recovered and preserved. In a good forensic insect collection, some maggots are preserved for later analysis and others are kept alive for rearing to adulthood for ease of identification, among other things. In some cases, though, particularly in older cases, immature insects have only been preserved. Later-stage maggots do have some morphological features that allow identification, but younger stages and eggs usually cannot be identified. In these situations, DNA is valuable and much work has been done to identify many forensically important species.

There is a second, much more exciting role for DNA in entomology, and that is using insects to retrieve DNA from the victim when the victim's DNA is no longer available. When maggots ingest tissue, the partially digested food is first stored in a crop or food storage organ, which is part of the foregut. It is eventually utilized and digested. The material inside this crop is an excellent source of the host DNA, that is, the deceased victim in which the insects are feeding. Although, in the early days of DNA, a relatively large sample of fresh tissue was required to get an identification, today a sample can be as small as a few cells, and extremely degraded tissue can be used to extract DNA. Even with highly decomposed human remains, it is still possible to obtain DNA, and as a result, the DNA in the insect's crop is often not needed.

Even more interesting, perhaps, is the analysis of DNA inside blood-feeding insects. Many insects blood feed, including mosquitoes, bed bugs, and fleas. The interesting thing about bed bugs is that they occupy a room rather than a person, so anyone entering that room might get bitten. A burglar may rob a bedroom and take great care not to leave fingerprints and DNA around in the usual fashion, but he may be bitten by bed bugs that are later analyzed to prove he was in the room.

8.4 REVIEW MATERIAL

8.4.1 KEY TERMS AND CONCEPTS

Carrion
Entomology
Instar
Puparial
Successional colonization

8.4.2 QUESTIONS

1. What is it about insect development that allows insects to be used to estimate elapsed time since death?
2. What is it about insect succession that allows insects to be used to estimate elapsed time since death?
3. When are blow fly eggs of most value?
4. What information is required to estimate elapsed time since death using maggot evidence?
5. What information is required to estimate elapsed time since death using insect succession?
6. How can insects be used as toxicological specimens?
7. How can insects be used to indicate wound sites?

8.4.3 ADVANCED QUESTIONS AND EXERCISES

1. Return to the example crime scene scenario presented in Chapter 3. How could entomology be used in the context of this case? What information could it provide?

BIBLIOGRAPHY AND FURTHER READING

BOOKS

Amendt, J., Goff, M. L., Campobasso, C. P., and Grassberger, M., Eds. *Current Concepts in Forensic Entomology*. Dordrecht: Springer, 2010. ISBN: 978-1402096839.

Gennard, D. *Forensic Entomology: An Introduction*. 2nd ed. Hoboken, NJ: John Wiley & Sons, 2012. ISBN: 978-0470689028.

Goff, M. L. *A Fly for the Prosecution: How Insect Evidence Helps Solve Crimes*. Cambridge, MA: Harvard University Press, 2012. ISBN: 978-0674007277.

ARTICLES

McIntosh, C. S., Dadour, I. R., and Voss, S. C. A Comparison of Carcass Decomposition and Associated Insect Succession onto Burnt and Unburnt Pig Carcasses. *International Journal of Legal Medicine* 131, no. 3 (May 2017): 835–845.

Richards, C. S., Simonsen, T. J., Abel, R. L., Hall, M. J. R., Schwyn, D. A., and Wicklein, M. Virtual Forensic Entomology: Improving Estimates of Minimum Post-Mortem Interval with 3D Micro-Computed Tomography. *Forensic Science International* 220, no. 1–3 (Jul 2012): 251–264.

Thompson, C. R., Brogan, R. S., Scheifele, L. Z., and Rivers, D. B. Bacterial Interactions with Necrophagous Flies. *Annals of the Entomological Society of America* 106, no. 6 (Nov 2013): 799–809.

Tomberlin, J. K., Mohr, R., Benbow, M. E., Tarone, A. M., and VanLaerhoven, S. A Roadmap for Bridging Basic and Applied Research in Forensic Entomology. *Annual Review of Entomology* 56 (2011): 401–421.

WEBSITES

Link	Description
www.forensic-entomology.com	Provides general information and an overview of forensic entomology
http://medent.usyd.edu.au/projects/maggott.htm	Fascinating site describing how maggots can be used in medical treatments
http://www.nafea.net/	North American Forensic Entomology Association

Biological Evidence

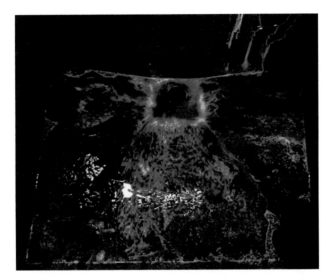

9.1 FINDING AND IDENTIFYING BIOLOGICAL EVIDENCE

Large bloodstains are easy to see, but not all crimes involve obvious stains and not all crimes involve blood alone. Other body fluids of forensic interest include saliva (oral fluid) and semen (seminal fluid). All are potential sources of DNA evidence, but they must first be located and second be identified as biological evidence before additional analysis can take place. As an example, suppose a bedsheet is delivered to a laboratory from a crime scene where a violent sexual assault took place. Clearly DNA testing will be done, but the question is where? The laboratory must first locate areas on the sheet to sample. Bloodstains may be obvious, but is it also possible that the stains are too small to see or that the sheet has been washed, leaving no visible stains. Biological evidence such as saliva or seminal fluid may not leave any visible stains at all. Thus, forensic scientists must apply specialized tests to the sheet to find the right area to sample.

Technologies such as DNA analysis have captured our attention for both the uniqueness involved and the certainty with which an individual may be identified. However, DNA is a relative newcomer to forensic science, not much more than 30 years old. What preceded DNA as a means of analysis of blood and body fluids? The discipline called forensic **serology**. A study of serology involves the examination and analysis of body fluids and, among those fluids, blood. Forensic serology deals not only with a variety of body fluids (blood, saliva, semen, and urine) but also, and more frequently, with samples that are in stain form and often degraded or deteriorated, making successful analysis more difficult.

9.2 BLOOD

Identification of a stain as blood is one of the most important preliminary tests performed on physical evidence and one that DNA protocols have not replaced. Because of the time, cost, and complexity of DNA typing, it is impractical to test every stain that appears to be blood. This is where screening or presumptive testing is most important. Typically, the flow of analysis of a stain suspected to be blood moves from more general and less specific testing to DNA typing, which under the best circumstances can link a stain to an individual with a high probability of certainty. The analysis begins with a careful visual examination of the item of evidence to locate any stains or material visibly characteristic of blood. Any stains that look promising are analyzed using a screening test; we will discuss the most common ones shortly. If the test is positive, then the species of origin of the stain may be tested, but this is not always the case. We will see why in Chapter 10. Finally, the blood is characterized using DNA, which will be discussed in detail in Chapter 10.

9.2.1 IDENTIFICATION OF BLOOD

We discussed the composition of blood in Chapter 4 in the context of bloodstain patterns, and a few points are worth highlighting. Blood has many functions and is a complex mixture of organic and inorganic materials, including electrolytes such as sodium, proteins, and several kinds of cells. The characteristic red of blood comes from the complex formed between **hemoglobin** in red blood cells (RBCs) and oxygen (Figure 9.1). This is important here because hemoglobin is the substance that the most common types of screening test for as the blood target. Hemoglobin binds with carbon monoxide as well as oxygen, but for purposes of screening tests, it is the hemoglobin component that is critical.

The identification of blood usually employs a presumptive test, frequently followed by a confirmatory test to clearly establish the identification. A visual observation of the untested stain, coupled with positive chemical presumptive and confirmatory tests, then provides sound data to support the identification of blood. A presumptive test is one that, when positive, would lead the forensic examiner to strongly suspect blood is present in the tested sample. However, the results are not absolute; further testing is always required to confirm the results. Therefore, the term *presumptive* is used to describe the test. When the results are negative, the stains likely need no further consideration. In the event of a positive test, when sufficient sample remains, further action to confirm the presence of blood is usually taken, since no single test is specific for blood. Presumptive tests may be recognized as those that produce a visible color reaction or those that result in a release of light. Both types rely on the catalytic properties of blood to drive the reaction.

9.2.2 CATALYTIC COLOR TESTS

Catalytic tests employ the chemical **oxidation** of a **chromogenic substance** (one capable of generating a colored species) by an oxidizing agent catalyzed by the presence of hemoglobin. Those

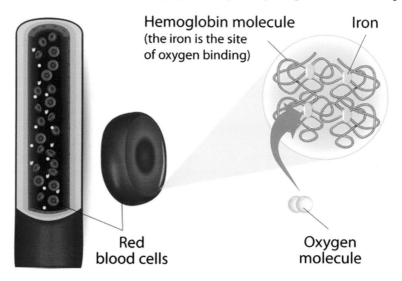

FIGURE 9.1   Hemoglobin and RBCs. The hemoglobin in the cells binds with oxygen and carries it to the tissues where it is released. Image courtesy of the National Library of Medicine, NIH, Genetics Home Reference.

tests that produce color reactions are usually carried out by first applying a solution of the chromogen to a sample of the suspected material/stain, followed by addition of the oxidizing agent (often hydrogen peroxide in a 3% solution). The catalyst is the peroxidase-like activity of the **heme group** of hemoglobin (structure containing the iron atom) present in the RBCs (Figure 9.2). A rapidly developing color, characteristic of the chromogen used, constitutes a positive test. Some methods employ the chromogen and the oxidant in a single solution. There is a potential disadvantage in this, however, as the order of addition of the specific reagents can be important. A nonblood sample capable of producing a color reaction will normally do so without the addition of the oxidizing agent so that when the color reagent is added first there is a reaction (without the addition of the oxidant). With a single solution, a reaction due to a nonblood material might be incorrectly interpreted.

Because these tests are presumptive in nature, inevitably there are instances of **false-positive** results (e.g., a positive result from a substance other than blood) and **false-negative** results (a negative result even when blood is present). Misleading results usually can be attributed to chemical oxidants (often producing a reaction before application of the peroxide), plant materials (vegetable peroxidases are thermolabile and can be destroyed by heat), or materials of animal origin (to include human) that are not blood but may contain contaminating traces of blood. The proper use of controls is needed to minimize false positives and false negatives. For example, a positive control for such a test would be a known blood sample, while a negative control might be plain water.

One method of applying the presumptive test involves sampling a questioned stain with a clean, moistened cotton swab and adding a drop of the color reagent solution, followed by a similar amount of hydrogen peroxide (H_2O_2). You may be familiar with hydrogen peroxide being used to treat wounds and prevent infection. When H_2O_2 is dropped onto a wound with blood present, it reacts with an enzyme and converts the peroxide to water and oxygen; the bubbling you see is the oxygen gas. Peroxides are also used as whitening agents for teeth. These are both oxidation reactions. Bleach is another example of a substance that is involved in oxidation reactions.

With this procedure (sample on a swab), the immediate development of the color typical of the reagent used indicates the presence of blood in the test sample. Alternatively, the evidence could be sampled by removal of a thread or fragment of dried material and testing it with the above reagents in a spot plate. Color development would then be observed in the spot plate as well. Immediate (within a few seconds) reading and recording of results is an important aspect of test result interpretation. A clearly negative result may appear positive several minutes after the test

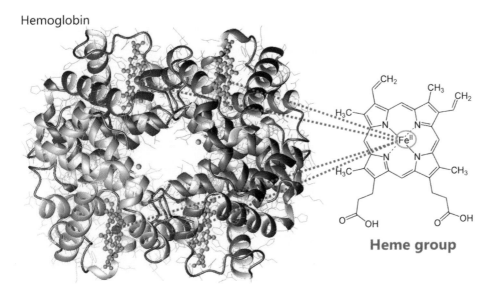

FIGURE 9.2 Hemoglobin with the heme groups shown. There are four such groups in each hemoglobin molecule. (The image of the hemoglobin is from Fermi, G., and Perutz, M. F., visualization author, user: Astrojan, GFDL [http://www.gnu.org/copyleft/fdl.html] or CC BY 3.0 [http://creativecommons.org/licenses/by/3.0], via Wikimedia Commons.)

is completed due to a slow oxidation that often occurs in air. These tests are not usually affected by the age of the stain.

Many substances have been studied as chromogens for presumptive testing for blood over the years, and many abandoned for assorted reasons, such as hazardous (carcinogenic) reagents. The two most commonly used presumptive tests today are the **Kastle–Meyer** and **Hemastix®** tests. The Hemastix test has been adopted for field use by many laboratories, particularly at crime scenes when containers of solutions can be hazardous or, at best, inconvenient. The test itself consists of a plastic strip with a reagent-treated filter paper tab at one end. Testing a bloodstain may be accomplished by moistening a cotton swab with distilled water, sampling the stain, and touching the swab sample to the reagent tab on the strip. The reagent tab is originally yellow, and a normally immediate color change to green or blue-green indicates the presence of blood (Figure 9.3). Interestingly, these test strips were originally designed (and are still sold) for detection of blood in urine, but the community was quick to adapt them for forensic applications.

Perhaps the most common presumptive test for blood currently used is based on the simple acid–base indicator **phenolphthalein**. Phenolphthalein produces a bright pink color when used as above in testing suspected blood (Figure 9.4). The reagent consists of reduced phenolphthalein (right side of Figure 9.4), which is oxidized by peroxide in the presence of hemoglobin in blood to form a pink color (left side of Figure 9.4). Notice that the only change in the indicator molecule is the loss of one hydrogen atom, but that is enough to cause the pink color to form. As with any of the catalytic tests, the result is read immediately, and a positive result a minute or more after the test is performed is usually not considered reliable.

MYTHS OF FORENSIC SCIENCE

Field tests only work on blood.

Field tests, also called presumptive or screening tests, are not specific for blood. Different reagents will react with some other substances, and results from these kinds of test should never be presented as definitive. The number of substances that can yield a false positive or create a false negative is relatively small, but they do exist. The role of a field test is to provide information needed to decide what should be done next. If a speck of reddish material is found on a table, it makes sense to test it; there is no reason to perform DNA analysis on ketchup. DNA analysis, as we will see in Chapter 10, is specific to human DNA, and even if a spot of ketchup was submitted to the lab, the DNA analysis would show that it was not a body fluid from a human being. Field tests help direct collection of evidence and, in the lab, help determine how a sample is to be handled, not to definitively identify anything.

9.2.3 TESTS USING CHEMILUMINESCENCE AND FLUORESCENCE

Often, the presence of blood is suspected, based on witness information or because it would be expected in a location, but under normal lighting and viewing little is to be seen. A drag pattern across a floor that has been cleaned up or a washed spatter pattern on a wall might be typical examples. At this point, a sensitive test such as the luminol test may be used, which involves spraying a chemical mixture on a suspected bloodstained area, usually *in situ*, and observing the result, often in a darkened area. The most well known of these is **luminol**. When luminol reacts with blood, light is produced that often enables the observer to determine the limits, shape, and some degree of detail in the original bloodstained area, often including an enhancement of blood

FIGURE 9.3   Hemastix test strips. Negative result on the left, positive on the right.

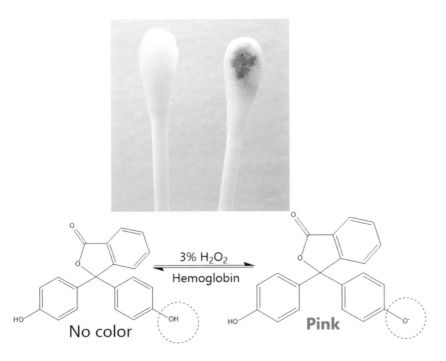

FIGURE 9.4   Simplified depiction of the Kastle–Meyer test reaction.

patterns already present. The picture at the beginning of this chapter shows luminol in a sink after blood had been rinsed down the drain.

The presence of blood and patterns displayed by blood can provide information of value, and one should be alert to both. Further, the nature of these tests makes them potential sources of contamination of the blood. Thus, if the stain can be seen and collected, these tests probably should not be used. Because these tests do add material to the tested area, their use should be carefully considered, often as a last resort. These tests are of more value in locating and defining blood than in specifically identifying it.

Luminol reacts in a fashion like that of the color tests discussed above, wherein luminol and an oxidizer are applied to a bloodstain. The catalytic activity of the heme group then accelerates the oxidation of the luminol, producing a blue-white to yellowish-green light (depending on the reagent preparation) where blood is present. The forensic application of luminol at crime scenes involves spraying a mixture of luminol and a suitable oxidant in aqueous solution over the area thought to have traces of blood present. A resultant blue-white to yellow-green glow will indicate the presence of blood. Outlines and details are often visible for up to 30 seconds before additional spraying is required. Luminol is one of the most sensitive of the presumptive tests (Figure 9.5), and as shown in the figure, even blood diluted 20,000 times shows a positive response. Interestingly, luminol can also react with blood that has been painted over, such as on walls.

9.3 SEMINAL FLUID

Next to blood, the body fluid most often seen in forensic cases is **semen** or **seminal fluid**. Semen is produced by postpubescent males and ejaculated following sexual stimulation. It is a semi-fluid mixture of cells, amino acids, sugars, salts, ions, and other organic and inorganic materials. Ejaculate volumes of human males range from 2 to 6 mL and typically contain between 100 and 150 million sperm cells per milliliter. Certain disease states, genetic conditions, excessive abuse of alcohol or drugs, prolonged exposure to certain chemicals, and elective surgery procedures may result in a drastically reduced sperm count or complete absence of sperm cells from semen.

9.3.1 SPERM CELLS

The principal cellular component of semen is the **spermatozoa**, or sperm cell, which has a characteristic tail and is approximately 55 μm in length. This flagellated tail is attached to the head via a short midpiece and accounts for about 90% of the total length of a sperm cell. A common method of detecting semen is to stain the sperm cells to allow for visual detection using a microscope. The condition of **azoospermia** (semen lacking spermatozoa) requires that other tests besides the microscopic identification of sperm be used.

9.3.2 ACID PHOSPHATASE

Acid phosphatases (APs) are a class of enzymes that can catalyze chemical reactions of certain organic phosphates. Seminal acid phosphatase (SAP) is a phosphatase found in human semen at uniquely elevated levels compared with other body fluids and plant tissues. In males, puberty

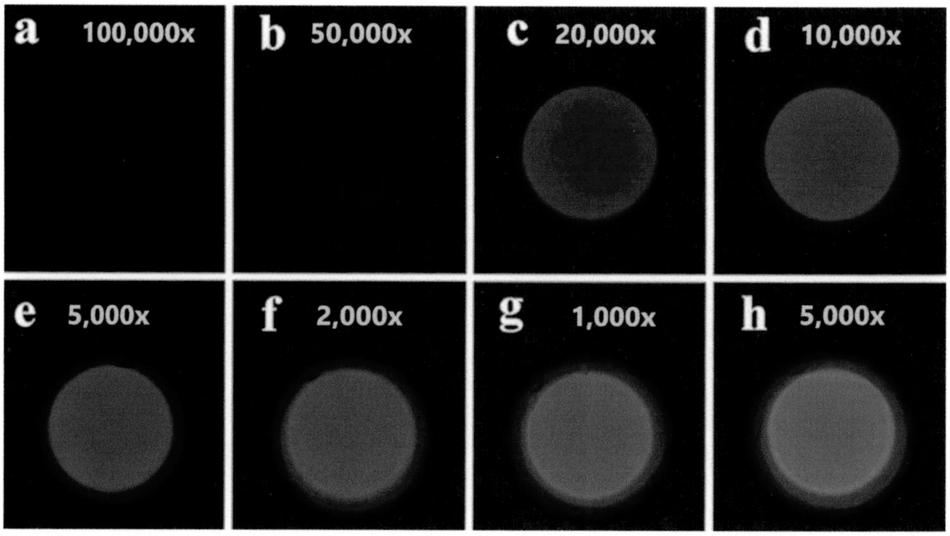

FIGURE 9.5   A version of the luminol test conducted with different dilutions of blood. (Adapted from Gao, W. Y., et al., Artemisinin-Luminol Chemiluminescence for Forensic Bloodstain Detection Using a Smart Phone as a Detector, *Analytical Chemistry* 89, no. 11 [Jun 2017]: 6161–6166. With permission from the American Chemical Society 2017.)

stimulates the large-scale synthesis of SAP by cells that line the prostate gland. SAP levels remain high until the age of about 40, after which they gradually decline.

Because SAP is present in high amounts in seminal fluids, it is the target of forensic screening tests. The current test involves a reaction that generates a color by combining a substrate (here, something that will react with the phosphatase enzyme) with a color developer. These two components are prepared in separate solutions and then combined to create a working solution or reagent that is sensitive enough to produce a positive result with semen diluted 500 times. This test is especially practical because it does not require that a suspected semen stain first be localized visually before applying it. Application of the test is a very simple two-step procedure. Because SAP is readily soluble in water, a piece of absorbent paper (filter paper is ideal since it comes in a variety of sizes) or cotton swab is moistened with sterile water and applied to the questioned stain. The reagent is added to the paper or swab and development of an intense purple color is noted (Figure 9.6). A strong reaction within 30 seconds is diagnostic for semen presence, but not definitive.

9.3.3 CONFIRMATORY TESTS FOR SEMEN

Microscopic identification of sperm cells provides unambiguous proof that a stain contains semen. It is unusual for a forensic scientist to examine semen in which sperm cells are motile since motility is lost within 3–6 hours of ejaculation; however, established staining techniques greatly assist the trained eye to easily distinguish sperm cells from extraneous material, such as epithelial cells. One of the most commonly encountered staining techniques uses specialized dyes that stain the nucleus of the sperm cell a red color and the tail portion a green color. Not surprisingly, the test is referred to as the **Christmas tree stain**, and it was developed specifically for sperm cell visualization. Stained sperm cells can be identified using microscopy. Sperm cell heads are ovoid and exhibit characteristic differential staining. The front portion is pink and the back is dark red or purple and often appears shiny (Figure 9.7). There are other methods of staining to visualize sperm cells besides the Christmas tree method, but generally these tests work in the same way by targeting compounds found in the cell. Finding sperm cells in forensic cases is often challenging because the sample (e.g., a swab or stain) is a complex mixture of material often including other cellular materials in addition to the sperm. An example is shown in Figure 9.8.

Other methods and types of staining have been developed and used to help locate sperm cells and seminal stains. Some take advantage of **fluorescence** of compounds in the cells. Many compounds can fluoresce under the right circumstances. It occurs when a compound is exposed to wavelengths of light that are absorbed by the compound, followed by the emission of light at a lower energy. For example, some compounds may absorb ultraviolet light (relatively high-energy light that we cannot see) and immediately emit light in the visible range (lower energy) that we can see. This is the basis of **alternate light sources** (ALSs) that are used to visualize stained areas on large surfaces, such as bedsheets or even at crime scenes. These tools will be described in more detail in a later section.

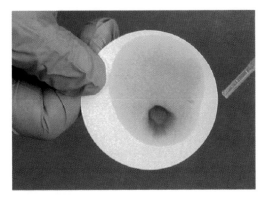

FIGURE 9.6   A positive test for SAP.

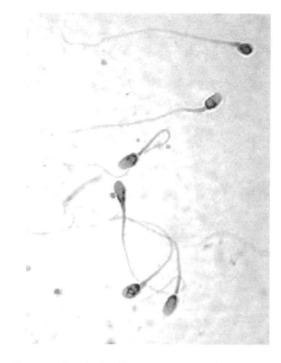

FIGURE 9.7   Sperm cells stained with the Christmas tree method.

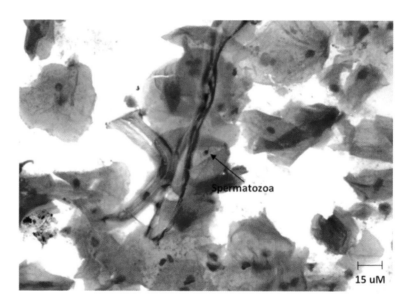

FIGURE 9.8   Stained sample showing sperm cell among other debris. (Reproduced from Westring, C. G., et al., Sperm Hy-Liter™ for the Identification of Spermatozoa from Sexual Assault Evidence, *Forensic Science International—Genetics* 12 [Sep 2014]: 161–167. With permission from Elsevier 2014.)

Finding sperm cells intact shows that seminal fluid is present, but confirmation is required. There are also cases of azoospermia in which seminal fluid is present but not sperm cells. You may have heard of the **prostate-specific antigen** (PSA, p30) test, which is used to detect prostate cancer in men. This antigen has also proven to be useful for identifying a stain as semen. PSA is found in elevated levels in the serum of males with prostate cancer, but because a positive reaction is usually required in tandem with PSA detection for semen to be unambiguously identified, this too is of little consequence forensically. Current tests for PSA are based on the appearance of a blue color, as shown in Figure 9.9.

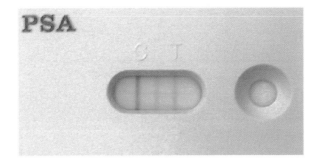

FIGURE 9.9   A positive result for PSA.

9.4 SALIVA

Saliva is a slightly alkaline secretion composed of water, mucus, proteins, salts, and enzymes found in the mouth. Humans produce 1–1.5 L of saliva per day. Its primary purpose is to aid in the initial stages of digestion by lubricating food masses for easier swallowing and initiating the digestion of starches using the enzyme **amylase**. No test is specific for saliva. Amylases are common enzymes found in both animals and plants. They are responsible for catalysis of the components of starch into smaller, less complex sugars. Although alpha-amylase is found in many body fluids and tissues, it still serves as a good marker because it is found at levels 50 times higher in saliva than in most other body fluids and is relatively stable. Some activity can be detected for up to 28 months.

9.4.1 STARCH–IODINE TEST

In the presence of iodine, starch appears blue. As amylase acts on starch to break it down, the color changes and subsides. This test has several drawbacks; for example, proteins that are present, particularly albumin and gamma-globulin originating in other body fluids (e.g., blood, semen), compete with starch for iodine and produce false-positive results. This test is also difficult to use as a locator test for stains on items. One version of this method is still in use as the radial diffusion test. An agar gel containing a known concentration of starch is prepared, and iodine is poured over the gel. An extract of the questioned sample is added to the wells in the gel. As the starch diffuses, it breaks down, leaving a circular void area proportional to the amount of amylase present.

9.4.2 PHADEBAS® REAGENT

More common is the use of commercial products in which starch is linked to a dye molecule to form an insoluble complex. When starch is cleaved from the dye by amylase, the dye molecule becomes soluble, producing a colored product that can be measured with a spectrophotometer. The degree of coloration is proportional to the amount of amylase in the sample. This procedure is commonly referred to as a *tube test*. Alternatively, the reagent can be dissolved and applied to large sheets of filter paper, which can then be placed on an item to map the location of amylase-containing stains for further analysis. This procedure is referred to as a **press test**, and **Phadebas** reagents are the most commonly used. An example is shown in Figure 9.10.

9.5 ALTERNATE LIGHT SOURCES

We mentioned the use of ALS devices previously, and this technique has become invaluable in forensic labs as well as at crime scenes. An ALS system typically has a light source and goggles, each with different filters to allow for one type of light to be applied to the sample and another to

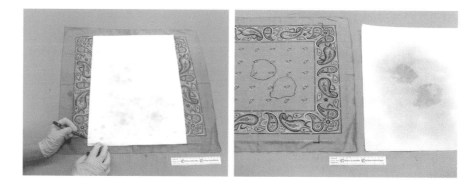

FIGURE 9.10   A positive result indicating the presence of saliva using a press test. The blue color is the starch–iodine complex.

be seen by the person wearing the goggles. These tools are particularly valuable for finding areas stained with biological fluids, including semen, saliva, urine, and vaginal fluids. Figure 9.11 illustrates how a generic ALS works. A light source is pointed toward the surface where a stain may be present. The light may be specialized, or it may be filtered to a specific range of wavelengths. When this light interacts with compounds that can fluoresce, light is emitted at a lower energy that is then either directly visible or visible through another type of filter. An example of an ALS in use is shown in Figure 9.12. Notice that the light is blue and the goggles are orange.

ALS systems can be used for many kinds of stains. While visualizing a stain with an ALS doesn't tell the forensic scientists what that stain is, it does provide critical information on what locations on the surface (generically referred to as the substrate) should be tested further. An example of a such a system, coupled to a camera, is shown in Figure 9.13. The camera system has a filter to illuminate the samples. In the top right frame, the top row shows colored cards in normal light and how they appear when exposed to the ALS light. Note how the orange card looks yellow; this is because of the filter system used. The stain is semen. In the bottom series of images, the top row shows semen and the bottom shows urine stains.

9.6 RETURN TO THE SCENE OF THE CRIME

The crime scene scenario laid out in Chapter 3 would require the use of many of the techniques described in this chapter. In the house, there is no lack of blood evidence, so an important consideration for the crime scene investigators (CSIs) will be the collection of representative samples. The evidence shown in Figure 3.13 (footwear impression and droplet) should be screened and the evidence collected for DNA analysis given that who the blood belongs to will be an

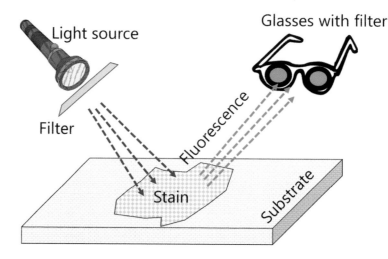

FIGURE 9.11   Simplified illustration of the principles of an ALS.

FIGURE 9.12   Investigator using an ALS. (Image from U.S. Army Criminal Investigation Division [CID].)

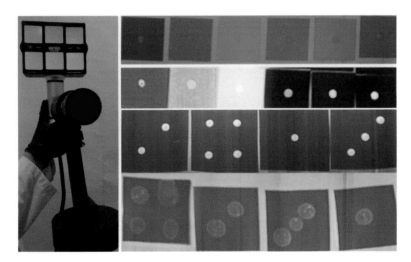

FIGURE 9.13   An ALS system used with various stains and substrates. (Reproduced from Sheppard, K., et al., The Adaptation of a 360 Degrees Camera Utilizing an Alternate Light Source (ALS) for the Detection of Biological Fluids at Crime Scenes, *Science and Justice* 57, no. 4 [Jul 2017]: 239–249. With permission from Elsevier 2017.)

important piece of information in determining what transpired at the scene. The room in shown in Figure 3.14 would require less in the way of screening tests for blood. For example, it would not be necessary to test each speck of red shown on the wall in the arterial spurt pattern; the cloth shown in Figure 3.15 would probably be tested at the lab using one of the techniques described above. In this scenario, a detailed search of the truck for traces of blood would be more challenging. There appears to be blood on the door area; whose blood would be important. Also, the bed of the truck would be examined. If the body was transported in the truck, finding the wife's blood there would be important and significant to the case.

9.7 REVIEW MATERIAL: KEY CONCEPTS AND QUESTIONS

9.7.1 KEY TERMS AND CONCEPTS

Acid phosphatase
Amylase
Azoospermia

Christmas tree stain
Chromogenic substance
False negative
False positive
Fluorescence
Hemastix
Heme group
Hemoglobin
Kastle–Meyer test
Luminol
Oxidation
Phadebas
Phenolphthalein
Prostate-specific antigen
Press test
Presumptive test
Screening test
Semen/seminal fluid
Serology
Spermatozoa

9.7.2 REVIEW QUESTIONS

1. An interesting procedure that can be used in conjunction with fluorescent staining for sperm cells is called laser dissection. The cell is located under a microscope and then cut from the sample. Why would this be useful in a case from a suspected sexual assault?
2. In what cases would it be possible to obtain a positive result for a PSA test but not to find sperm cells?
3. Why might it be of value to find a urine stain?
4. If luminol is such a sensitive reagent, why isn't it used on all samples instead of tests like the Kastle–Meyer test?
5. What is the main purpose of screening tests for blood and seminal fluid?
6. Can you tell what a stain is just by use of an ALS? Why or why not?

9.7.3 ADVANCED QUESTIONS AND EXERCISES

1. Suppose that the results of the processing of the truck indicate that blood is present in the cab but not in the bed. What does this suggest in regard to the crime scene scenario? List several possibilities.
2. Share your ideas from the previous question in a group. The narrative in Chapter 3 provided information about how the crime was committed. Did your responses to the previous question reflect confirmation bias? Explain.
3. Since DNA analysis is specific to human body fluids such as blood, why do CSIs even bother with doing field tests for biological fluids? Why not submit all samples that might have any kind of biological fluid to the lab?

BIBLIOGRAPHY AND FURTHER READING

BOOKS

Gaensslen, R. E., and Lee, H. C. *Forensic Science: An Introduction to Criminalistics.* 2nd ed. New York: McGraw-Hill, 2007. ISBN: 978-0072988482.

Li, R. Forensic Serology, in *Forensic Chemistry Handbook* (ed L. Kobilinsky), Hoboken, NJ, USA: John Wiley & Sons, Inc., 2011. ISBN 978-0471739548.

Li, R. *Forensic Biology*, 2nd ed. Boca Raton: CRC Press, 2015. ISBN: 978-1439889701.

ARTICLES

De Moors, A., Georgalis, T., Armstrong, G., Modler, J., and Fregeau, C. J. Sperm Hy-Liter™: An Effective Tool for the Detection of Spermatozoa in Sexual Assault Exhibits. *Forensic Science International—Genetics* 7, no. 3 (May 2013): 367–379.

Miller, K. W. P., Old, J., Fischer, B. R., Schweers, B., Stipinaite, S., and Reich, K. Developmental Validation of the Sperm Hy-Liter™ Kit for the Identification of Human Spermatozoa in Forensic Samples. *Journal of Forensic Sciences* 56, no. 4 (Jul 2011): 853–865.

Myers, J. R., and Adkins, W. K. Comparison of Modern Techniques for Saliva Screening. *Journal of Forensic Sciences* 53, no. 4 (Jul 2008): 862–867.

Sheppard, K., Cassella, J. P., Fieldhouse, S., and King, R. The Adaptation of a 360 Degrees Camera Utilising an Alternate Light Source (ALS) for the Detection of Biological Fluids at Crime Scenes. *Science and Justice* 57, no. 4 (Jul 2017): 239–249.

Vitiello, A., Di Nunzio, C., Garofano, L., Saliva, M., Ricci, P., and Acampora, G. Bloodstain Pattern Analysis as Optimisation Problem. *Forensic Science International* 266 (Sep 2016): E79–E85.

Westring, C. G., Wiuf, M., Nielsen, S. J., Fogleman, J. C., Old, J. B., Lenz, C., Reich, K. A., and Morling, N. Sperm Hy-Liter™ for the Identification of Spermatozoa from Sexual Assault Evidence. *Forensic Science International—Genetics* 12 (Sep 2014): 161–167.

WEBSITES

Link	Description
http://www.exploreforensics.co.uk/serology.html	Overview of testing of biological fluids

DNA Typing

10.1 CELLS, CHROMOSOMES, AND DNA

DNA (deoxyribonucleic acid) is found in cells. There are two types of DNA that can be typed using forensic methods—**nuclear DNA** (found in the nucleus) and mitochondrial DNA (**mtDNA**; found in the mitochondria). Aside from red blood cells, other cells in blood, skin cells, and cells found in other body fluids do have nuclear DNA. We begin by discussing this type of DNA, which carries our genetic information on structures called chromosomes. The relationship between cells, chromosomes, and DNA is shown in Figure 10.1.

The DNA molecule found in chromosomes is the currency of genetics. Recall that DNA is a long double-stranded molecule that in the cell is found in a twisted ladder shape referred to as a double helix. Nuclear DNA is found on structures called **chromosomes**. Normally, humans have 23 pairs of chromosomes, including the gender chromosomes called the X and Y chromosomes. Why pairs? For each chromosome, we inherit one from our mother and one from our father. If we are female, we will inherit an X from our mother and an X from our father. If we are male, we will inherit one X chromosome from our mother and one Y chromosome from our father. Human chromosomes are shown in detail in Figure 10.2. The differences in male and female chromosomes and DNA are used as part of the DNA typing procedure to determine the sex of a person who is the source of a biological sample.

The characteristics that we inherit from our parents are encoded in regions of the chromosomes referred to as a **gene**. A gene is thus a region on a chromosome that is composed of DNA that provides instructions on how to produce a particular protein. One gene codes for one protein, but not necessarily one trait. For example, eye color is a **polygenetic** trait because many genes are involved in determining it. The relationship between chromosomes and genes

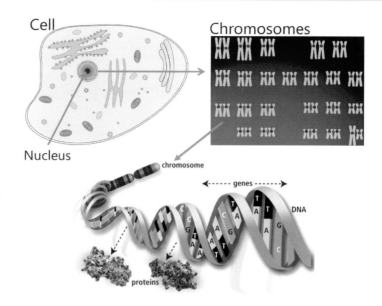

FIGURE 10.1   Cells to proteins. Nuclear DNA is found in the nucleus on 23 pairs of chromosomes. The DNA in the chromosomes provides instructions for making proteins. (Adapted in part from graphics archived by the Human Genome Project, National Institutes of Health.)

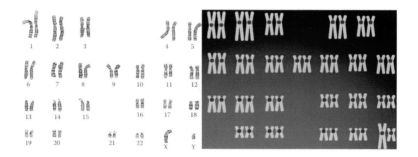

FIGURE 10.2   Human chromosomes. Left: Microscopic images of actual chromosomes. Right: Diagram form.

is shown in Figure 10.3. In this example, the chromosome map at the bottom of the figure corresponds to chromosome 9; the chromosome contains DNA that encodes for a protein made by the gene.

The molecule at the heart of chromosomes and genes is DNA (Figure 10.1), which is structured as a double helix in which the two strands are mirror images of each other. This means that having one strand of DNA automatically defines what the other strand is made of because of the complementary nature of the building blocks. The intact DNA molecule is composed of adenine (A), thymine (T), cytosine (C), and guanine (G) **nucleotides** (Figure 10.4). A nucleotide is defined as the unit consisting of the base (A, C, G, or T) connected to a sugar molecule and phosphate group. The sugar and phosphate group is the backbone of the helix. The complementarity of the two halves of the molecule or the two strands depends on the fact that A only pairs with T, and C only pairs with G, except during mutation. This leads to the term **base pair** (bp), which refers to the one base and its complement on the complementary backbone of DNA. Human chromosomes (Figure 10.2) each contain ~50 million to 300 million bp each, and the current estimate of the number of genes in human chromosomes is about 19,000.

The base pair sequence in DNA is the order in which the bases are found on the DNA strand. As we will see in the section on sample preparation, the fact that A pairs with T (and not G or C)

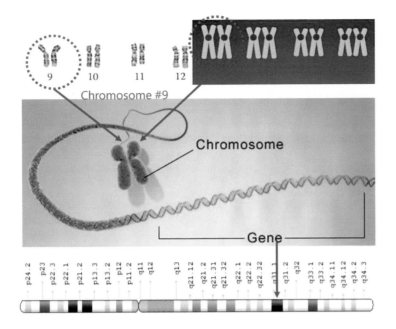

FIGURE 10.3. Genes are defined regions of DNA on a chromosome. (Adapted from images obtained from the National Library of Medicine and archive by the Human Genome Project, National Institutes of Health.)

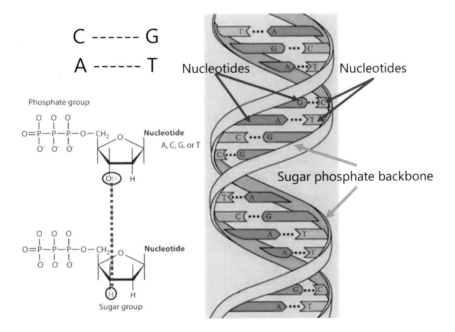

FIGURE 10.4 Base pairs are attached to the sugar–phosphate backbone of DNA.

and that C pairs with G (and not A or T) is invaluable because the sequence on one of the sugar–phosphate backbones automatically dictates the sequence on the other strand. If you "unzip" the double helix and separate it into two strands, you can recreate the original double-helix pattern from either strand due to the complementary nature of the bases (AT and CG; Figure 10.5). Strand 1 (red), when untwisted, has the sequence of bases shown in red letters. The complementary sequence is on the other sugar–phosphate backbone. Either strand could be used to recreate the complementary one. The first base pair in this example strand would be the TA pair and the base pair sequence would be shown as TGACTGGTCAA.

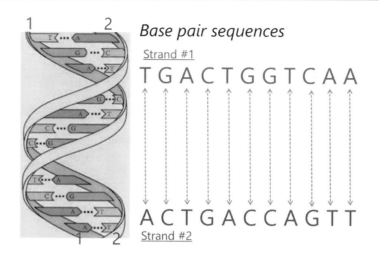

FIGURE 10.5   The base pair sequence in the DNA segment shown here and in Figure 10.4.

MYTHS OF FORENSIC SCIENCE

DNA evidence is infallible.

There is no such thing as infallible evidence. DNA is powerful evidence, but it is not infallible. In mixture cases, where there is more than one person's contribution, it can be challenging to interpret and identify which type belongs to which person. Touch DNA cases, where the tiniest bits of DNA are present, are equally difficult, as are samples that are old and degraded. Suppose a sample was taken from the handle of the door to your school. DNA methods have become so sensitive that the results would be a mixture of possibly hundreds of contributors. There is currently a significant debate and discussion occurring in the forensic DNA community regarding how DNA mixture data should be interpreted and why types of numerical and statistical analyses provide the most reliable information.

10.2 DNA TYPING

The number of genes in the human genome does not account for all the DNA present—far from it. Current estimates are that 90% or more of the DNA in chromosomes is noncoding DNA. Much of this noncoding DNA sequence is repetitive; the same sequence is repeated over and over, either in a tandem array or dispersed over the chromosomes. Interspersed repeats are distributed throughout the genome and are found between and sometimes in the coding regions. These repeats account for about 50% of the human genome and are what are targeted in forensic DNA typing methods. Specifically, we will focus on **short tandem repeats** (STRs). STRs are small, in that most have less than 40 repeats of a given sequence; most sequences contain 4 or 5 bp. Including their adjacent **flanking regions**, the STR regions are only 300–400 bp long. The forensic value of STRs is that the sequence itself does not vary, but the *number* of repeats does. The term used to describe this is **polymorphic** (more than one form), and the number of repeats is an inherited characteristic. For example, suppose that at a specific location on a chromosome, the sequence contributed by your mother has four repeats of a sequence, such as ATTA, and that contributed by your father has two repeats of that same sequence. Your type at that location on the chromosome would be referred to as "4,2," and this would be described as **heterozygous**.

If both parents have four repeats, you would be a type 4 and described as **homozygous** at that location on the chromosome.

10.2.1 LOCI AND STRS IN FORENSIC DNA TYPING

Currently, most forensic DNA typing methods focus on 13 locations (loci) with polymorphic STRs. These are not the only known loci that can be typed, but they represent the ones most frequently used in casework. These were selected by a task force in the late 1990s and this list was used as the standard to develop a database of DNA results. This database is called **CODIS** (Combined DNA Index System); it is discussed in a later section. The 13 loci are shown in Figure 10.6; refer to Figure 10.2 for chromosome information. Notice the X and Y chromosome in the lower right of the figure, which, as noted above, tells the sex of the donor. The amelogenin gene (AMEL) is found on both chromosomes, but there is a specific section of that gene that is different on X and Y. By targeting this region, it is possible to tell male from female donors because a female with be homozygous at that loci while a male will be heterozygous.

10.2.2 SEQUENCE OF ANALYSIS

The flow of a DNA typing analysis is shown in Figure 10.7. The form of the sample can be almost anything with a biological material on it. For example, a common method of collecting DNA samples is to swab the inside of a person's mouth (called a **buccal swab**), a process that collects cells and debris that contains nuclear DNA. Other common samples include blood (liquid and dried stains), semen, vaginal swabs, hair with the root attached, bone, tissue, and urine. The amount of DNA present in samples is usually measured in units of nanograms per microliter. A typical can of soda (12 ounces) contains about 350 mL. A nanogram is 1 one-billionth of

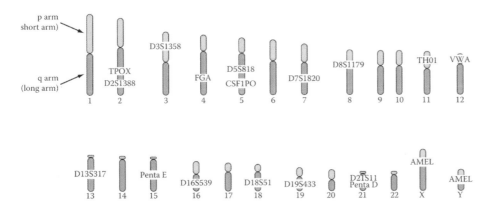

FIGURE 10.6   The 13 STR loci used in DNA typing and the chromosomes where they are located.

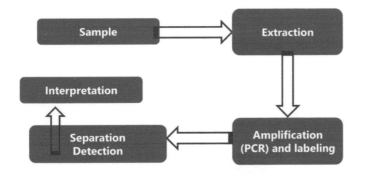

FIGURE 10.7   Flow of a DNA typing analysis.

a gram; a penny weighs about 2.5 g, or 2,500,000,000 ng. Thus, even fresh whole blood, which contains on the order of 30,000 ng/mL DNA, still has an astonishingly small amount of the material.

The first step in the analysis is the extraction of the DNA from the sample, such as a piece of clothing, which is usually accomplished by swabbing. The swab is then extracted to capture the DNA. There are several methods that can be used, ranging from a simple solvent extraction to automated procedures. Associated with the extraction, there is usually a quantitation to determine how much DNA is present in the extract. This is important as there is a range of DNA concentration that will yield the best analytical result; too little DNA can be as bad as too much. In the latter case, the sample can be diluted as needed, but if too little is present, the next step (amplification) can be altered to increase the amount present.

HOW DNA BECAME A FORENSIC TOOL

Sir Alec Jeffreys.

In November 1983, a 15-year-old schoolgirl, Lynda Mann, was raped and strangled on the Black Pad footpath in Narborough, Leicestershire. Despite a massive manhunt, the murder went unsolved. A few years later, on July 31, 1986, another 15-year-old schoolgirl, Dawn Ashworth, was raped and strangled nearby on Ten Pound Lane near a psychiatric hospital. A cook at the psychiatric hospital eventually confessed to the second murder. Police collected more than 4500 voluntary samples from local men between the ages of 16 and 34 from three villages. **Sir Alec Jeffreys**, a geneticist at Leicester University, had recently developed DNA testing (Figure 9.5). Jeffreys performed DNA tests at Leicester University on the evidence samples and found that the same man had committed both murders and that man was not the cook.

Colin Pitchfork, a local baker and convicted flasher, married with children, had avoided taking the voluntary blood test by having a coworker take it for him. No matches were found between the sperm taken from both victims and the 4500 men tested. Then, in September 1987, another coworker tipped off police. The police immediately arrested Pitchfork. When they asked him why he had done it, he stated because they were there. In January 1988, Pitchfork became the first criminal caught with DNA evidence and was sentenced to two life sentences. The cook who had confessed to the second murder became the first suspect exonerated with DNA.

10.2.3 PCR AND DNA AMPLIFICATION

Because of the way nucleotides pair (AT and CG), if you have one strand of DNA, you automatically know the base pair sequence is on the other. Thus, each single strand of DNA can create two exact copies of the original. This property is exploited to copy and amplify DNA using a technique called the **polymerase chain reaction** (PCR). The first step in the process is to separate or unzip the DNA double helix, a process that is also referred to as denaturing or melting, at elevated temperatures around 90°C. To reassemble the strands (rezip), the sample must be cooled and an enzyme added (a **DNA polymerase**).

Most DNA polymerases function at normal body temperature, 37°C (98.6°F), but not at elevated temperatures. Thus, the heat needed to unzip the DNA also made it impossible to rezip the molecule into the double helix. The solution to this came from scientists working with Dr. Kary Mullis at Cetus Corporation when they decided to check the stability of DNA polymerases from bacteria that live in hot springs. Dr. Mullis was awarded the 1993 Nobel Prize in Chemistry for "contributions to the developments of methods within DNA-based chemistry," specifically for his invention of the PCR process. Unlike other known polymerases, the hot springs polymerases that his group studied could withstand the 95°C heat required to split DNA into its two strands. This was a critical discovery because it made it possible to unzip and rezip DNA strands using heat. This also meant that the process of opening and closing the molecule could be automated using repeated heating and cooling cycles (**thermocycling**). The process is outlined in Figures 10.8 and 10.9.

Figure 10.8 shows what occurs during a single cycle. The sample is heated to 94°C, which causes the DNA strands to separate (melt or denature). The sample is then cooled, and a mixture of **primers** are added. These primers are sequences from 6 to more than 39 bps that bind to the flanking region of the target (the STR loci to be copied). Once the strands open, the primers find the portion of the DNA that is on either side of the STR that is to be copied. A mixture of nucleotides (A, T, C, and G) are added and bind to the two separate strands based on their complementary nature in a process called **annealing**. The DNA polymerase is added, the strands zip back together, and one strand of DNA has become two. Each cycle only takes a few minutes and the entire PCR process (20–30 cycles) typically requires about 3 hours and produces millions of copies of the DNA sections of interest. These copies are known as **amplicons**.

10.2.4 ANALYSIS

DNA samples are analyzed using an instrument that combines a separation unit and a detector. Conceptually, this is like the instruments we discussed in Chapter 6 (see Figure 6.15). The instrument is designed to separate the different amplified DNA segments and then to detect and record the findings. The separation is accomplished using a technique called **capillary electrophoresis**

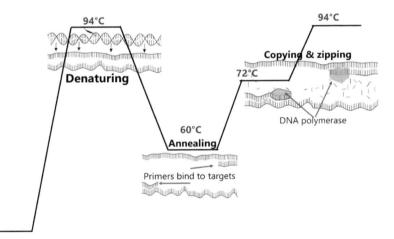

FIGURE 10.8   A single PCR thermocycle.

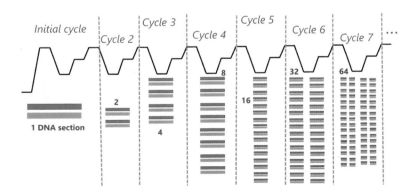

FIGURE 10.9   Multiple PCR cycles and resulting DNA copies.

(CE). The capillary (or in some instruments, capillaries) is a small tube filled with a buffer solution containing a polymer mixture. The buffer is designed to keep the DNA fragments negatively charged. The polymer is a gel-like material that slows the DNA as it is drawn through the capillary; the larger the DNA fragment, the slower it moves through the gel. Electrical potentials are used to drive the flow; the negatively charged DNA fragments are drawn to the oppositely charged end of the tube. A schematic of the instrument is shown in Figure 10.10.

The fragments are detected using a process called **fluorescence**. Fluorescence is the process of light emission that, in this case, is stimulated by exposing the DNA fragments to an intense laser. During the amplification process (PCR), a fluorescent molecular probe is attached to the amplified regions of the DNA. These probes are designed to emit light of different colors, which further helps identify the DNA fragment as it passes in front of the laser and emits an intense fluorescent signal. The processed output, called an **electropherogram**, is shown in the lower part of Figure 10.10. The labels below the peaks correspond to STR loci, as shown in Figure 10.6.

At each locus shown in the electropherogram, there will be one or two peaks, assuming the sample is from a single donor. If there are two peaks, the donor is heterozygous, meaning that the person inherited two different repeat lengths from their parents. If there is only one peak, the person is homozygous, and the number of repeats inherited from each parent is the same.

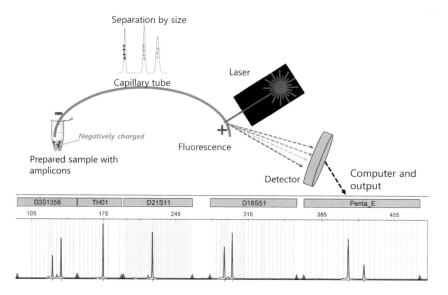

FIGURE 10.10   Separation and detection of amplified STRs. DNA is drawn toward the oppositely charged end of the tube, where a laser illuminates a small window; the molecular probes fluoresce and provide a signal. An example output is shown in the lower part of the frame.

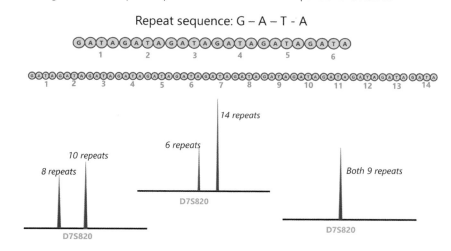

FIGURE 10.11   How repeating units appear in an electropherogram.

A detailed example is shown in Figure 10.11. In this example, the sequence of base pairs that repeats is GATA, and it corresponds to the loci called D7S820. The donor of the sample inherited 6 repeats from one parent (top of the figure) and 14 repeats from the other parent. When the DNA typing is performed, the electropherogram will show the pattern seen in the middle of the figure and would be reported as a 6,14. If both parents contributed nine repeats, one peak would result (right of figure); an 8,10 type is shown at the left.

All DNA typing results include an **allelic ladder**. It is created by analyzing a standard with known numbers of base pairs, and it is used as a reference and to ensure that the amplified samples are appearing at the proper time in the analytical run. The ladder is the lower part of the output shown in Figure 10.12, and the numbers refer to the number of base pairs; the larger the number, the larger the fragment and the longer it takes to travel through the capillary column. This figure also shows how the sex of the donor can be determined using DNA typing. Notice at the upper left of the figure that there are peaks labeled X and Y; this means that the donor of this sample was male. A sample from a female would have only one peak at X.

10.2.5 CODIS

We discussed earlier how 13 loci were selected for routine DNA STR typing in the United States. The database used to store and access the results from DNA typing was established in the early 1990s through the efforts of the FBI and other agencies. In January 2017, the number of loci was increased to 20. Results can be added to the database as long as the analysis meets strict quality standards. CODIS is a system of linked databases that start at the local level and move through the state level to the national level. The data is also grouped into indices such as a convicted offender index and a missing persons index; collectively, this is referred to as the NDIS (National DNA Index System).

There is a common misconception that personal data is stored with DNA profiles in the CODIS system, but that is not the case except for missing persons. Rather, the profile is stored with information regarding the agency that submitted it. Once a submitted profile is found to match one in the database, the submitting agency is put in contact with the agency that placed the profile in

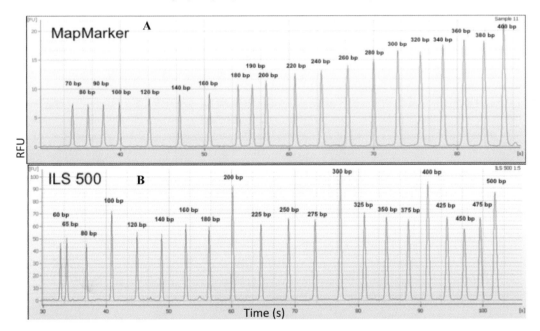

FIGURE 10.12 Example allelic ladders. The label at left, RFU, refers to the intensity of the fluorescent signal. The peaks are labeled by number of base pairs (bp). Two different methods are shown. (Reproduced from Aboud, MJ., Gassman, M., McCord, B., "Ultrafast STR Separations on Short-Channel Microfluidcs Systems for Forensic Screening and Genotyping, *Journal of Forensic Sciences* 60, no. 5 [September 2015]: 1163–1170. With permission from Wiley 2015.)

the database originally. The CODIS system has grown dramatically. At its inception, it held about 120,000 profiles; as of October 2017, it contained about 13 million offender profiles, 2.8 million arrestee profiles, and 800,000 forensic (casework) profiles.

A (COLD) CASE EXAMPLE

One of the most valuable uses of DNA typing and DNA databases is in solving cases that are years or even decades old. As long as evidence was collected and preserved, this is possible. One example is from a vicious sexual assault that occurred more than 30 years ago in the early 1980s, before DNA typing methods were available. The victim was attacked, saw her friend shot in the head and killed, and was then shot multiple times herself and left for dead. Remarkably, the victim survived and consented to the collection of evidence related to the sexual assault. Decades later, when the evidence was examined using DNA and the results compared with DNA databases, a matching profile was found and the suspect, now in his late 50s, was already in prison for another offense and was convicted and given 30 years without parole.

Source: Connery, S. A., Three Decade Old Cold Case Murder Solved with Evidence from a Sexual Assault Kit, *Journal of Forensic and Legal Medicine* 20, no. 4 (May 2013): 355–356.

10.3 MITOCHONDRIAL DNA

Mitochondrial DNA is located outside the nucleus of the cell in the energy-producing organelles known as **mitochondria** (Figure 10.13). The forensic advantage of this type of DNA analysis is the large number of mitochondria per cell and DNA molecules per mitochondrion. The mitochondrial genome may be present in a cell, from 100 copies to several thousand. A single hair root can be easily typed using mitochondrial analysis, as can hair shafts, bones, teeth, ancient samples of tissue, saliva, blood, semen, and any sample that is amenable to DNA sequencing. One feature of mtDNA inheritance is that human relatedness may be evaluated over several generations. All members of a **maternal lineage** will share the same mtDNA sequence. Your mtDNA comes from your mother, which came from your grandmother, which came from your great grandmother, and so on. Thus, standards for comparison may be collected from any maternal relative. In general, if one compares the mutation rate of mtDNA versus nuclear DNA, mtDNA mutates at a much higher rate, sometimes estimated at one order of magnitude, or 10 times that of single-copy DNA.

Mitochondrial DNA is a relatively small molecule; however, some regions of the molecule are very polymorphic, such as **hypervariable regions** 1, 2, and 3 (HVI, HVII, and HVIII) (Figure 10.14). HVI and HVII are sufficiently polymorphic to allow two samples to be differentiated quite easily, but polymorphism in the third region is currently being evaluated and will increase the power of mtDNA analysis. Full mtDNA analysis is currently being explored as a source for further polymorphisms, which will increase the power of this test. This type of DNA is useful when skeletal remains are found or when other DNA is not usable. An example is shown in Figures 10.15 and 10.16. In this case, a paper towel was found burned during the investigation of a suspected arson. Parts of the towel were treated with ninhydrin to help visualize any fingerprints (Figure 10.15). This chemical creates a purple color and can hamper DNA recovery. DNA typing using

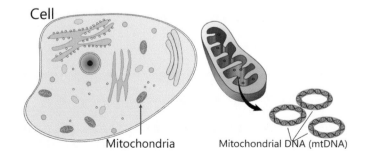

Cell

Mitochondria

Mitochondrial DNA (mtDNA)

FIGURE 10.13 Mitochondria in the cell and mtDNA.

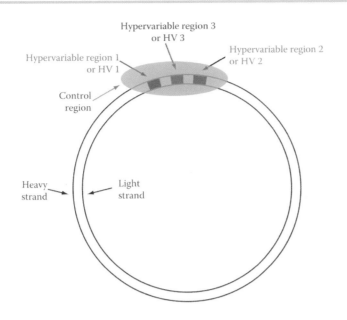

FIGURE 10.14   Variable regions of mtDNA.

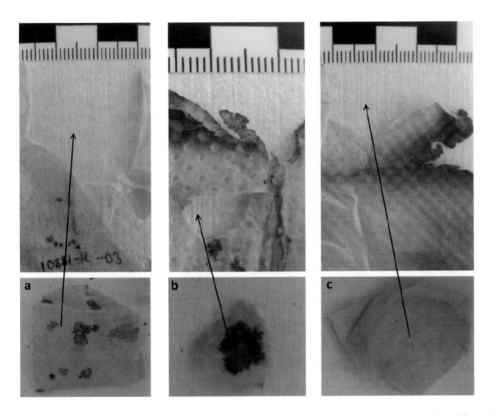

FIGURE 10.15   Sections of a burned paper towel sampled for DNA analysis. (Reproduced from Bus, M. M., et al., Analysis of Mitochondrial DNA from a Burned, Ninhydrin-Treated Paper Towel, *Journal of Forensic Sciences* 61, no. 3 [May 2016]: 828–832. With permission from Wiley 2016.)

STRs was attempted and resulted in mixtures, a topic we address in the next section. Analysis of the mtDNA also produced mixtures, but two samples (sample 10, section c in Figure 10.15 and sample 13, section b in Figure 10.16). were found to be from a single donor. Notice that the typing of the mtDNA is done by sequencing the base pairs; Figure 10.16 shows the base at the top of each frame (G, C, T, and A). The numbers at the bottom indicate the position (location) with the hypervariable regions that were sequenced, HVI and HVII in this case.

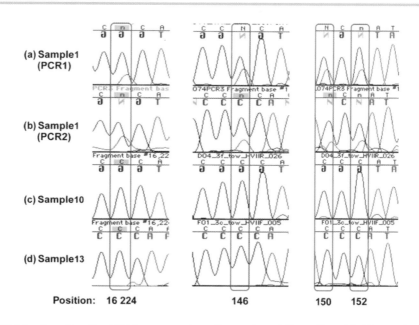

Position: 16 224   146   150 152

FIGURE 10.16 Results of mtDNA sequencing. (Reproduced from Bus, M. M., et al., Analysis of Mitochondrial DNA from a Burned, Ninhydrin-Treated Paper Towel, *Journal of Forensic Sciences* 61, no. 3 [May 2016]: 828–832. With permission from Wiley 2016.)

10.4 CURRENT ISSUES IN DNA TYPING

10.4.1 MIXTURES AND TOUCH DNA

In many sexual assault cases, it would be beneficial to be able to distinguish male DNA from female DNA. Several approaches have been developed to deal with these types of mixtures. The first is based on **differential extractions**, in which an attempt is made to separate cellular material from the female from the sperm cells of the male. Typically, the cells from the woman are broken open with an enzyme that does not work on sperm cells. The DNA from the female cells are then separated out for analysis and the sperm cells treated with a different enzyme to open the cells and free the nuclear DNA. There are other methods that use lasers and high magnification to physically select and separate the sperm cells from the mix. This approach is called **laser-capture microdissection** (LCM). Look back at Figure 9.8, which shows a mixture of cells; you can see how exacting this process must be to separate the sperm cells from everything else. An interesting approach using magnetic beads that are treated to specifically attract sperm cells is shown in Figure 10.17. Once attached, the beads and cells are separated using a magnet.

Another method applied to mixed samples focuses on the sex chromosomes (X and Y). A set of STR markers associated with just the Y chromosome has recently been developed. Recall that within our 23 pairs of chromosomes, women have an XX and men have an XY pair. Unlike conventional STRs, where two alleles per locus is the norm, a man only has one Y in his XY gender chromosome. And because women do not have Y chromosomes as part of their genetic material, Y-STR typing is the only test that can unambiguously determine what a male has contributed to

FIGURE 10.17   Sperm cell attached to bead. (Reproduced from Zhao, X. C., et al., Isolating Sperm from Cell Mixtures Using Magnetic Beads Coupled with an Anti-Ph-20 Antibody for Forensic DNA Analysis, *PLoS One* 11, no. 7 [Jul 2016] [open source].)

a mixed sample, like those collected as part of most rape investigations. Y-STRs are especially invaluable when very few sperm are detected in the sample, when the rapist has had a vasectomy or is sterile, when previous STRs show no Y signal at the amelogenin (gender) locus, when differential extraction is unsuccessful, when there are multiple semen donors, or when the ratio of female to male DNA is so large that the female DNA masks the male DNA. For the same reason, Y-STRs are advantageous in analyzing fingernail scrapings from a female assault victim, separating out the male skin cells from a ligature used in the strangulation of a female victim, or testing microscope slides from cold case rape kits.

A different version of a mixture problem arises from improvements in DNA typing methods. The techniques used in DNA typing have become so sensitive that usable data can be retrieved from samples that are not visible. This adds power to the technique but also can cause complications. For example, suppose a DNA sample is taken from a doorknob or a steering wheel of a family car. It is likely that many people have touched such objects and interpreting the results can be difficult because of the mixture issues as well as determining which profiles are of forensic interest and which are not. The small residuals left by contact with an object are referred to as **touch DNA**. Because DNA amplification methods have become so efficient and effective, DNA has acquired some of the characteristics of trace evidence, such as secondary transfer. Thus, interpretation can become more difficult as DNA may be found on an exhibit of evidence that did not get there through any event linked to a crime. An example study of this type of transfer is shown in Figure 10.18. In this experiment, the researchers simulated a stabbing in which only one person handled the knife. You can see in several locations where contributions from other people are evident. The researchers suggest that the extraneous DNA could have come from secondary transfer from other people in the lab onto the knife.

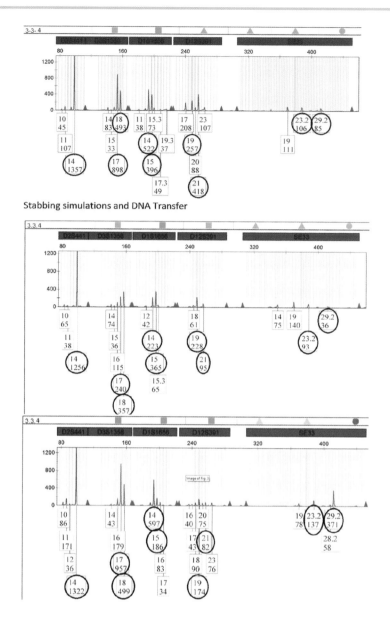

FIGURE 10.18 Study results for a simulated stabbing. The circled data is associated with the simulated stabber, the one who handled the knife. The three electropherograms are three replicates of one sample obtained from the knife handle. (Reproduced from Samie, L., et al., Stabbing Simulations and DNA Transfer, *Forensic Science International—Genetics* 22 [May 2016]: 73–80. With permission from Elsevier 2016.)

10.4.2 RAPID DNA AND PORTABLE SYSTEMS

The process of DNA typing from extraction to analysis is amenable to increased automation and miniaturization, and there has been much research and development in this area over the past decade. An example of one design is shown in Figure 10.19, and the output of this small device

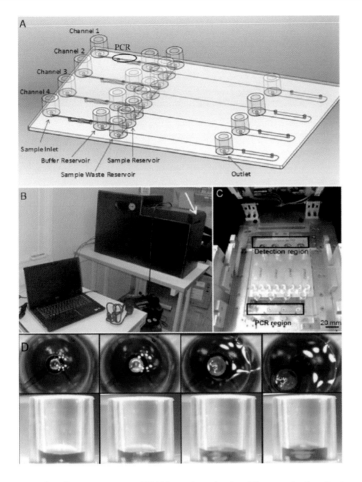

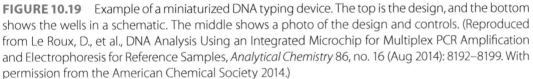

FIGURE 10.19   Example of a miniaturized DNA typing device. The top is the design, and the bottom shows the wells in a schematic. The middle shows a photo of the design and controls. (Reproduced from Le Roux, D., et al., DNA Analysis Using an Integrated Microchip for Multiplex PCR Amplification and Electrophoresis for Reference Samples, *Analytical Chemistry* 86, no. 16 (Aug 2014): 8192–8199. With permission from the American Chemical Society 2014.)

is shown in Figure 10.20. The device is about 5×3 inches and is designed to analyze four samples simultaneously in about 90 minutes. Figure 10.19 presents an example output for samples obtained from the same person. The top (A) is from the device and it is compared with the middle (B), which was obtained using standard DNA typing methodology. The results are comparable. As of 2017, at least one commercial instrument designed to perform rapid DNA is commercially available, although the results cannot be added to CODIS given the rigorous quality assurance that is required. However, progress in the field is rapid and exciting, and it may not be long before automated methods or completely innovative approaches are developed and used in forensic laboratories.

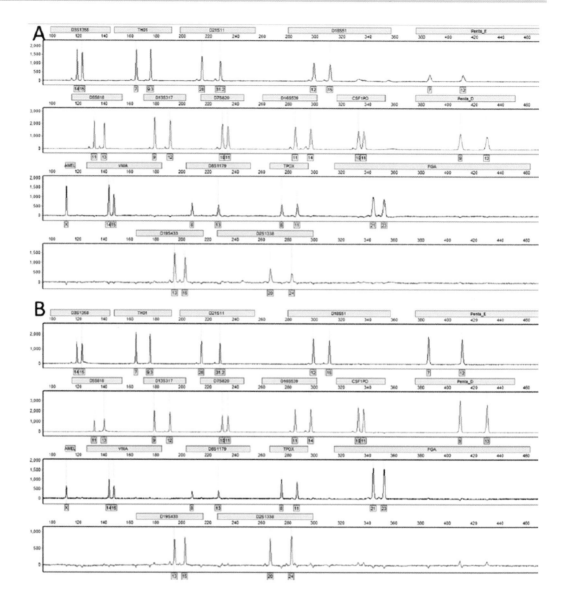

FIGURE 10.20   Output compared with typical STR methods. (Reproduced from Le Roux, D., et al., DNA Analysis Using an Integrated Microchip for Multiplex PCR Amplification and Electrophoresis for Reference Samples, *Analytical Chemistry* 86, no. 16 (Aug 2014): 8192–8199. With permission from American Chemical Society 2014.)

10.5 RETURN TO THE SCENE OF THE CRIME

DNA will be integral in the investigation set forth in the crime scene scenario from Chapter 3. It will be vital to determine if the blood at the house is that of the wife and if any of it, such as the tiny drip in the garage, belongs to the husband. The same would be true of the blood that appears on the door of the truck. The crime scene investigators (CSIs) that are processing the scene will be tasked with collecting enough of the important samples without collecting so many that the sample load overwhelms the laboratory. For example, it would not be necessary to sample every one of the drops on the wall that appear to be associated with the arterial spurt. Knowing what to collect is one of the key skills of the CSIs since collecting too much evidence can be as damaging to a case as not collecting enough.

10.6 REVIEW MATERIAL: KEY CONCEPTS AND QUESTIONS

10.6.1 KEY TERMS AND CONCEPTS

Allelic ladder
Amplicons
Annealing
Base pairs
Buccal swab
Chromosomes
Differential extractions
DNA polymerase
Electropherogram
Flanking region
Fluorescence
Gene
Heterozygous
Homozygous
Hypervariable regions
Mitochondria
Nuclear DNA
Nucleotide
Polygenetic
Polymerase chain reaction
Polymorphic
Primers
Touch DNA

10.6.2 REVIEW QUESTIONS

1. Why must a hair have its root attached for STR DNA typing methods to work?
2. Go through Figure 10.10 and list which loci are homozygous and which are heterozygous.
3. Suppose a DNA is sample is taken from a sheet found at a crime scene. The electropherogram shows that there are three loci with three peaks and one locus with four peaks. What does this mean?
4. How does touch DNA demonstrate Locard's principle?
5. There is usually much more mtDNA in a typical forensic sample than nuclear DNA. Why is STR typing the preferred method?

10.6.3 ADVANCED QUESTIONS AND EXERCISES

1. Suppose the DNA analysis of samples from the crime scene presented in Chapter 3 were as follows:
   - Towel recovered from the house: Mixture of husband's and wife's blood
   - Blood inside the truck cab: Husband
   - Blood traces from the bed of the truck: Mixture
   - Blood on the tarp: Mixture
   - Blood spot in the garage: Husband

   What scenarios could explain this? Do your best to avoid confirmation bias. There are many scenarios imaginable. List what other evidence would be needed to prove or disprove your idea.

2. Suppose that a third person's DNA is found in one of the stains on the inside door of the truck and that is the only instance in which it is found. The DNA type is not found in CODIS. How might this come about?
3. The CSI working the truck does not swab the steering wheel for DNA. Why?

BIBLIOGRAPHY AND FURTHER READING

BOOKS

Butler, J. M. *Advanced Topics in Forensic DNA Typing: Interpretation.* San Diego: Academic Press, 2015.
Butler, J. M. *Advanced Topics in Forensic DNA Typing: Methodology.* San Diego: Academic Press, 2012.
Butler, J. M. *Fundamentals of Forensic DNA Typing.* San Diego: Elsevier, 2010.

ARTICLES

Arenas, M., Pereira, F., Oliveira, M., Pinto, N., Lopes, A. M., Gomes, V., Carracedo, A., and Amorim, A. Forensic Genetics and Genomics: Much More than Just a Human Affair. *PLoS Genetics* 13, no. 9 (Sep 2017): e1006960.
Kayser, M. Forensic Use of Y-Chromosome DNA: A General Overview. *Human Genetics* 136, no. 5 (May 2017): 621–635.
Kim, Y. T., Heo, H. Y., Oh, S. H., Lee, S. H., Kim, D. H., and Seo, T. S. Microchip-Based Forensic Short Tandem Repeat Genotyping. *Electrophoresis* 36, no. 15 (Aug 2015): 1728–1737.
Mapes, A. A., Kloosterman, A. D., de Poot, C. J., and van Marion, V. Objective Data on DNA Success Rates Can Aid the Selection Process of Crime Samples for Analysis by Rapid Mobile DNA Technologies. *Forensic Science International* 264 (Jul 2016): 28–33.
Mapes, A. A., Kloosterman, A. D., van Marion, V., and depot, C. J. Knowledge on DNA Success Rates to Optimize the DNA Analysis Process: From Crime Scene to Laboratory. *Journal of Forensic Sciences* 61, no. 4 (Jul 2016): 1055–1061.
Prahlow, J. A., Cameron, T., Arendt, A., Cornelis, K., Bontrager, A., Suth, M. S., Black, L., et al. DNA Testing in Homicide Investigations. *Medicine Science and the Law* 57, no. 4 (Oct 2017): 179–191.
Romsos, E. L., and Vallone, P. M. Rapid PCR of STR Markers: Applications to Human Identification. *Forensic Science International—Genetics* 18 (Sep 2015): 90–99.
Salceda, S., Barican, A., Buscaino, J., Goldman, B., Klevenberg, J., Kuhn, M., Lehto, D., et al. Validation of a Rapid DNA Process with the Rapidhit® ID System Using Globalfiler® Express Chemistry, a Platform Optimized for Decentralized Testing Environments. *Forensic Science International—Genetics* 28 (May 2017): 21–34.

WEBSITES

Link	Description
http://strbase.nist.gov/	Database of STR sites maintained by the National Institute of Standards and Technology
https://www.swgdam.org/	Scientific Working Group on DNA Analysis Methods
https://www.fbi.gov/services/laboratory/biometric-analysis/codis	CODIS and NDIS overview

Drugs and Poisons

11.1 FORENSIC CHEMISTRY

The next four chapters cover topics in forensic chemistry. As with forensic biology, there are many forensic applications of chemistry, including seized drugs and poisons, forensic toxicology (introduced in Chapter 6, along with death investigation), arson and fire investigation, and explosives. Many aspects of trace evidence analysis (Chapter 17) also integrate chemistry and chemical analysis. Even DNA typing (Chapter 10) relies on basic chemistry and chemical analysis; the capillary electrophoresis (CE) instrument used for short tandem repeat (STR) typing is a chemical analysis instrument.

Our focus in this chapter is on drugs and poisons and how they integrate in forensic science. In Chapter 6, we discussed one aspect of forensic toxicology, postmortem toxicology. We learned how drugs and poisons can be ingested and how they can cause or contribute to death. Postmortem toxicology is one of many types of forensic toxicology. In this chapter, we focus on the ingested substances themselves and how they become forensic evidence before and after the ingestion process.

A **drug** is a substance that causes a physiological response. For example, aspirin is a drug but baking soda is not. Taking an aspirin initiates processes and pathways in the human body that result in responses such as decreased swelling and a reduction of fever. Ingesting baking soda as part of a cookie or cake does start any such processes.

A **poison** also invokes a physiological effect, but one that is harmful or has fatal effects. Any drug can be a poison depending on the dose. Aspirin, a safe drug, can also be a poison if too much is ingested at once. The saying "the dose makes the poison" is true, and the dose can differentiate between what is a therapeutic dose and a toxic or lethal dose. Even water can be a poison; refer to Chapter 6, Table 6.1 and Figure 6.6.

In this chapter, we focus on drugs and poisons that cause physiological effects that range from therapeutic and beneficial to toxic, causing impairment or intoxication. Because the type of effect is usually dependent on dose, we can think of the effects as a continuum (Figure 11.1), from no noticeable effect through to death. As an example, consider ethanol (the alcohol in beer, wine, and spirits). Ingesting a single shot of whiskey would cause a noticeable effect in most people,

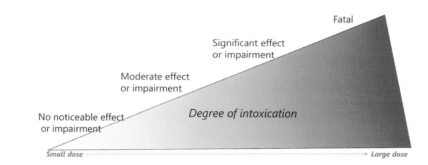

FIGURE 11.1   Continuum of intoxication.

while two or more shots could cause intoxication and impairment. More shots would mean more intoxication, and for a typical male, drinking 12–13 shots would be fatal. The **degree of intoxication** is the critical issue for the toxicologist and for law enforcement because above a certain level of intoxication, the person is incapable of driving and of making rational decisions.

There are two forensic disciplines that deal with drugs and poisons. Forensic toxicology was introduced in Chapter 6, as it relates to postmortem toxicology. Forensic toxicologists also work with drugs, poisons, and metabolites in other contexts, such as in driving under the influence (DUI) cases, workplace drug monitoring, and performance-enhancing drugs in sports. The common thread is the study of drugs and poisons that have been ingested. In contrast, the forensic discipline of seized drugs deals with these same substances as physical evidence. This includes exhibits such as seizures of drugs, pills, **drug paraphernalia** (items associated with drug use, such as syringes or pipes), and plant matter. **Seized drug** chemists do not focus on what happens to a substance when it is ingested; rather, their focus is on identifying the substance as physical evidence.

11.2 DRUGS AND POISONS

11.2.1 CLASSIFICATION

Because any drug can be toxic (a poison), there are many ways in which these substances are described and classified. We will examine the areas that are most relevant to forensic chemistry. Historically, one of the first delineations made was between plant-based drugs and mineral-based ones. Aspirin (acetylsalicylic acid) was first synthesized in the 1800s by the Bayer Company, but the active ingredient is found in willow leaves and bark. Ancient civilizations such as the Egyptians and Greeks knew that ingesting extracts from the plant relieved pain and reduced inflammation. Cocaine is found in coca leaves, which when chewed produce a stimulant effect. Marijuana has also been used for centuries across cultures and as treatments for many conditions and maladies. All are examples of plant-based drugs.

The most common plant-based drugs are chemically classified as **alkaloids**. These substances were originally called "vegetable bases" since they were extracted from plants and the extracts were found to be basic or alkaline. A basic substance tastes bitter and, when dissolved in water, will cause the pH to rise about 7. Examples of alkaloids include nicotine, morphine, cocaine, atropine, and thebaine. Opium poppies are the source of morphine and related drugs, such as codeine and heroin.

Perhaps the most famous mineral-based toxin is arsenic, which was (and still is) used in paint pigments and as a poison. Arsenic is abundant in the natural environment and thus easily obtained even with primitive technology. The element arsenic is found in many different chemical forms, such as As_2O_3 (arsenic trioxide or white arsenic), AsH_3 (arsine gas), and As_2S_3 (arsenic sulfide or King's yellow). White arsenic and King's yellow are examples of minerals, which is where the term originated. The colors of different forms of arsenic led them to be used as pigments in early paints and colorants, but the toxic effects were also well known. White arsenic was at one time called "inheritance powder" because it was so commonly used as a poison. It was so common that much of the early pioneering work in forensic toxicology was focused on developing methods to detect it in food and human tissues. Other well-known mineral-based (metal-based) poisons are lead, thallium, and chromium.

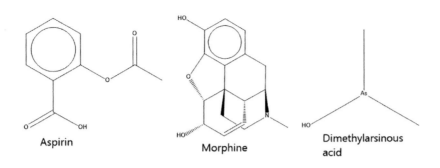

FIGURE 11.2   Example structures. All are organic.

Substances are also classified as **organic** or **inorganic**. Here, organic refers to a compound that contains carbon, while an inorganic compound does not. Nicotine, morphine, and aspirin are examples of organic compounds, while arsenic is inorganic. There are a few exceptions; for example, the notorious poison cyanide contains carbon but is classified as an inorganic. Interestingly, arsenic can exist in organic forms that can arise from ingestion and metabolism. Once in the body, arsenic is converted to a compound called dimethyl arsinous acid, which is organic because it includes carbon. See Figure 11.2 for some example chemical structures. Note that in this figure, carbons are not shown explicitly, but every place that two lines contact each other, there is a carbon atom, unless a different letter, such as N (nitrogen), is shown. On the arsinous acid structure, there are carbons at the end of each of the three lines: C–OH on the one with the –OH, and CH$_3$ on the other two.

Physical form is another way to classify substances. Most are solids, but there are liquid toxins, such as ethanol (the alcohol in beverages such as wine), methanol (wood alcohol), and the element mercury, which is a liquid. Cyanide (HCN) is a gas, as is the arsine form of arsenic (AsH$_3$) and carbon monoxide (CO). Carbon monoxide is a by-product of fire and flames and is a common cause of accidental poisoning and death in fires. We talk about gases as poisons in a later section.

11.3 SEIZED DRUG ANALYSIS

11.3.1 CONTROLLED SUBSTANCES ACT

The concept of an illegal drug or controlled substance is relatively recent. Regulation of drugs in the United States dates to the early part of the 20th century, a time during which heroin and cocaine were legal and used in commercial products. The modern form of regulation began in the 1970s with passage of the **Controlled Substances Act** (CSA) in 1970. The CSA recognizes that many of the substances have a useful, legitimate medical purpose. Therefore, it has divided the drugs into regulated schedules based on their medical use and potential for abuse. Many state and local laws concerning the possession of controlled substances are based on these federal regulations. Others have taken the list of controlled substances and simply made their possession illegal without regard to legitimate use or abuse potential.

The CSA classifies drugs into one of five schedules (indicated with a Roman numeral) based on accepted medical use and the danger of physical and psychological dependence. Each schedule also specifies penalties and how the drug can be obtained, as well as whether a permit is necessary. **Precursor** chemicals (the starting chemicals that are chemically altered to generate a controlled substance) used in clandestine labs are also listed. In 1984, an amendment was added as part of the **Comprehensive Crime Control Act** that permitted the **Drug Enforcement Administration** (DEA) to temporarily add substances to a schedule without having to wait for a more formal and lengthy procedure to take place. This allows the DEA to address newly discovered precursors and designer drugs synthesized by clandestine laboratories. In the Anti-Drug Abuse Act of 1986, a provision was added to address **analog** drugs (drugs similar but not identical to existing controlled drugs) synthesized by clandestine labs to act similarly to other controlled substances. Substances have been added at a rapid pace since the sharp increase in synthetic cannabinoids and opioids that began about 2010. As an example of this, in November 2017, the DEA was authorized to list all fentanyl analogs as Schedule I substances.

Drugs are placed into schedules based on medical use, risk of addiction, and danger to the user. Schedule I drugs are those that have a high potential for abuse and no accepted medical applications. Heroin, LSD, and marijuana are listed on this schedule. Schedule II drugs (morphine, cocaine, and methamphetamine) are similar but have accepted medical uses; abuse can lead to severe addiction and dependence. Schedule III drugs have less potential for abuse and addiction and include anabolic steroids and some codeine and barbiturate preparations. Schedules IV and V drugs reflect decreasing risk and increasing legitimate uses; drugs such as Valium® are on Schedule IV, and **over-the-counter** (OTC) cough medicines with codeine are on Schedule V. The current list of scheduled drugs is available at the DEA website.

11.3.2 EXAMPLE DRUGS AND DRUG FAMILIES

11.3.2.1 OPIATES

Opiates constitute a large class of drugs distinguished by their ability to cause profound euphoria as well as to relieve pain. They were introduced in Chapter 6. Codeine, which we just discussed, is an example of an opiate. Many possess high potency as pain relievers and are derived from or related to morphine, which comes from the opium poppy, a plant that grows in copious quantities in Southeast Asia and several other areas of the world. **Morphine** was the first compound extracted from the plant and used for pain relief; codeine is also directly extractable. Heroin is easily prepared from morphine through simple chemical reactions. Many other synthetic opiates (**opioids**) have been made to find stronger pain relievers, although the stronger these become, the more dangerous they become. Opiates have also been used as poisons and as murder weapons.

HAROLD SHIPMAN

Dr. Harold Shipman, a physician who lived in England, is thought to have killed more than 200 people, mostly elderly women, from about 1970 until he was arrested in 1998. He poisoned his victims with fatal doses of heroin, which was legal for doctors to have and use in the United Kingdom at that time. He was eventually caught after he forged a will of a prominent citizen, a woman who despite her age was in otherwise good health. Many of the victims were cremated, but once he was suspected, some of the buried victims were exhumed and tested to confirm that they had been poisoned. Shipman was imprisoned and committed suicide in 2004.

Semisynthetic opiates are those made by a simple modification of the morphine or codeine molecule, as shown in Figure 11.3. These include hydromorphone, hydrocodone, oxymorphone, and oxycodone. OxyContin® is a sustained form of oxycodone that has caused many deaths. It was designed to be taken orally and allow patients who suffer from cancer and related pain to experience long-term relief. Abusers crushed the tablet and injected or smoked the residues to defeat the time-release design and many deaths occurred. Fentanyl (Figure 11.4) is another example of a synthetic opiate that is not based directly on the morphine skeleton. Opiates are classified as **depressants**. Accordingly, they produce, in addition to the initial euphoria, reduced muscle activity, depressed respiration and heartbeat, and an inclination to sleep. In overdose, they cause death, usually by paralysis of the respiratory center.

11.3.2.2 AMPHETAMINES

In contrast to opiates, amphetamine and its analogs, such as methamphetamine, cause elevations of heart rate, blood pressure, and respiration. They are classified as **stimulants**, and their use provokes intense euphoria. Methamphetamine can be synthesized easily by clandestine laboratories, starting with ephedrine. To prevent the illicit production of methamphetamine, governments have recently passed legislation that limits the availability of ephedrine. The image at the start of this chapter shows crystal methamphetamine, which can vary from this nearly clear

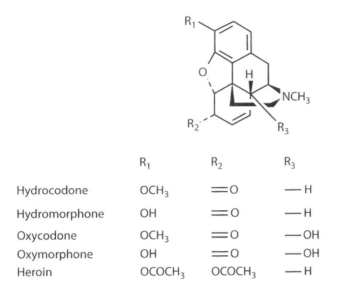

	R₁	R₂	R₃
Hydrocodone	OCH₃	=O	—H
Hydromorphone	OH	=O	—H
Oxycodone	OCH₃	=O	—OH
Oxymorphone	OH	=O	—OH
Heroin	OCOCH₃	OCOCH₃	—H

FIGURE 11.3 Morphine skeleton with additions that generate other opiates. R is generic for a substitution.

FIGURE 11.4 Forms of fentanyl as evidence. (Images courtesy of the DEA.)

form to off-yellow. It is also sold as pills and powder. Many compounds resemble amphetamine both structurally and pharmacologically. These compounds are sold by prescription or over the counter and have decongestant, anti-insomniac, and **anorexic** (appetite suppressant) actions. These compounds are numerous and include ephedrine, phenylephrine, and phenmetrazine, among others. Ephedrine and pseudoephedrine are often used as precursors for the clandestine synthesis of methamphetamine; as a result, many states now have controls to limit purchases of these drugs.

11.3.2.3 COCAINE

Cocaine is a stimulant that resembles amphetamine in its abuse potential and pharmacological responses. Unlike amphetamine, however, cocaine is a natural product found in the coca leaf. It was widely (and legally) used in the United States early in the 20th century as a medicine and simulant (Figure 11.5). It was an ingredient in early recipes for Coca Cola®. *Erythroxylon coca*, the source of cocaine, grows in damp, mountainous regions, especially the Andes range of South America. Cocaine is extracted by a straightforward process from the plant material. Because cocaine is alkaline in nature and is usually extracted with hydrochloric acid, the substance produced is cocaine hydrochloride, in which the hydrochloric acid is bonded to the nitrogen atom of cocaine.

Cocaine hydrochloride may be treated with a base and extracted into an organic solvent such as ether. This additional treatment produces "free base" or "crack" cocaine (examples are shown

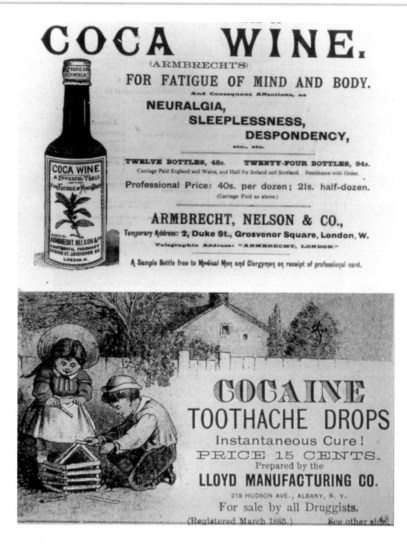

FIGURE 11.5 Advertisements for products containing cocaine.

in Figure 11.6). In actuality, crack cocaine and free base are chemically the same. The two names refer to slight differences in the manner of preparation. Free base and crack cocaine have much lower boiling points than cocaine hydrochloride. This difference in physical properties is very important because it makes it possible to smoke cocaine. An attempt to smoke cocaine hydrochloride would simply burn the drug. When a drug is smoked, the large surface area of the lungs is available for drug absorption. The result is that larger amounts may be absorbed per unit of time and a greater drug effect is experienced. Free base and crack cocaine were introduced into the United States during the 1980s.

11.3.2.4 CANNABINOIDS

Marijuana is a name that applies to parts of the *Cannabis sativa* plant. Many related psychoactive compounds come from this plant, and the collective term **cannabinoids** is often applied to them. **Tetrahydrocannabinol** (THC) is the major active agent and is present to the extent of 2% and higher by weight in cannabis. An oily extract of the plant, hashish, has a much higher THC content (12%) and accordingly produces a greater psychoactive response when used.

The forensic aspect of marijuana has changed dramatically in the last few years. In 2012, Colorado and Washington State legalized the recreational use of marijuana by adults; several states followed. Prior to that date, many states had decriminalized its use or permitted medicinal uses of the plant and its extracts. However, as of 2018, marijuana remained on Schedule I of the CSA, which complicates enforcement. The legalization and decriminalization has led to the

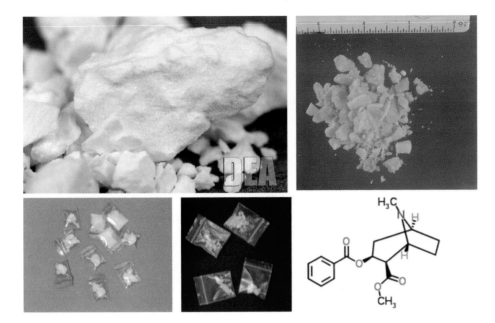

FIGURE 11.6   Various forms of cocaine. (Images courtesy of the DEA.)

creation of legal analytical laboratories that determine the concentration of THC and related compounds in marijuana and its derived products.

Another challenge that has arisen as a result is the need to determine if someone is driving under the influence by THC. Unlike alcohol, which can be detected in the breath in a way that can be related to the blood concentration, THC does not get into the exhaled breath except in tiny amounts that are currently difficult to detect. THC intoxication is also more difficult to gauge than drunkenness, even when a blood sample is obtained. The degree of impairment (Figure 11.1) can vary widely between people, as can the speed of metabolism. Thus, significant research is being directed toward developing a test for screening purposes. In most cases, a police officer must use their best judgment regarding impairment and the need to arrest someone and get a blood test.

11.3.3 NOVEL PSYCHOACTIVE SUBSTANCES

Novel psychoactive substances (NPSs) are compounds that have emerged as an unprecedented threat in seized drug analysis, public health, and toxicology, in terms of both intoxication and fatalities. Also referred to as **designer drugs**, these substances are clandestinely manufactured to produce similar effects to other illegal substances while avoiding legal controls, such as listing on the CSA. Most of these compounds are manufactured on a large scale in laboratories in Asia and smuggled into the country where they are consumed. Currently, it is estimated that a new NPS is identified roughly every 7–10 days, making it impossible for forensic chemists, medical examiners, and law enforcement to keep pace.

One of the first groups of compounds to appear in the clandestine market was the synthetic cannabinoids, which were and are marketed as safe and legal alternatives to marijuana. An overview is presented in Figure 11.7. They are called cannabinoids because the drugs are designed to cause the same type of physiological response as marijuana. However, these drugs have proven to be much more dangerous and unpredictable than the compounds found in marijuana. There are seven families of synthetic cannabinoids, which are grouped based on their chemical structure, with dozens of examples in each category and with new variants being identified at a weekly to monthly rate. These drugs are usually dissolved in a solvent and sprayed on plant material that is then smoked as a means of ingestion. Early packaging referred to the substances as "spice" or "incense." Figure 11.7 shows examples of packaging and plant matter with cannabinoids.

Cannabinoids are chemically complex; the current naming convention is based on four parts of the molecule—core, tail, link, and ring. An example, JWH-018, is shown in the figure, along

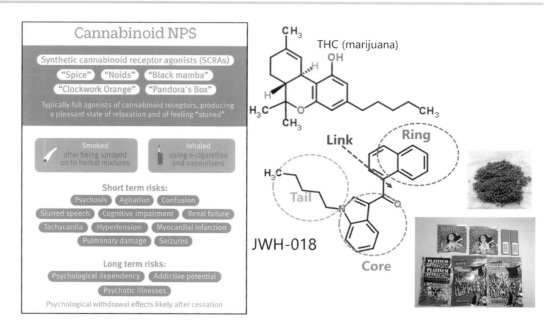

FIGURE 11.7   Overview of synthetic cannabinoids. (Evidence images courtesy of the DEA. Poster reproduced from Tracy, D. K., et al., Novel Psychoactive Substances: Types, Mechanisms of Action, and Effects, *BMJ* 356 [Jan 2017]. With permission.)

with THC, which is the active ingredient in marijuana. Any of these four structures can be altered, which leads to a seemingly endless flow of new drugs. The poster in the left of the figure was generated for public education and shows some of the short- and long-term risks associated with use of these drugs. Because so little is understood about these compounds, each variant is a new form of human experimentation. As of 2017, 26 of these substances were listed on Schedule I of the CSA. The CSA also now defines substances as analogs and allows these substances to be controlled.

A second category of NPSs is the synthetic cathinones, a group that includes stimulants; informally, the term **bath salts** is used because this is how samples are sometimes labeled. The compound on which these substances are based is **cathinone**, which is the active ingredient in **khat leaves**. Khat leaves are chewed to get the stimulant effect of the substance.

As shown in Figure 11.8, there are many points in the molecule that can be altered and thus lead to the regular appearance of new variants. These drugs are chemically like the amphetamine/methamphetamine family and cause many of the same effects.

The latest group of NPSs to appear on the illicit market are synthetic opiates (opioids), which we discussed in Chapter 6. These drugs are related to morphine, heroin, and fentanyl. As with cathinones and cannabinoids, there are many locations on the core molecule that can be altered, leading to the same types of concerns.

MYTHS OF FORENSIC SCIENCE

The best way to test a drug is to taste it.

This is such a bad idea for so many reasons that it is hard to list them all. While it remains a staple in action movies, it is not done; as we have learned in this chapter, a few tiny particles of some drugs are lethal. It is true that early chemists knew that taste could help distinguish some compounds; basic compounds tend to taste bitter and acid compounds tend to taste sour. This is clearly subjective. Also, since most illegal drugs are basic, the majority are likely to taste bitter. Arsenic powder also tastes bitter, so the test is not very selective. Drugs in the cocaine family are local anesthetics and numb the tongue, but so do compounds like benzocaine. Thus, tasting is extraordinarily dangerous and provides very little useful data (but great drama!).

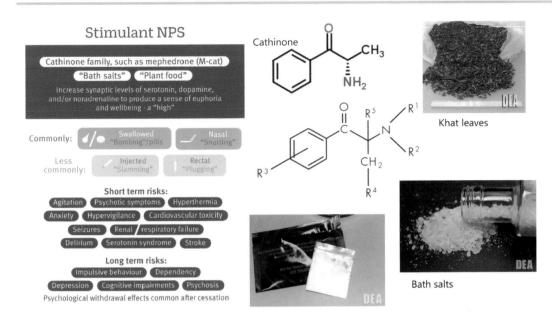

FIGURE 11.8   Synthetic cathinones. (Evidence images courtesy of the DEA. Poster reproduced from Tracy, D. K., et al., Novel Psychoactive Substances: Types, Mechanisms of Action, and Effects, *BMJ* 356 [Jan 2017]. With permission.)

11.3.4 CLANDESTINE LABORATORIES

Clandestine drug laboratories are hidden or secret locations that manufacture controlled substances. The types and numbers of labs seized reflect national and regional trends concerning the types and amounts of illicit substances that are being manufactured, trafficked, and abused. They have been found in remote locations, urban and suburban neighborhoods, hotels and motels, industrial complexes, and academic and industrial laboratories. They range in size from tabletop setups used to produce gram quantities to large multiple-location operations that generate kilograms of final product. In each of these, toxic and explosive fumes can pose a significant threat to the health and safety of anyone living or working nearby.

The sophistication of clandestine labs varies widely, and their investigation is one of the most challenging efforts of law enforcement. An example small-scale laboratory is shown in Figure 11.9. Although we are focusing on clandestine drug labs in this chapter, a clandestine lab can manufacture other types of materials, including explosives, chemical warfare agents, and biological agents. The identification, investigation, and prosecution of a clandestine lab is a collaboration of the efforts of law enforcement, forensic experts, scientists, and criminal prosecutors to present a case that definitively demonstrates how a group of items with legitimate uses are being used to manufacture an illegal controlled substance.

Processing a clandestine lab scene is more complicated than the traditional crime scene search. Because of the chemicals involved, the site of a clandestine lab is a hazardous materials incident and necessitates invoking different protocols for crime scene processing. People processing the scene must wear full protective gear, as shown in Figure 11.10. A complete forensic laboratory analysis is a critical element of a clandestine lab investigation. The analysis of a reaction mixture is more complex than simply identifying the controlled substance it contains. The identification of precursor and reagent chemicals, as well as reaction by-products, is necessary to establish the manufacturing method used. The identity of unique chemical components within a sample can be used as an investigative tool to connect the clandestine lab under investigation to other illegal activity.

Up until the appearance of the designer drugs and NPSs around 2009, clandestine methamphetamine laboratories were one of the greatest concerns related to drugs of abuse. The problem has diminished but has not gone away; the opioid crisis has overshadowed methamphetamine labs. Clandestine labs that produce synthetic opioids are especially dangerous to first responders, law enforcement, and forensic chemists, not only due to the unsafe nature of clan labs but also

FIGURE 11.9   A small clandestine seizure. (Image courtesy of the DEA.)

FIGURE 11.10   Responders to a clandestine lab scene. Notice that the suits are rinsed, and the water collected as this is also hazardous. (Image courtesy of the DEA.)

because of the inherent toxicity of the products being made. Full protective gear is essential when these scenes are processed.

11.4 METHODS OF ANALYSIS

The flow of a typical forensic analysis began with screening or presumptive testing. In seized drug analysis, the screening test does not use immunoassay but rather relies on simple color changes when a reagent is added to a small portion of the seized material. These simple tests are designed to give the analyst an idea of what might be present but alone can never identify a substance. This information is used in turn to direct further testing that leads to identification of the substance(s) present in the sample. The instrument most often used for confirmation is gas chromatography linked to mass spectrometry (GC/MS). Refer to Chapter 6 for a refresher of how GC/MS works.

Until recently, it was common for police officers to conduct a form of presumptive test in the field, but these tests are much less common now due to the danger presented by the NPSs.

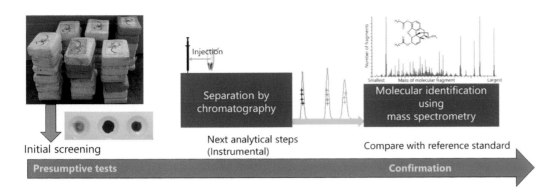

FIGURE 11.11   Flow of a typical GC/MS analysis. (Image of evidence courtesy of the DEA.)

Even opening packets of powders presents a significant hazard to the officers and even drug-sniffing dogs. In addition, color test reagents do not work with many of the NPSs, and as such, presumptive tests, if they are done, are more often done in the lab than in the field.

On overview of the flow of an analysis is shown in Figure 11.11. In most forensic applications of GC, a sample is prepared by dissolving it in a solvent, and the solution is injected into the instrument using a syringe. For example, to analyze a white powder suspected of being cocaine, a small portion is weighed out and dissolved in a solvent such as methanol. A tiny portion of the sample is then drawn up into a syringe and injected into the heated injector port of the instrument. The mobile-phase gas (carrier gas) also enters the injector port, picking up the volatilized sample and introducing it into the column where the separation process occurs. If the sample contains cocaine, it will emerge from the column at a given time (the retention time), which can be compared to the retention time of a known standard sample of cocaine. The retention time in conjunction with information obtained from the detector is used to positively identify the compound as cocaine, if indeed it is present.

For specific identification, more than retention time is needed. That additional assurance is provided by coupling a GC with a detector that can provide a specific identification of a substance (in most instances). Such a detector is a mass spectrometer, and the combination is referred to as GC/MS, which was introduced in Chapter 6. In some cases, mass spectra are not unique. Some molecules are resistant to fragmentation and produce fragments in low abundance that strongly resemble mass spectra of other substances. In such cases, additional analytical information is required for definitive identification.

How are substances emerging from chromatographic columns identified? This is usually left to a computer search, in which the unknown mass spectrum is matched with spectra of known toxins. The **National Institute of Standards and Technology** (NIST) has a library of more than 140,000 compounds that can be searched for matches. Other libraries can also be employed, but most are smaller than the NIST library. In some circumstances, such as searching specific areas, they may have advantages over the NIST library. To obtain definitive identification, a reliable standard must be analyzed using the same GC/MS conditions to ensure that the retention time and the mass spectrum are the same as for the sample.

11.5 TOXICOLOGY

We introduced toxicology in Chapter 6 as it relates to death investigation (postmortem toxicology). All the concepts presented in that chapter (doses, effective dose, lethal dose, etc.) apply to toxicology and intoxication in the living; the difference is that the focus is on intoxication and impairment rather than fatalities. There are several types of forensic toxicology of the living, including blood and breath alcohol (related primarily to DUI cases), workplace drug monitoring, and performance enhancement, such as doping in sports. There are also interesting substances, such as cyanide and carbon monoxide, that are particularly important in toxicology.

TYLENOL MURDERS WITH CYANIDE

Cyanide was used in a series of murders that occurred in 1982 in the Chicago area. The first death was that of a 12-year-old girl who was given one Tylenol for symptoms of a cold. The same day, a man died of a what was at first thought to be a heart attack. His brother and sister-in-law, overcome with grief, developed severe headaches and each took a Tylenol tablet from the same bottle that the victim had; within 2 days, both were also dead. The investigation and analysis revealed that the Tylenol capsules taken by all victims (eventually seven) had been opened and potassium cyanide (KCN) added. Millions of bottles were taken off the shelves and tested. More tainted bottles were found. An investigation showed that the tampering had occurred after the bottles left the factories, but the person or persons responsible were never caught. The incident led to the development of tamper-proof capsules, bottles, and containers that are used with OTC medicines today.

11.5.1 BLOOD AND BREATH ALCOHOL

Ethanol is beverage alcohol and is the compound referred to informally as "alcohol." There are other compounds classified as alcohols that are seen in forensic toxicology as poisons, such as methanol (wood alcohol) and isopropanol. Methanol, for example, is found in homemade distilled beverages and in sufficient quantities can cause blindness or death. The toxic effect is due primarily to the metabolites rather than the methanol parent compound. Methanol is converted into formaldehyde and formic acid principally by the liver, and both substances are toxic; treatment of methanol overdose is best accomplished by preventing the conversion of methanol into its metabolites.

Beverage alcohol enters the blood mainly from the small intestine. The intoxicating effects of ethanol are due to the ethanol itself. Within the liver, about 90% of the average dose of ethanol is converted into acetaldehyde and acetic acid (the acid in vinegar). The remainder is eliminated via sweat or urine. Ninety minutes after ethanol ingestion is the approximate time that peak blood levels are reached. This figure is certainly highly variable and is based on the interaction of many factors. As a general guide, one 12-ounce can of beer or one cocktail (1.5 ounces of 100-proof alcohol) raises the blood concentration of an average size individual by 0.02%. This value is referred to as the **blood alcohol concentration** (BAC). Alcohol is also metabolized while it is in the blood, so that effect must be considered when judging the level of intoxication. The average **clearance rate** (rate of removal of ethanol by metabolism) is about one drink per hour. These approximations are very inexact because an individual's handling of alcohol depends on many factors, one of which is the person's drinking experience. Figure 11.12 shows examples of what is considered one drink.

As an example of how intake and clearance are balanced, consider a person (male in this case) who consumes four drinks within an hour. He stays at the bar for another 2 hours before leaving. Would you expect him to be intoxicated? Four drinks would result in a BAC of ~45×0.02%, or 0.08%. However, he waits 2 hours, so the amount cleared from his system is 0.02×2, or ~ 0.04%. When he leaves that bar, his BAC is estimated at 0.04%. In the United States, the legal limit for driving in the United States is 0.08% BAC. If a person's BAC is higher, that person is legally intoxicated. Thus, in this simple example, the man would not be legally intoxicated. Other countries have different legal limits, such as 0.00% in predominantly Muslim countries such as Pakistan, 0.03% in China, and 0.05% in many European and Nordic countries.

The toxicity of beverage alcohol is well known. Acute toxicity correlates with dose and blood level. Alcohol contributes to numerous disorders because of chronic abuse. The liver is the organ that is most vulnerable, and it shows pathological response to alcohol, ranging from fatty accumulation up to liver cancer. The brain may also be attacked with very significant injury, including several psychosis-like syndromes. BACs greater than ~0.35% are often fatal. The right side of Figure 11.12 relates physiological effects with BAC.

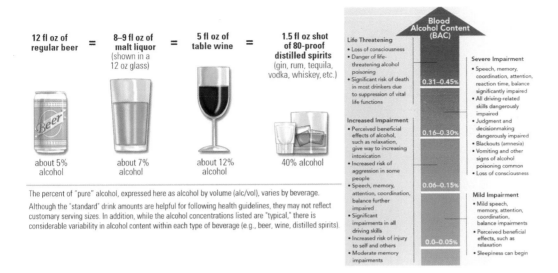

FIGURE 11.12 Drink equivalents and degrees of impairment. (From the National Institute on Alcohol Abuse and Alcoholism.)

One of the interesting aspects of alcohol is that it can be detected in the breath at a concentration that can be linked to that in the blood. Equilibrium exists between alcohol in the bloodstream and alcohol in the lung such that, on average, the concentration of blood alcohol is about 2300 times greater than the concentration of **breath alcohol** (BrAC). Since this equilibrium exists, one can measure the BrAC and infer the corresponding alcohol concentration in the bloodstream. Since it is the BAC that determines intoxication, measuring BrAC provides an indirect measurement of BAC. The obvious advantage of breath testing is that it may be carried out without requiring blood collection with its hazards and associated difficulties. Further, unlike blood testing, the result is immediately available so that an impaired driver can be removed from the highway right away. In the past, there were objections to the specificity limits of breath testing. Modern methods that are based on infrared analysis provide specificity that approaches that of blood testing by sophisticated instruments. These devices are called **breathalyzers** (Figure 11.13). It is important to realize that these devices are used to estimate BAC; a blood test is required to determine the actual concentration.

The method used to measure blood alcohol levels relies on GC/MS, although the sample is introduced differently. Because alcohol evaporates at a relatively low temperature (indeed, that is why it can be detected in breath), a blood sample can be directly heated to a stable constant temperature in an enclosed vial. Once the concentration in the headspace (empty part of the vial) is balanced with the concentration in the blood sample, the headspace is sampled and injected into the GC system where it is detected, and the amount of ethanol can be accurately measured.

11.5.2 CYANIDE

Cyanide is a toxic substance that is present in myriad forms in nature. The fastest-acting form is the gas hydrogen cyanide. Salts such as sodium cyanide are highly poisonous, but their onset of action is somewhat slower than that of hydrogen cyanide gas. Many industrial chemicals, such as acetonitrile, are metabolized to cyanide and produce the same symptoms as cyanide, although to a lesser degree. Many plant materials, such as amygdalin and linamarin, are poisonous to humans because their ingestion leads to the production of cyanide in the body.

Cyanide is dangerous because it binds to iron ions in a critical enzyme, which causes an interruption of the central pathway for energy generation in human biochemistry. Death occurs quickly. Inhalation of large amounts of hydrogen cyanide is fatal in less than 1 minute. Cyanide in the form of HCN (hydrocyanic acid or prussic acid) is an inhaled poison, but HCN can also be produced by swallowing a solid such as sodium cyanide (NaCN) or potassium cyanide (KCN).

FIGURE 11.13   An example of a breathalyzer in use.

The acidic environment in the stomach leads to formation of the deadly HCN. HCN is often produced in fires in which man-made materials such as plastics are burning, and frequently contributes to fire-related death. Forensic laboratories can test for cyanide in whole blood, and its concentration correlates well with severity of poisoning. The normal level is less than 40 ng/mL. Levels greater than 1000 ng/mL are associated with stupor. Amounts above 2500 ng/mL are usually fatal.

11.5.3 CARBON MONOXIDE

Some studies suggest that carbon monoxide (CO) causes more deaths than any other toxic substance. This may be true, because CO is present in fires and, together with other poison gases that result from combustion, causes more fire deaths than thermal injury. Most carbonaceous materials in the world are reduced forms of carbon (carbon bonded to hydrogen). Heat converts them to oxidized forms of carbon (carbon monoxide and carbon dioxide). In the presence of adequate heat and oxygen, reduced carbon is fully converted to relatively harmless carbon dioxide. In the presence of inadequate oxygen, however, the product of carbon oxidation is CO. Faulty heaters, indoor fires, and other situations in which carbon is not fully oxidized are the usual scenarios in which excessive amounts of CO are produced.

Carbon monoxide is toxic for many reasons. It binds hemoglobin much more tightly than oxygen, so the hemoglobin is unable to fulfill its normal function of transporting oxygen to tissue. CO also causes a left shift in the hemoglobin dissociation curve. This means that hemoglobin binds oxygen more tightly at any given partial pressure of oxygen. This is dangerous because the oxygen, tightly bound to hemoglobin, cannot be transferred to its intended destination—cells in need of oxygen.

Carbon monoxide testing is commonly conducted in clinical laboratories and in postmortem evaluation. Persons whose blood **carboxyhemoglobin** (hemoglobin bound to CO instead of O_2) levels exceed 60% are at elevated risk of death. Lesser degrees of hemoglobin–CO binding are associated with lesser danger of death, but it is important to be cautious in interpreting carboxyhemoglobin levels in the blood because removal from the source of CO immediately starts the process of unloading CO from hemoglobin. If testing is delayed, a carboxyhemoglobin level will underestimate the degree to which a patient was exposed to CO.

A case example of an accidental death due to CO poisoning is shown in Figure 11.14. In this case, the victim was using a washer powered by a gasoline engine. Even though he was working in a large space with open doors, there was not sufficient ventilation where he was working and, as a result, he collapsed and died.

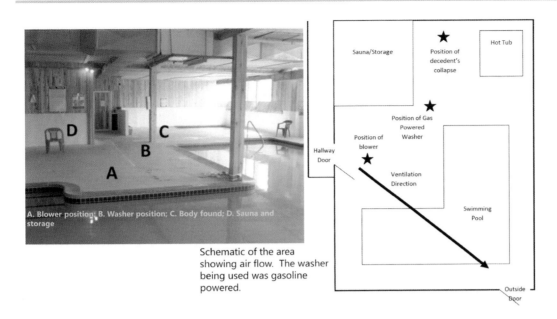

A. Blower position; B. Washer position; C. Body found; D. Sauna and storage

Schematic of the area showing air flow. The washer being used was gasoline powered.

FIGURE 11.14   Case example of CO poisoning. (Reproduced from Huston, B., et al., Carbon Monoxide Poisoning and Death in a Large Enclosed Ventilated Area, *Journal of Forensic Sciences* 58, no. 6 [Nov 2013]: 1651–1653. With permission from Wiley.)

11.6 REVIEW MATERIAL: KEY CONCEPTS AND QUESTIONS

11.6.1 KEY TERMS AND CONCEPTS

Alkaloids
Analogs
Anorexic
Anti-Drug Abuse Act
Bath salts
Blood alcohol concentration
Breath alcohol
Breathalyzer
Cannabinoids
Carboxyhemoglobin
Cathinone
Clearance rate
Comprehensive Crime Control Act
Controlled Substances Act
Degree of intoxication
Depressants
Designer drugs
Drug
Drug paraphernalia
Inorganic substance
Khat
Morphine
Novel psychoactive substances
Opiates
Opioids
Organic substance
Over-the-counter drugs

Poison
Precursor
Seized drugs
Stimulants
Tetrahydrocannabinol

11.6.2 REVIEW QUESTIONS

1. Using Figure 11.1 as a guide, explain how aspirin can act as a drug or a poison.
2. Can a substance be just a drug and not a poison?
3. Is caffeine a drug? If so, why isn't it listed as a controlled substance?
4. Suppose a person drinks two beers over an hour, waits another hour, and leaves in her car. Is she legally intoxicated?
5. How many drinks would someone have to ingest in a short period of time to drive their BAC over the fatal level?
6. If a person has a BAC of 0.2%, estimate what the BrAC would be.
7. Why do accidental deaths due to CO increase significantly in the winter?
8. Why is it so critical to use BrAC to estimate the concentration in the blood? Why not just rely on breath concentrations?

11.6.3 ADVANCED QUESTIONS AND EXERCISES

1. Criminal acts associated with seized drugs, such as possession, are based on laws that reflect social norms rather than acts that are intrinsically evil. For example, most people agree that murder is wrong, but not all agree on whether marijuana should be legal. What factors do you think are or should be used to draft drug legislation? What behaviors are drug laws trying to control or limit?
2. How do you think drugs such as fentanyl have changed the way that seized drug chemists work in the lab?
3. Refer to the crime scenario presented in Chapter 3. Can you identify items of evidence that might be submitted for seized drug analysis? Suppose the items are submitted and found to contain illegal drugs. How would this influence the outcome of the case? Does this finding support one version of events over the other?

BIBLIOGRAPHY AND FURTHER READING

BOOKS

Bell, S. *Forensic Chemistry*. 2nd ed. Upper Saddle River, NJ: Prentice Hall, 2011.
Dargan, P. I., and Wood, D. M., Eds. *Novel Psychoactive Substances: Classification, Pharmacology, and Toxicology*. London: Academic Press, 2013.
Karch, S. B., Ed. *Drug Abuse Handbook*. 2nd ed. Boca Raton, FL: CRC Press, 2011.
Klaassen, C., Ed. *Casarett and Doull's Toxicology: The Basic Science of Poisons*. 8th ed. New York: McGraw-Hill, 2013.
Levine, B. *Principles of Forensic Toxicology*. 4th ed. Washington, DC: American Association for Clinical Chemistry Press, 2013.
Moffat, A. C., Osselton, M. D, Widdop, B., and Watts, J., Eds. *Clarke's Analysis of Drugs and Poisons*. 4th ed. London: Pharmaceutical Press, 2011.
Powers, R. H., and Dean, D. E. *Forensic Toxicology: Mechanisms and Pathology*. Boca Raton, FL: CRC Press, 2015.

ARTICLES

Tracy, D. K., Wood, D. M., and Baumeister, D. Novel Psychoactive Substances: Types, Mechanisms of Action, and Effects. *BMJ* 356 (Jan 2017).

WEBSITES

Link	Description
https://www.deadiversion.usdoj.gov/schedules/orangebook/e_cs_sched.pdf	Current list of controlled substances, also called the "Orange List"
https://www.cdc.gov/drugoverdose/opioids/index.html	Information on the opioid overdose problem
http://www.emcdda.europa.eu/drug-profiles	European website with resources on NPSs and other drugs
http://www.emcdda.europa.eu/topics/pods/synthetic-cannabinoids	Website that shows how synthetic cannabinoids are structured and named
https://pillbox.nlm.nih.gov/pillimage/search.php	Site maintained by the National Library of Medicine used to identify tablets based on color, size, and other features
https://www.nlm.nih.gov/visibleproofs/galleries/index.html	Site with several entries and cases of poisoning
https://pubs.niaaa.nih.gov/publications/alcoholoverdosefactsheet/overdosefact.htm	Facts about alcohol and BAC

Arson and Fire Investigation

12.1 ASPECTS OF FIRE INVESTIGATION

The evaluation of physical evidence recovered from fire scenes is considered part of forensic chemistry. In addition to laboratory analysis, fire investigation also incorporates elements of forensic engineering and crime scene analysis. In most states, fire investigation falls under the jurisdiction of a fire marshal's office, which is tasked with safety and investigation authority. The insurance industry is also involved in fire investigation given that insurance money is a common motive for intentionally setting a fire (**arson**).

Fire investigation has been controversial. In 2011, the Texas Forensic Science Commission issued a report regarding two cases in which men were convicted of murder associated with arson fires. The cases of Cameron Todd Willingham and Ernest Willis occurred in 1986 and 1992, at a time where there were few standard practices used in fire investigation and little in the way of training outside of working with senior investigators. Willingham was executed in 2004, while in the same year the case against Willis was dismissed. The report made clear that Mr. Willingham was convicted and executed based on faulty investigation, incorrect evaluations of burn patterns, and subsequent testimony. The report helped to push reforms in fire investigation, training, and interpretation of burn patterns. One of the core documents used by fire investigators is published by the National Fire Prevention Association (NFPA), entitled *NFPA 921: Guide for Fire and Explosion Investigation*, the most current version being published in 2017; the first edition was published in 1992. This guide was added to the Organization of Scientific Area Committees (OSAC) registry of approved standards in 2017, which indicates that progress is being made in fire investigation and linking the forensic aspects of such investigations to current scientific findings.

In the forensic laboratory, the forensic chemical analysis focuses on determining if an **accelerant** was used in setting the fire. The classic example of an accelerant is gasoline. Generally, an accelerant is a substance or material used to ignite or sustain a fire. The term ***ignitable liquid***

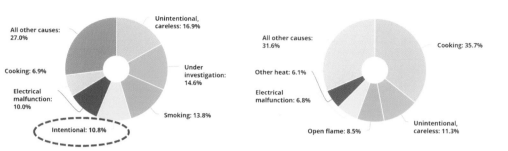

Causes of fatal residential building fires (2015)

Causes of residential building fires resulting in injuries (2015)

FIGURE 12.1 Causes of residential fires, 2015. Notice that the majority of fires are not intentionally set. (From https://www.usfa.fema.gov/data/statistics/.)

is used for substances such as gasoline that are often used as accelerants. After a fire, it is often possible to detect residues of an accelerant using methods that we have seen before—gas chromatography linked to mass spectrometry (GC/MS) or similar techniques; the challenge arises from the matrix and other debris that can make interpreting the results more challenging. Thus, practices and procedures at the scene and proper collection of evidence are critical. Crime scene practices that we introduced earlier would apply to a fire scene investigation as much as to the scene of a homicide. Sadly, sometimes a fire scene is also a homicide scene, as we will see later in this chapter.

According to a report published by the NFPA for the period 2010–2014, an average of more than 260,000 intentional fires were reported per year and resulted in the deaths of 440 people. About three-quarters of these fires were outside, about 20% were structural (buildings), and about 6% were vehicles. More than 80% of the deaths were associated with structural fires. The estimated damage was about a billion dollars. A breakdown of the fires and sources is given in Figures 12.1 and 12.2, data from 2015. When a fire occurs, it is the job of the fire investigator to determine the cause of the fire, and as the figures show, the percentage of fires that are intentionally set is relatively small.

MYTHS OF FORENSIC SCIENCE

Setting fire to a crime scene is a great way to destroy evidence.

While it certainly can complicate the investigation, this tactic is not as effective as movies, books, and TV would have you believe. Fires are usually reported and extinguished relatively quickly, and it is rare that a fire so damages a body that an autopsy cannot provide critical information. Not every part of a scene will burn completely, and evidence such as fingerprints and even blood and DNA can often be recovered.

12.2 CHEMISTRY OF COMBUSTION

You may have seen a **fire triangle** (Figure 12.3) that shows three necessary ingredients for fire—a fuel, oxygen, and heat. If one of these elements is removed or exhausted, fire is no longer supported. Although correct, this is a simple depiction of a complex process that is better represented

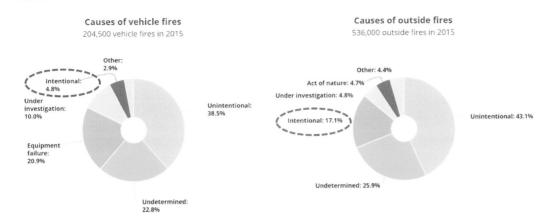

FIGURE 12.2 Causes of car fires and outside fires, 2015. Outside fires, including wildfires, were intentionally set more often than residential fires. (From https://www.usfa.fema.gov/data/statistics.)

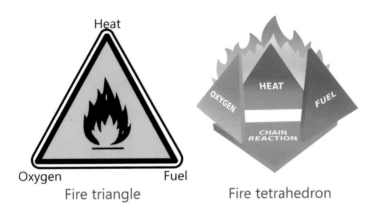

FIGURE 12.3 Depiction of components of fire.

by a **fire tetrahedron** with four components—fuel, heat, oxygen, and chemical chain reactions that are combustible and generate what we call a flame. The fuel can be anything that can burn, but that fuel must be in the gas phase for the reaction to occur. Wood doesn't burn; the vapors driven off by the heat of the fire are what ignite and produce a flame. We will come back to this point, as the concentration of fuel vapors is an important consideration in fire investigation.

The second ingredient needed is an oxidizer; here we will assume that this is oxygen from the atmosphere. Normal atmospheric air contains about 20% oxygen. Typically, flaming combustion occurs above 15% oxygen. A smoldering fire (e.g., glowing logs in a campfire) may continue at much lower percentages.

Heat is a necessary ingredient because it drives evaporation of components of fuel into the gas phase where they can burn. Once a fire is established, the heat needed comes from the reaction itself, but some form of energy is needed to start the combustion process. If you have gas burners at home, there is a pilot light that ignites them; to start a campfire, a match is used. This energy is sufficient to start the fire, which is a complex series of chain reactions that will continue until one of the key components is gone. In the open air for a typical fire, oxygen does not run out, so the loss of fuel or the removal of heat causes combustion to cease. However, oxygen can become the limiting ingredient within confined spaces. For example, if a fire starts inside a room with no open windows or doors, the oxygen in that space can be consumed quickly. The chain reactions can continue but flames cannot be supported. If one of the windows breaks, oxygen can rush back into the room, causing an eruption of fire that can be very dangerous to firefighters, which is referred to as **backdraft**.

DEATH BY FIRE?

On occasion, murderers will set incendiary fires to destroy evidence related to the death. When a body is found after a fire, the skills of both fire and death investigators are needed. One such case occurred in 2005 when a fire erupted at the home of a nurse practitioner; evidence uncovered in the investigation showed that the woman had secretly left work at a hospital shortly before the fire broke out. After the fire was extinguished, the body of her husband was found, badly burned. An autopsy and postmortem toxicology showed that the levels of CO in the blood were not elevated, which was evidence that the victim was not breathing during the fire. Toxicological analysis also revealed the presence of a drug called rocuronium at elevated levels in the victim's body, while crime scene processing turned up a charred needle cap in the fire debris. The drug is used during surgical procedures to relax and paralyze muscles; thus, the victim was paralyzed and died of suffocation. The wife was convicted in 2007 of arson and murder.

Source: Johnstone, R. E., et al., Homicides Using Muscle Relaxants, Opioids, and Anesthetic Drugs: Anesthesiologist Assistance in Their Investigation and Prosecution, *Anesthesiology* 114, no. 3 (Mar 2011): 713–716.

12.2.1 CRITICAL CONDITIONS AND POINTS

The fire triangle shows what must be present for fires to occur and continue. However, the presence of all three does not guarantee that combustion will occur. Fuel and oxygen must be present within specific concentration ranges near an ignition source for the fire to start. As an example, suppose a would-be arsonist wishes to burn down a failing business to collect insurance money (a common motive for arson). He obtains gasoline and empties a large can onto a wooden floor and immediately drops a lit match into the gasoline puddle. Likely, nothing happens, and the match is snuffed out in the liquid gas. Recall that only gases burn, and this is true of gasoline in the liquid phase. Gasoline is a complex mixture of compounds called hydrocarbons, many of which evaporate quickly and easily (i.e., they are **volatile**). Thus, if gasoline is poured onto a surface, it takes time (less if warm, more if cold) for enough vapor to be formed in the air where it can mix with oxygen to form a flammable mixture. Even if this mixture is present, an ignition source (energy) must be provided in the right place to start the reaction.

Fire investigation relies on tools and tables to determine if the conditions existed for a fire to develop. When considering the fuel source, it is important to determine how likely it is under a given set of circumstances to support a fire. Because a liquid's boiling point indicates imminent vaporization, it is one measure of the **volatility** of a liquid fuel. The lower the boiling point of a liquid, the more volatile it is. Two other indicators are **flash point** and **fire point**. The flash point is the temperature at which a liquid gives off sufficient vapors to form an ignitable mixture at its surface (gasoline is −45°F and kerosene is 100°F). The fire point is the temperature at which a liquid produces vapors that will sustain combustion. This is generally several degrees higher than the flash point (e.g., gasoline is 495°F and kerosene is 110 °F). The generally accepted temperature used to distinguish flammable liquids from combustible liquids is defined as 100°F; temperatures below 100°F classify flammable liquids, and temperatures greater than 100°F classify combustible liquids. Finally, the relative concentrations of air to fuel needed to support combustion are given by the **upper and lower flammability limits** (UFL and LFL), which are usually expressed as a percentage of fuel within a volume of space. The term *explosive limit* is also used. As an example, the **flammable range** of methane (natural gas) is between 5% by volume (LFL) and 15% (UFL).

Figure 12.4 shows the relationship of the concentration of fuel in air to flammability. When the amount of fuel present is too low to support combustion (lean mixture), combustion cannot be supported. When the concentration reaches the LFL, combustion is possible if sufficient energy is present to start it. Combustion will be supported at increasing fuel concentrations until that

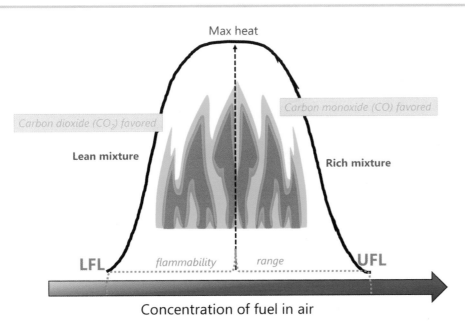

Max heat

Carbon dioxide (CO₂) favored

Carbon monoxide (CO) favored

Lean mixture

Rich mixture

LFL   *flammability*   *range*   UFL

Concentration of fuel in air

FIGURE 12.4   A simplified depiction of the flammability range of a substance such as natural gas.

amount reaches the UFL. Beyond that concentration, the mixture cannot support combustion. Within the flammability range, the amount of heat produced and the composition of the by-products will change as a function of the **fuel-to-air ratio** (F/A ratio). On the lean side of optimal balance, the formation of carbon monoxide (CO_2) is favored, and on the rich side (lower relative oxygen concentration), carbon monoxide (CO) formation is favored. Recall from Chapter 12 that CO poisoning can be deadly and often plays a role in deaths related to fires.

Suppose a natural gas leak is occurring in an enclosed basement room that is relatively air-tight. For ease of example, assume that the space has a volume of 100 cubic feet. The mixture will not become flammable until 5 cubic feet of that volume consists of methane and the remaining 95 cubic feet is air. If there is a spark or other ignition source in the right place for the right amount of time, a fire can occur. If it does not and the leak continues, when the volume of natural gas exceeds 15 cubic feet, then the mixture is no longer flammable. When there is an imbalance in concentrations in which there is too much fuel and too little oxygen, the mixture is too rich for combustion, and when there is too little fuel and too much oxygen, the mixture is referred to as lean. Not surprisingly, the larger the flammability range (LFL–UFL), the more dangerous the substance is; hydrogen gas is flammable from 4% to 75% concentrations by volume.

Other important descriptors are the **ignition temperature** (the minimum temperature at which a substance can ignite) and the **ignition time**, which is the time from putting an energy source in contact with a flammable mixture that is needed to create a self-sustaining chain of chemical reactions. Intentionally set fires (arson, also called **incendiary fires**) must be started somehow; the device used is called the **incendiary device**. Finding this is critical to proving that a fire was intentionally set. However, these devices must have sufficient energy delivered to the right place (where a flammable mixture exists) at the right time and for long enough to start the fire. This is not as simple as it is often perceived to be. Common incendiary devices include matches, cigarettes, and road flares. The remains of these devices are critical physical evidence in fire investigation. Examples are shown in Figure 12.5.

12.2.2 HEAT

Heat is needed to drive flammable materials into the gas phase to sustain a fire and ignite it. Since combustion reactions generate copious amounts of heat, the reaction can be self-sustaining, as we have already discussed. There are four ways in which heat is typically produced related to combustion and fire investigation: **chemical**, **mechanical**, **electrical**, and **compression**:

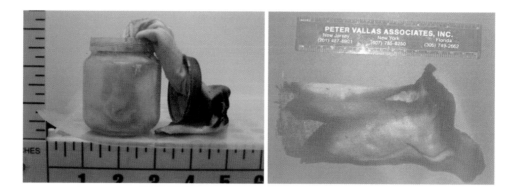

FIGURE 12.5 Two examples of incendiary devices. Left: A jar with a rag and gasoline (called a "Molotov cocktail"). Right: A diaper soaked in paint thinner. (Image at left courtesy of the Minnesota Department of Public Safety.)

Chemical—Chemically produced heat is the result of rapid oxidation. The speed of the oxidative reaction is a key factor. Rust is a product of oxidation, but a very slow one.

Mechanical—Mechanical heat is the product of friction. Internal metal components of machinery can overheat due to lubricant breakdown or ball-bearing failure and cause ignition of available combustibles.

Electrical—Electrical heat is the product of arcing, shorting, or other electrical malfunction. Poor wire connections, too much resistance, a loose ground, and too much current flow through an improperly sized wire are other sources of electrical heat.

Compressed gas—When a gas is compressed, its molecular activity is greatly increased. Consider the operation of a diesel engine. The gaseous fuel is compressed within the cylinder, increasing its molecular activity. The heat generated by this activity eventually reaches the ignition temperature of the fuel itself. The resulting contained explosion forces the piston back to the bottom of the cylinder, and the process repeats.

Once heat is produced, it is transferred by different mechanisms. Three of the most common forms of heat transfer are **conduction**, **convection**, and **radiation**:

Conduction is the transfer of heat through direct contact. If you touch a hot stove, the pain you first feel is the result of conducted heat passing from the stove directly to your hand. In a structural fire, superheated pipes, steel girders, and other structural members, such as walls and floors, may conduct enough heat to initiate fires in other areas of the structure. Heat transfer by conduction is responsible for the spread of fire in almost every structure.

Convection entails the transfer of heat by a circulating medium, usually air or liquid. The superheated gases evolved from a fire are lighter than air and consequently rise. As they travel and collect in the upper levels of the structure, they can and do initiate additional damage.

Radiation is **radiated heat** moving in invisible waves and rising much the same as sunlight or x-rays. Radiated heat travels at the same speed as visible light (186 miles/s); infrared (IR) energy is heat energy. It is primarily responsible for the exposure hazards that develop and exist during a fire. Radiant heat travels in a direct or straight line from the source until it strikes an object.

12.2.3 COMBUSTION PRODUCTS

Fire is an extraordinarily complex process both physically and chemically. The chain reactions that drive it produce a mix of by-products that vary depending on the fuel, amount of oxygen, F/A ratio (Figure 12.4), temperature, location, and so forth. **Smoke** is a generic term

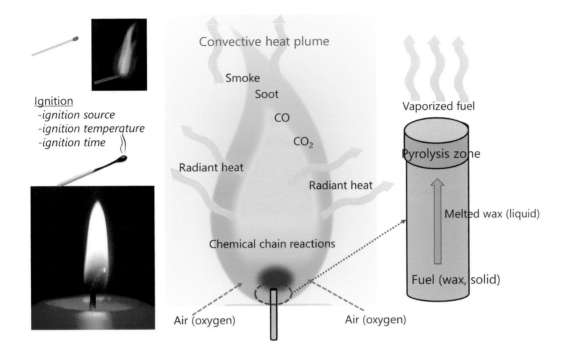

FIGURE 12.6 Combustion in a candle, as described in the text.

for by-products of fire, but many of the compounds produced are not visible but still deadly. Smoke is visible in part because it contains tiny particles and droplets of combustion by-products. **Soot** is defined specifically as particles of black carbon. We noted in Chapter 11 that cyanide can also be produced in fires from materials that contain the element nitrogen. Many modern plastics and polymers contain it, so the danger of cyanide poisoning in fires is also present.

12.2.4 INTEGRATED EXAMPLE

Let's go through a simple example of a combustion event to show how these terms and descriptors can be used in the context of a fire. Figure 12.6 shows a candle. To light the candle (ignition), a match is used. When you push the match across the rough strip of a matchbook, this generates friction and mechanical heat that ignites the material in the match head. The lit match is the ignition source that you place against the wick for a few seconds (ignition time). While the match is touching the wick, conductive heat is transferred into the wick. The heat emitted by the burning match head contacts the wick, melts some of the wax, vaporizes it, and if held long enough in place, ignites the vapor. Once the candle flame is burning hot enough to melt wax, vaporize the components, and support the reaction, the candle will continue to burn. Because the chemical reactions generate heat (chemical heat), the solid fuel (wax) will melt and be drawn up the wick to a zone where the components of the wax are driven into the gas phase and start to degrade. This area is referred to as the **pyrolysis zone**. Each of the compounds in the wax will have different LFL values, but in these conditions, all are quickly exceeded, and combustion is supported. Because the heated air rises away from the flame, the UFL won't be reached under typical conditions.

Once these vapors ignite, the reactions take place, generating heat and light. As long as there is fuel, air, and heat (the fire triangle), the complex chemical reactions continue (the fire quadrangle). A complex mix of products will be produced by the flame, including CO, CO_2, smoke, and soot. Because hot air is less dense than cooler air, it rises and creates a convective heat plume. The candle will continue to burn until the wax is depleted (runs out of fuel), oxygen is cut off (e.g., you cover the candle with a drinking glass or grab the wick with wet fingers), or heat is cut off (the wick is cooled with wet fingers). Although greatly simplified, the basic principles shown in this example translate to fire investigation practices and procedures.

MYTHS OF FORENSIC SCIENCE

It is easy to tell if a fire was arson.

As with all forensic investigations, all the information uncovered during an investigation must be internally consistent to reach that conclusion. Intentionally set fires require that some source of ignition be used (an incendiary device), a combustible source of materials must exist (accelerants), and these elements must be purposely placed in the right place at the right time for a fire to start. The detection of residual accelerants is not always easy. Gasoline is commonly used to start fires, but gasoline is a petroleum product, as are many of the plastics and synthetic materials found in a home or other building. Sometimes finding residuals is straightforward, but often it is not. All the evidence recovered through forensic analysis, fire investigation, and law enforcement investigations should tell the same story before a conclusion is reached.

12.3 FIRE INVESTIGATION

When a suspicious fire is identified, the forensic investigation focuses on identifying the cause. This task is analogous to identifying the cause of death when a questioned death arises (Chapter 5). Fires can be classified as accidental, natural (e.g., caused by lightning), intentionally set, or in some cases, undetermined. Fire investigation combines engineering with laboratory analysis to determine the cause and if a crime has been committed. Two key questions for investigators to answer are:

1. Where did the fire start (what is the **point of origin**?). This is essential information as this will also be the point where ignition occurred. If more than one point of origin is found, these raise concerns that the fire might have been intentionally set. Finding the point of origin is also important for finding any incendiary devices if the fire was intentionally set. As an example, suppose that a fire burns in an older home and the point of origin is found to be at a wall socket. The source of ignition could be an electrical short that ignited wallpaper or another flammable in the area close to the outlet. However, if the point of origin is found to be in the middle of the room with no electrical wiring or other obvious ignition source, then the likelihood of an intentionally set fire increases and the investigator would turn attention to question 2.
2. Was something used to start and support the fire (gasoline, paper, matches, etc.)?

The first question requires study of the fire scene and knowledge of fire behavior. The second requires laboratory work and analysis of physical evidence.

12.3.1 ACCELERANTS AND INCENDIARY DEVICES

An accelerant is the flammable material that is used to start the fire. Accelerants can be solids, liquids, or gases, with gasoline being the most commonly used. Solid accelerants include paper, fireworks, highway flares, and black powder. Butane (cigarette lighter fuel), propane, and natural

FIGURE 12.7   Unlined paint cans used to collect fire debris evidence. The stopper is removed during laboratory analysis.

gas are examples of gaseous accelerants, which do not leave any residue at a fire scene. However, gases must be contained and transported, so severed gas lines or spent containers serve as critical physical evidence in such cases. Liquid accelerants fall into two broad categories: petroleum distillates, which include gasoline and other petroleum products, and nonpetroleum products, such as methanol, acetone (used in nail polish remover), and turpentine. **Petroleum distillates** are derived from crude oil and are also called **hydrocarbons** or petroleum hydrocarbons. In crude oil, the volatility of the individual components ranges from extremely volatile substances, such as propane (a gas at room temperature), to asphalt, which remains solid even at elevated temperatures. Petroleum distillates such as gasoline and kerosene are not single hydrocarbons but mixtures of different components with similar volatilities. The volatility of an accelerant is an important consideration in the combustion process, determining how much residue will be left and how quickly it will evaporate after the fire is out.

Fire debris evidence is typically collected in metal cans that are paint cans without any lining material (Figure 12.7). The debris is scooped into the can and the can is sealed and transported to the laboratory. At the scene, it is also critical to collect control and background samples for comparison. This is because many materials are made from hydrocarbons or synthetic materials derived from petroleum products. Examples include flooring, carpeting, and plastics. Note that the cans have stoppers inserted—these are removed during laboratory analysis; however, the cans, if properly sealed, are airtight and will preserve vapors that arise from any accelerants present.

12.3.2 LABORATORY ANALYSIS

The instrument most often used in fire debris analysis is GC/MS (see Chapters 6 and 11). Typically, the debris in the can is gently heated to drive any residual volatiles into the headspace. Conceptually this is the same approach that is used to test blood alcohol (see Chapter 11). A sample of the headspace vapor can be drawn out and injected directly, but many methods involve an additional step of concentrating the vapors. One way to do this it to place a charcoal strip in the container while it is being heated; the charcoal efficiently absorbs the vapors and traps them. This is the same way a charcoal filter works for drinking water. The charcoal strip would then be placed in a small amount of solvent to extract the volatile materials, and this extraction is injected into the GC/MS. Accelerants such as gasoline contain many different individual

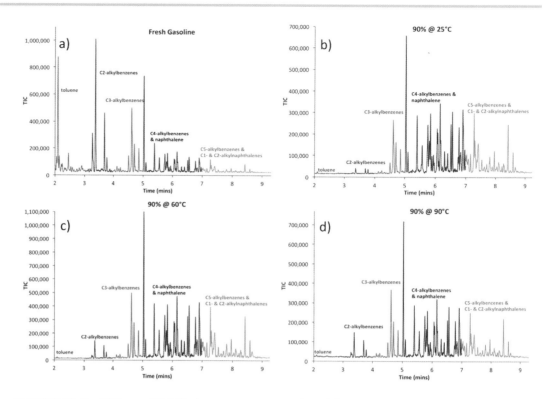

FIGURE 12.8 Different GC/MS chromatograms produced from gasoline. (Reproduced from Birks, H. L., et al., The Surprising Effect of Temperature on the Weathering of Gasoline, *Forensic Chemistry* 4 [Jun 2017]: 32–40. With permission from Elsevier.)

compounds, and as a result, there will be many peaks in the resulting chromatogram if gasoline or a similar accelerant is present. Other accelerants may have fewer peaks, such as if pure alcohol or acetone is present. The pattern of the peaks is an important piece of information, but not the only one needed to determine if an accelerant is present. See Figure 12.8 for examples of gasoline patterns.

In the upper left frame (a), a chromatogram from a fresh gasoline sample is shown. The x-axis shows the time that has elapsed since the injection (the **retention time**) and the y-axis shows the intensity of the peak. The chromatogram can be divided into zones (colored) based on the types of compounds that have retention times in that area. For example, toluene is a relatively small molecule and, as such, has a short retention time on the column. As you go farther to the right, the molecular weight of the compounds is generally increasing and the volatility of the compounds is generally decreasing. You can see that there are dozens of peaks in the chromatogram because there are many individual compounds found in gasoline.

In the experiments reported in the research shown, the gasoline has been allowed to **weather** to mimic what might occur to residual gasoline at a fire scene. Consider a case where a fire is intentionally set with gasoline; it burns for some time until firefighters arrive and extinguish it. It may be several hours or days before samples are collected at the scene, and during the delay, compounds will continue to evaporate; the more volatile compounds (the ones that come out earliest in the chromatogram) will generally be affected the most. In Figure 12.8b, the overall intensity of peaks has decreased (see the numbers on the y-axis; decreased from ~1 million to ~700,000) and the most volatile compounds (in blue) are all but gone. Depending on the degree of weathering and temperature, slight differences in the pattern can be seen. Thus, the possibility of weathering must be considered as part of data interpretation. This data is also coupled to the MS data to provide additional information useful for identifying the presence of any likely accelerant. An example of the chromatographic pattern for diesel fuel is shown in the following textbox; note that it is dramatically different from that for gasoline.

ACCELERANTS IN BLOOD AND WATER

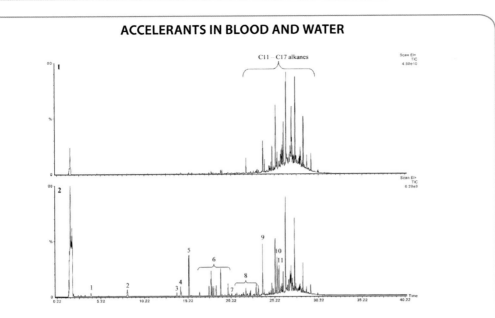

Case 1. GC/MS data from water samples. The top frame shows a sample taken from the top of the well, and the lower frame shows the results from water taken from a faucet. (Reproduced from Borusiewicz, R., Substrate Interferences in Identifying Flammable Liquids in Food, Environmental and Biological Samples: Case Studies, Science and Justice *55, no. 3 [May 2015]: 176–180. With permission from Elsevier.)*

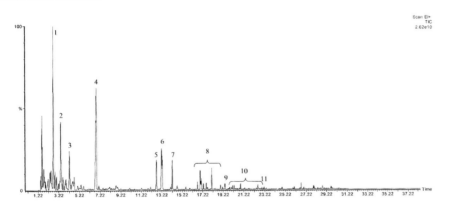

Case 2. GC/MS data from a blood sample. (Reproduced from Borusiewicz, R., Substrate Interferences in Identifying Flammable Liquids in Food, Environmental and Biological Samples: Case Studies, Science and Justice *55, no. 3 [May 2015]: 176–180. With permission from Elsevier.)*

Common accelerants such as gasoline and diesel fuel are also potential environmental contaminants and poisons. Two examples are shown in which analytical methods similar to that used for fire debris were used to detect petroleum hydrocarbons in water and blood. In the first case, a well was suspected of being contaminated with kerosene, a mixture of hydrocarbons that are generally heavier than those in gasoline (Figure 12.8). The sample that produced the result in the top frame of the figure was drawn from the top of the well. Since kerosene (like gasoline) floats on water, the top of the well was essentially pure kerosene. The sample from the tap shows less evidence of kerosene because kerosene is not soluble to any significant degree in water. In the second case, a blood sample was collected from a victim found near a gasoline vapor source. The chromatogram shows the components of gasoline but is different than would be expected based on typical gasoline patterns, as shown in Figure 12.8. The blood sample contained relatively more of the lighter and more water-soluble portions of gasoline that would be seen in gasoline alone; this could be because the vapors were inhaled and the substances that were most soluble in the blood were the ones seen at the highest concentrations.

12.3.3 CASE EXAMPLE

As we saw in Figures 12.1 and 12.2, only a fraction of fires are intentionally set, yet fire investigators still must determine causes and make recommendations, particularly about public safety. One of the worst fatal fires in the United States occurred in 2003 at a nightclub called the Station in Rhode Island. One hundred people were killed and more than 200 injured, many severely. The NFPA lists this fire as the 8th deadliest in the category of public assembly in the United States and 10th worldwide. A large investigation was undertaken to understand what had happened, why the fire spread so quickly, and what steps could be taken to avoid similar disasters in crowded spaces. A comprehensive report was issued by the National Institute of Standards and Technology (NIST) in 2005 describing research and findings, and it provides an example of large-scale fire investigations.

The fire began during a rock concert at the nightclub when pyrotechnics (similar to fireworks) were ignited. At the time, the audience was packed into the area in front of the stage, which added to the death toll. The sparks from the device impacted a soundproofing foam on the wall of the stage and within a few seconds (ignition time), the foam began to burn. The fire spread incredibly fast; within less than a minute, smoke was coming out of the windows and within 5 minutes, the fire was burning through the roof. The building did not have an automatic sprinkler system, so there was nothing to stop the spread of the fire. Figure 12.9 shows a map of the nightclub with the points of ignition. The right side of this figure shows where bodies were recovered. Once people inside the club realized the danger, they pushed toward the main exit, which quickly became blocked, trapping many people as the flames, heat, and smoke enveloped the building. The NIST report noted that within about 90 seconds after ignition, anyone remaining in the club would have been severely incapacitated or dead (Figure 12.10).

The speed of the spread of the fire and the combustibility of the foam were the focus of the investigation. The area around the stage had the foam material, but there was also interest in ceiling tiles and wood paneling in the building. One test result is shown in Figure 12.11. On the left is the visible image, and on the right, an IR (heat) image. The area where burning started is seen clearly in the IR image; this is an example of the concepts of ignition time and temperature. The reports also describe how the materials involved were tested for combustion products, including gaseous cyanide.

Finally, computer modeling was used to estimate the temperature and oxygen concentrations in different areas of the club during the fire. Figure 12.12 shows the temperatures about 2 feet above the floor as a function of time after the foam ignited. Notice the color shading and coding; red color indicates an air temperature of nearly 1000°F. Similarly, Figure 12.13 shows the oxygen

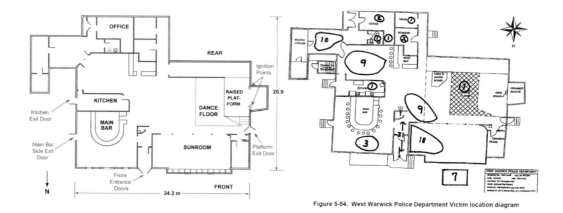

Figure 5-54. West Warwick Police Department Victim location diagram

FIGURE 12.9 The nightclub floorplan. The ignition points (points of origin) are shown in the left frame behind the stage. On the right is the police map of the scene showing where bodies were recovered. (Reproduced from NIST NCSTAR 2: Vol. 1, Report of the Technical Investigation of the Station Nightclub Fire, National Institute of Standards and Technology, Fire Research Division, June 2005. https://www.nist.gov/el/station-nightclub-fire-2003.)

FIGURE 12.10   Recreation of the first seconds of the fire. Notice the black smoke on the ceiling appearing in less than 1 minute. (Reproduced from NIST NCSTAR 2: Vol. 1, Report of the Technical Investigation of the Station Nightclub Fire, National Institute of Standards and Technology, Fire Research Division, June 2005. https://www.nist.gov/el/station-nightclub-fire-2003.)

concentrations on the same timescale. Blue color corresponds to the normal ~20% oxygen by volume in air, while red indicates no oxygen at all. The rapid depletion of oxygen shows how fast the fire was burning in the early stages—too fast for air to replace it.

The investigation concluded that many factors contributed to the large loss of life, including overcrowding, lack of a sprinkler system, and use of dangerously flammable materials. Since 2003, there have not been any larger fatal nightclub fires in the United States, but sadly other large-fatality fires have occurred worldwide. Anytime crowds gather in enclosed spaces, fire will always be a danger, and fire investigation is invaluable in determining causes and contributing factors and making recommendations to prevent future incidents.

WILDFIRE INVESTIGATIONS

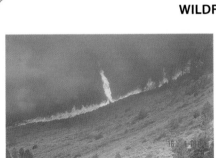

Wildfire near Casper, Wyoming. (Image courtesy of the National Oceanic and Atmospheric Administration.)

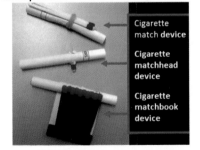

Example of cigarette-based incendiary devices. (Image from National Wildfire Coordinating Group, Guide to Wildland Fire Origin and Cause Determination, PMS 12 NFES 1874, April 2016. https://www.nwcg.gov/ publications/412.)

Every year, thousands of acres of forest and wildlands are destroyed by incendiary fires.

A government group, the National Wildfire Coordinating Group, provides a clearinghouse for information and coordination of agencies that are involved in fighting and investigating fires. The also publish the *Guide to Wildland Fire Origin and Cause Determination*, which assists wildland fire investigators with determining if a fire was intentionally set. With large fires, finding the origin can be a challenging task that combines witness statements with fire behavior and burn patterns. Meticulous grid or lane searching, as described in Chapter 3, may be used to find and collect evidence. Common causes of wildland fires include lightning, campfires, smoking, and sparks from passing vehicles and trains. Incendiary devices can be as simple as a match, cigarette, or even a combination device that is designed to burn slowly down and ignite the fire long after the person setting it has left.

Source: National Wildfire Coordinating Group, *Guide to Wildland Fire Origin and Cause Determination*, PMS 12 NFES 1874, April 2016. https:// www.nwcg.gov/publications/412.

FIGURE 12.11 Tests of a foam with the pyrotechnic device. (Reproduced from NIST NCSTAR 2: Vol. 1, Report of the Technical Investigation of the Station Nightclub Fire, National Institute of Standards and Technology, Fire Research Division, June 2005. https://www.nist.gov/el/station-nightclub-fire-2003.)

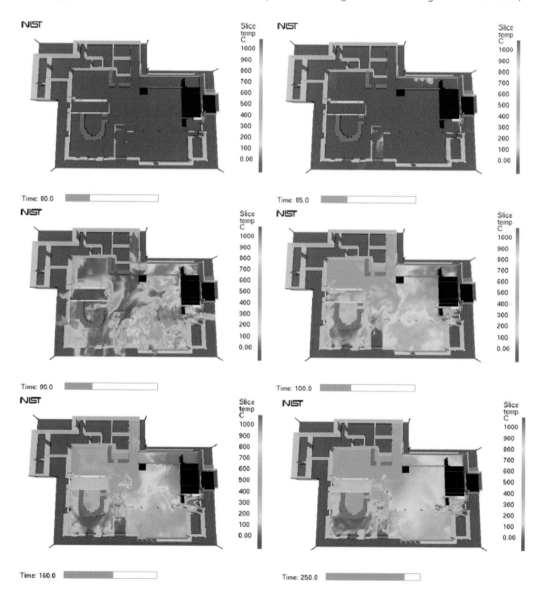

FIGURE 12.12 Temperatures in the fire zone. (Reproduced from NIST NCSTAR 2: Vol. 1, Report of the Technical Investigation of the Station Nightclub Fire, National Institute of Standards and Technology, Fire Research Division, June 2005. https://www.nist.gov/el/station-nightclub-fire-2003.)

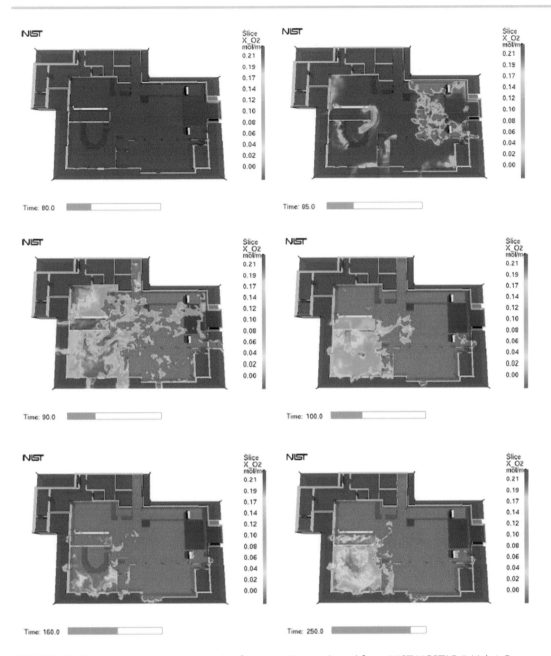

FIGURE 12.13   Oxygen depletion in the fire zone. (Reproduced from NIST NCSTAR 2: Vol. 1, Report of the Technical Investigation of the Station Nightclub Fire, National Institute of Standards and Technology, Fire Research Division, June 2005. https://www.nist.gov/el/station-nightclub-fire-2003.)

12.4 REVIEW MATERIAL: KEY CONCEPTS AND QUESTIONS

12.4.1 KEY TERMS AND CONCEPTS

Accelerant
Arson
Backdraft
Chemical heat
Compression heat
Conduction
Convection

Electrical heat
Fire point
Fire tetrahedron
Flammability range
Flash point
Fuel-to-air ratio
Hydrocarbon
Ignitable liquid
Ignition temperature
Ignition time
Incendiary device
Incendiary fire
Lower flammability limit
Mechanical heat
Petroleum distillates
Petroleum hydrocarbons
Point of origin
Pyrolysis zone
Radiation
Retention time
Smoke
Soot
Upper flammability limit
Volatile/volatility
Weather/weathering

12.4.2 REVIEW QUESTIONS

1. What are the three conditions necessary for combustion?
2. Suppose a tank of hydrogen is leaking into a small, well-sealed laboratory room with a volume of 200 cubic feet. How many cubic feet of hydrogen would have to leak in to allow a combustible mixture to exist?
3. Look at the picture at the start of the chapter. Notice the flames coming out of the window and down the side of the door. Describe what is happening in a way like the candle example (Figure 12.6).
4. If you put a drinking glass over a burning candle, the candle flame dwindles and dies, and the amount of soot increases. Describe what happened and why (hint: UFL/LFL).
5. Identify where flash point and fire point data come into play in the nightclub case study.

12.4.3 ADVANCED QUESTIONS AND EXERCISES

1. Revisit the crime scene scenario laid out in Chapter 3. If you were the investigator, would you request any analyses at the site of the clandestine grave? If so, why? What evidence would you be seeking. If not, why not?
2. Suppose a body is found at a fire scene, as was the case described in the insert. What types of information could be provided by an autopsy and by postmortem toxicology that could be used to determine if the victim was alive at the time of the fire or already dead?
3. At many arson scenes, gasoline is easily detectable. If so, why is it so challenging to prosecute arson cases?

BIBLIOGRAPHY AND FURTHER READING

BOOKS

Icove, D. J., and Haynes, G. A. *Kirk's Fire Investigation*. 8th ed. New York: Pearson Education, 2018.

National Fire Protection Association. *NFPA 921: Guide for Fire and Explosion Investigations*. 2017.

Quintiere, J. G. *Principles of Fire Behavior*. 2nd ed. Boca Raton, FL: CRC Press, 2017.

ARTICLES AND REPORTS

Report of the Texas Forensic Science Commission: Willingham/Willis Investigation, April 15, 2011. http://www.fsc.state.tx.us/documents/FINAL.pdf.

U.S. Fire Administration. Fire in the United States. 19th ed. https://www.usfa.fema.gov/downloads/pdf/publications/fius19th.pdf.

WEBSITES

Link	Description or Abbreviation
http://www.nfpa.org/News-and-Research/Resources/Fire-investigations	National Fire Prevention Association—information on fire investigations
https://www.nist.gov/el/fire-research-division-73300/national-fire-research-laboratory-73306/videos	NIST fire science videos
https://www.atf.gov/arson/atf-fire-research-laboratory	Bureau of Alcohol, Tobacco, Firearms and Explosives Fire Research Lab info and videos
https://www.usfa.fema.gov/data/	Federal Emergency Management Agency U.S. Fire Administration data site
https://www.nwcg.gov/	National Wildfire Coordinating Group
https://www.atf.gov/arson/atf-fire-research-laboratory-gasoline-flame-jetting-videos	Videos of gasoline jets from containers
http://ilrc.ucf.edu/	Ignitable liquids database

Explosives and Improvised Explosive Devices

13.1 DETONATION AND DEFINITIONS

13.1.1 BURNING AND PROPAGATION

In Chapter 12, we discussed fire investigation and combustion from the point of view of the fire triangle and tetrahedron (fuel, air, heat, and reactions) but did not go into detail about the composition of the fuel. We also assumed that air supplied all the oxygen that was needed to support a combustion reaction, which is evidenced by flame or smoldering. As we move into the topic of explosives, we need to be more specific about fuels, oxygen, where the oxygen comes from, and the speed of the reaction. Fire, flames, and explosions are related; a fuel and an oxidant are involved. In familiar fires and flames, the oxidant is oxygen from the air, but that is not the only way an oxidant can be supplied to a reaction. The flame you see on hot water heaters or gas burners on stoves is a type of combustion reaction that can be described by a balanced chemical equation:

$$CH_4(g) + 2O_2(g) \rightarrow CO_2(g) + 2H_2O(g) + heat$$

Each molecule of natural gas (methane, CH_4) reacts with two molecules of oxygen (O_2) to form one molecule of carbon dioxide (CO_2) and two molecules of water. The reaction produces heat that feeds back into the cycle to allow it to continue until either the fuel (methane) or oxygen is depleted. The reaction will not spontaneously start on its own; an initial input of energy is required. On a stove, this is accomplished by a pilot light or match, which is the source of ignition. The symbol (g) indicates that the reactions occur in the gas phase. The burning of methane is a combustion reaction involving a fuel and an oxidant.

There are other such reactions that are important in forensic science. Propellants such as gunpowder (Chapter 15) are also based on combustion; the fuel is a mixture of compounds, primarily nitrocellulose and nitroglycerin, which burn rapidly to produce the large volume of gases that propel the bullet down the barrel of the gun. Propellants are also used in rockets to generate thrust; in this case, the hot expanding gases are directed outward through the nozzle. When propellants burn, there is a progression of the burning and distinct zones of reacted and unreacted components. Another term used in these cases is a **propagation reaction** since the reaction has a direction and moves progressively.

Explosives are also based on fuel and oxidants. Key differences between simple flame combustion, propellants, and explosives are the speed of the reaction, the degree of confinement of the reaction, and the source of oxidizers. For a stove burner and typical flames and fire, the atmosphere can supply oxygen as the oxidizer. However, in other cases, such as with fast-burning propellants and explosives, oxygen must be supplied from other chemical compounds.

The linear burn rate of a fuel is one metric that can be used to characterize the speed of a reaction (Figure 13.1). Imagine that a thin-line gasoline (fuel) is poured on a surface and one end is ignited. The flame propagates down the line. By measuring the distance covered over a given time, the **linear burn rate** can be calculated as 5.0 m/s in this example. The way in which the flame propagates down the line of gasoline is shown in Figure 13.2. The burning process

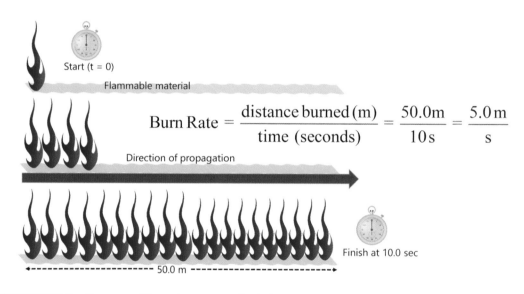

$$\text{Burn Rate} = \frac{\text{distance burned (m)}}{\text{time (seconds)}} = \frac{50.0\,\text{m}}{10\,\text{s}} = \frac{5.0\,\text{m}}{\text{s}}$$

FIGURE 13.1   Concept of linear burn rate calculation.

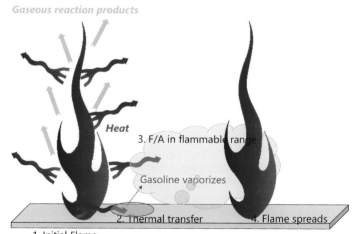

FIGURE 13.2   How a flame front propagates.

generates intense heat and hot expanding gases. In the open air, the gases dissipate, as does much of the heat. However, enough of the heat is transferred to the liquid gasoline in front of the flame (**thermal transfer,** 2 in Figure 13.2) to vaporize it and create a flammable mixture (based on the fuel-to-air ratio [F/A ratio]; 3). The heat of the flame is sufficient to ignite the vapors and move the flame front down the line of gasoline. Figure 13.3 shows how this burning progresses. If you were to watch the line of gasoline burn, you would observe the flame moving along the line of fuel; this is referred to as the propagation of the **flame front**.

13.1.2 DEFLAGRATION

Another term used to describe how flames spread is ***deflagration***. Deflagration is an oxidative decomposition (burning) in which the propagation speed is slower than the speed of sound (subsonic). The speed of sound in air is about 343 m/s (~3430 cm/s; ~760 miles/h or ~1100 ft/s); in our example from Figure 13.1, the propagation speed was 5 m/s and as such is deflagration. Another important characteristic of deflagration is that the direction of flow of the reaction products is opposite the direction of the reaction front (Figure 13.4). A rocket is an example of a deflagration in that the flame front is propagating in the opposite direction of the reaction products (the exhaust). In the case of the rocket, the deflagration reaction is contained in such a way that the reaction products are forced to exit through the nozzle at the rear. In the example of flames in open air (Figures 13.1 through 13.3), the reaction products will drift up and away.

MYTHS OF FORENSIC SCIENCE

Bombs always go off.

Fortunately, no. Although these devices can be very easy to make, making one that will go off as intended can be difficult. It is not unusual for bombers to have a place that they test their devices before planting them as part of their planned crime. Getting the right mix of compounds is difficult, as is developing an effective detonation or ignition device or procedure. A recent example of a failed attempt occurred in 2017 at a subway terminal in New York City. The simple device burned, but fortunately did not explode as intended and no one was killed. In December 2001, Richard Reid attempted to detonate a shoe bomb by lighting it with a match; he was stopped by alert passengers and crew.

13.1.3 EXPLOSIONS

Although the terms *explosion* and *detonation* are often used interchangeably, they are two distinct things. An **explosion** is a sudden, rapid, and often violent release of hot expanding gases.

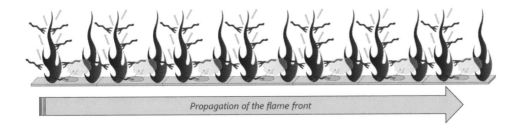

Propagation of the flame front

FIGURE 13.3 How a flame front propagates along the line of fuel in open air.

FIGURE 13.4   Propellant in a solid rocket deflagrates. (Image from NASA.)

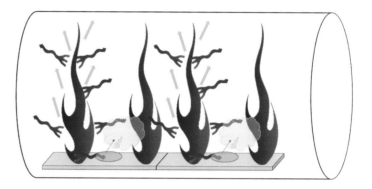

FIGURE 13.5   Deflagration in a sealed container.

Not all explosions are caused by explosives. Heating water in a confined space can cause a steam explosion; this type of accident was not unusual in the age of steam engines. Volcanos can also produce steam explosions. A buildup of pressure through confinement is integral to creating explosions. The fuel and oxidizers are typically trapped, if only for a fraction of a second, as the reaction is taking place. The hot expanding gases cannot escape, causing increases in the pressure until the confining vessel fails. The failure allows for the sudden release of gases and heat.

In Figure 13.1, we used the example of gasoline burning (deflagrating) in open air with no confinement. Now suppose gasoline is placed inside a sealed container such that a flammable mixture exists, and it is ignited at one end, as shown in Figure 13.5. The flame front will propagate as before, but now the reaction products and hot expanding gases are trapped inside the container. If there is sufficient fuel and oxidizer available, the reaction will continue and the pressure will build until the container ruptures and produces an explosion (Figure 13.6).

Once the container fails, the hot expanding gases drive outward and carry pieces of the container along with them (**shrapnel**). The escaping gases form a pressure wave that moves rapidly away and eventually will dissipate. However, this **pressure wave** (also called a **blast wave**) can do tremendous damage to structures and people. Blast waves are generally spherical; two examples are shown in Figure 13.7. Notice that in one case, the deflagration occurs within the barrel of a gun, while in the other case the explosion occurred in a small container obliterated in the blast. In the gun barrel, the space behind the projectile is the sealed space, and as the pressure builds, it pushes the projectile forward and out the end of the barrel. The barrel must be strong enough to contain the pressure, and the projectile must be tight in the barrel, but still movable. If not, the barrel will explode. Thus, the key elements of an explosion are confinement (even if only for a fraction of a second) and rapid buildup of hot expanding gas.

The pressure waves shown in Figure 13.7 (top) were captured by a type of photography called **Schlieren imaging**. This technique detects changes in the refractive index of gases. Refractive index is discussed in detail in Chapter 17, but for our purposes here, this means the image detects changes in density of the air. Pressure waves compact the air molecules in front of it, creating

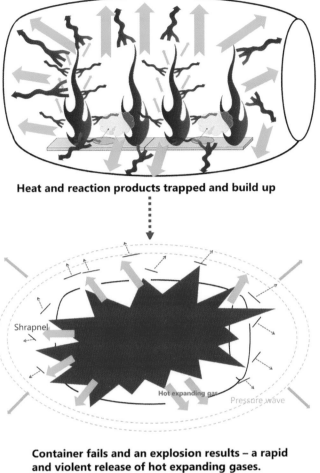

Heat and reaction products trapped and build up

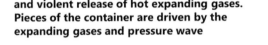

Container fails and an explosion results – a rapid and violent release of hot expanding gases. Pieces of the container are driven by the expanding gases and pressure wave

FIGURE 13.6   Deflagration in confined space leading to an explosion.

areas of much greater density than surrounding air, and thus make it possible to see pressure waves. In the image in the lower frame, the pressure waves are revealed by patterns in the soil. Since this is a normal photo and not a Schlieren image, the pressure sphere is not visible, only the damage that it is doing to the ground.

13.2 DETONATION

Suppose we set up an experiment in a tube filled with a flammable mixture and ignite it at one end (Figure 13.8). The burning will progress from left to right into the unreacted gases, producing a flame front in the tube. As the burning starts, heat is transferred to the reactants to keep the reaction going, as was shown in Figure 13.3. Since the reactants are now at a higher temperature, more heat is produced, and the second wave front moves faster since the expanding gas driving it is hotter than was the initial wave. This process continues as the speed of propagation increases. Notice in the bottom frame of Figure 13.8 that the distance between the fronts is decreasing because the one generated last is moving the fastest, while the first one generated is the slowest. An example in which a mixture of hydrogen and air is ignited in this same way is shown in Figure 13.9. High-speed imaging is essential to obtaining image series like this; the total time from first to last image is ~9.2 milliseconds or ~0.0092 seconds. The frame on the right shows the temperatures as per the scale; the mixture is hotter near the starting point (left) than at the end (right).

FIGURE 13.7   Examples of pressure waves.

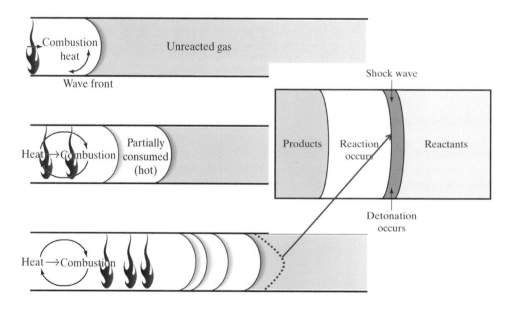

FIGURE 13.8   Wave fronts in tube of combustible mixture.

A **detonation** occurs when the speed of the reaction front exceeds the speed of sound in the media in which it is propagating. As noted above, this would be ~343 m/s in air. The speed of sound in water is around 1500 m/s, and in general the denser the material, the faster the wave will propagate. When a wave front exceeds the speed of sound, a shock wave is produced, which is an area in which the media is extremely compressed. In a detonation, it is the **shock wave** that causes the mixture to ignite and *not* the heat of the reaction. The process is shown in Figure 13.10. In this model, the red dotted wave generated last was so hot that the wave fronts

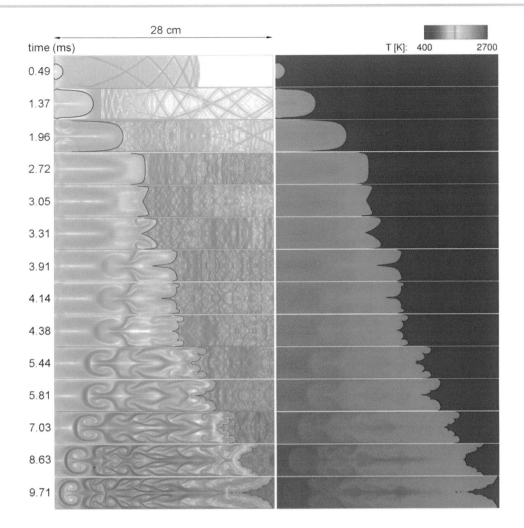

FIGURE 13.9   A sequence of Schlieren and temperature maps showing the flame front development in a mixture of hydrogen and air inside a 4×28 cm tube. (Reproduced from Xiao, H. H., et al., Premixed Flame Propagation in Hydrogen Explosions, *Renewable & Sustainable Energy Reviews* 81 [Jan 2018]: 1988–2001. With permission from Elsevier.)

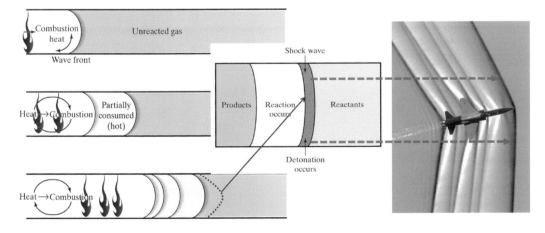

FIGURE 13.10   Detonation and shock wave.

it drove caught up to and passed the others and formed a shock wave. The mixture is no longer deflagrating, it is detonating. A sonic boom is the result of the shock wave (zone of highly compressed air) formed when an aircraft exceeds the speed of sound in air (Figure 13.10 right). Using Schlieren imaging, compression waves can be seen behind the plane, and the shock wave at the front and tail of the plane. If this plane flew over you, you would hear loud sharp booms as these waves reached your ears.

An actual detonation shock wave is shown in Figure 13.11. Two techniques are used to create the composite image on the right side. The term *OH PLIF* refers to a specialized imaging technique called planar laser-induced fluorescence (PLIF), and the OH refers to the hydroxide species OH. These specialized methods are required given how fast the reaction occurs. The detonation shock wave is easily seen in the figure. This is *not* an explosion because there is no rapid violent release of hot expanding gases. However, many explosives are designed to detonate and, by doing so, create an explosion.

There are many legitimate uses of explosives, such as in mining and demolition. Military forces use explosives to destroy structures and equipment. Criminal and terrorist bombs are designed to destroy and kill, and sadly, explosive devices can easily cause mass causalities. There are two general ways in which explosions injure and kill people. The first is the pressure of the **blast wave** (Figure 13.7; sometimes referred to as the **overpressure**), which can cause injuries such as ruptured eardrums, injuries to the lungs and heart, and other damage to internal organs. The blast wave can cause secondary blunt force injuries by launching the victim into objects, or objects such as glass or shrapnel into the victim. Those close to the bomb can also be burned. Another source of injury and death is flying shrapnel, which may travel beyond the range in which overpressure injuries occur. A **pipe bomb** (often filled with gun powder) detonation is shown in Figure 13.12, along with measurements of the dispersal and speed of shrapnel created from the pipe. The top shows a series of images from a high-speed camera starting with ignition. The container is heating and starting to expand at 600 microseconds (0.0006 seconds), and the first failure in the container can be seen at the cap at 900 microseconds. Most of the fragments are traveling faster than 500 ft/s and many faster than 1000 ft/s. At that speed and energy, it is easy to see how fragments can be found hundreds of feet from where a bomb is detonated.

Another recent study (Figure 13.13) detailed the performance characteristics of pipe bombs, which remain the most common type of explosive device encountered in forensic settings. The researchers examined bombs in a controlled setting with multiple measuring devices (top left).

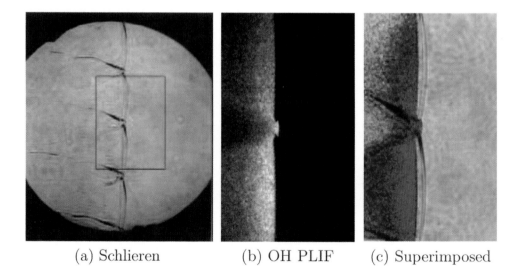

(a) Schlieren   (b) OH PLIF   (c) Superimposed

FIGURE 13.11 Detonation shock wave observed in a tube. (Reproduced from Mevel, R., et al., Application of a Laser Induced Fluorescence Model to the Numerical Simulation of Detonation Waves in Hydrogen-Oxygen-Diluent Mixtures, *International Journal of Hydrogen Energy* 39, no. 11 [Apr 2014]: 6044–6060. With permission from Elsevier.)

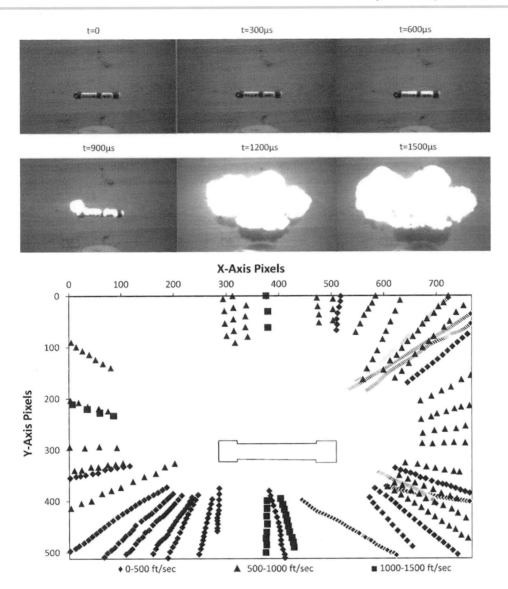

FIGURE 13.12   Dispersion of fragments from a pipe bomb explosion. (Reproduced from Bors, D., et al., The Anatomy of a Pipe Bomb Explosion: Measuring the Mass and Velocity Distributions of Container Fragments, *Journal of Forensic Sciences* 59, no. 1 [Jan 2014]: 42–51. With permission from Wiley.)

Bombs were constructed to generate explosions based on deflagration or detonation. Variables measured included fragment velocities and the mass of fragments recovered. In a forensic analysis, this type of data is useful in estimating the power of the bomb and how much of the energetic material was used in the construction. The portion of the figure at the upper right shows an example of fragments recovered on panels and where the fragments were from on the device. The bigger the sphere, the larger the fragment.

Recall earlier in the chapter that we discussed burn rate; a similar metric, called **detonation velocity**, can be calculated for detonation reactions. In this study, the average detonation velocity was ~4.4 mm/µs. Thus, if a 6-inch pipe bomb is filled with ~4 inches of powder, the detonation reaction would take less than a fraction of a second. This is also shown in the top part of Figure 13.13, where times are measured not in milliseconds (1/1000th of a second) but in microseconds (1/1000th of a millisecond). In the detonations in this study, the sides of the pipe failed first and fragments were ejected at speeds of ~1000 km/s (lower frame of the figure, a TNT-based detonation). When the explosion was the result of deflagration, the end cap failed first and fragment speeds were ~240 km/s. In these experiments, fragments were found as far as 300 m (nearly 1000 feet) from the blast site. Thus, bombings create enormous crime scenes.

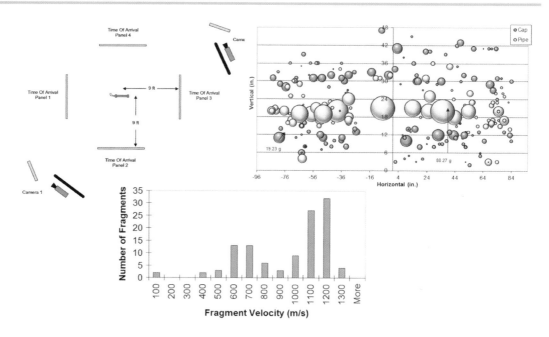

FIGURE 13.13   Power and performance of experimental pipe bombs. (Reproduced from Oxley, J. C., et al., Characterizing the Performance of Pipe Bombs, *Journal of Forensic Sciences* 63, no. 1 [Jan 2018]: 86–101. With permission from Wiley.)

THE UNABOMBER

Sketch of the Unabomber (left) and Kaczynski's mug shot.

Ted Kaczynski, known as the Unabomber, was a serial bomber arrested in 1996 after 17 years of bombings. His devices killed three people and injured many more. His bombs were sophisticated and made of parts that were commonly available but difficult to trace (a signature). Interestingly, he marked parts of the bombs with the initials FC, for "Freedom Club," which was a characteristic that linked the bombings. He made the bombs in a cabin in Montana that had no electricity, toilet, or running water. It was later learned that he chose victims at random, further complicating the investigation. He was finally caught due in large measure to a manifesto he sent to a newspaper that was published; Kaczynski's brother thought he recognized the writing style and reported it to authorities.

Source: https://www.fbi.gov/history/famous-cases/unabomber

13.3 TYPES OF EXPLOSIVES AND DEVICES

In a forensic setting, explosives are encountered most often as improvised explosive devices (IEDs), such as pipe bombs. IEDs are created by individuals or groups with the intent to cause damage and death. Everything from the fuel and oxidants through the mechanism of the device can be made by the bomber. Some of these contain commercially available materials (stolen or diverted) or military explosives. In other cases, gunpowder or synthesized compounds are used. The purpose of an IED is to cause damage and injure or kill people, which can be accomplished by the pressure wave and shrapnel. Some devices are designed to maximize one over the other, but both pressure and shrapnel can be lethal. Additional sources of shrapnel, such as nails or tacks, may be added to the bombs.

A pipe bomb is an example of a common type of IED; the pressure cooker bombs set off at the Boston Marathon in 2013 are another. An example of a diffused pipe bomb is shown in Figure 13.14. The IED is made from pipe secured at both ends by threaded caps to provide containment. Most pipe bombs, like this one, are filled with commercial smokeless gunpowder, which can be seen on the ground to the right. All IEDs require a means to initiate the reaction; in this case, a simple fuse placed through a hole in the cap was used. You can see the sealant that was used around the fuse next to the left end of the device. When the fuse burns into the powder, it ignites and burns rapidly while hot expanding gases are trapped. When the pressure exceeds the ability of the pipe or caps to contain it, the container ruptures. The result can range from one of the end caps being ejected (as in Figure 13.12) through shattering of the pipe and the creation of sharp hot shrapnel. In Boston Marathon bombing, pressure cookers stuffed with nails and ball bearings were used to create deadly shrapnel. Both types of devices can be made with materials that can be easily purchased in many places.

All IEDs must be initiated. The fuse noted above is simple and effective for gunpowder or other easily ignited materials, but for other materials, such as military or commercial explosives, a more sophisticated approach is needed. Military explosives, such as TNT or RDX (an ingredient in a formulation called C4), are very stable and do not burn or ignite; thus, more complex initiation mechanisms are needed to start the chemical reaction. Consider how a gun fires a bullet (which is discussed in detail in Chapter 15): when the trigger is pulled, the firing pin strikes the primer, which contains a shock-sensitive explosive mixture. The impact causes the primer to ignite, sending a hot flash into the propellant, which ignites it. On the other extreme are military explosives such a C4, which is a stable plastic material that requires a shock wave to detonate.

The way in which a bomber designs the ignition system can provide critical information. Many terrorist bombings utilize cell phones to detonate IEDs remotely, while others may use timers powered by batteries. When the signal is sent, or the timer reaches zero, typically the device generates an electrical spark or surge that initiates the explosive train or ignites the fuel. Timers require batteries, which can provide investigators with useful information if recovered. Detonation cord is another way to initiate the reaction and is widely used in for drilling, seismic studies, and demolition; blasting caps are also used. Because these items have many legitimate applications, they can be obtained through theft or misappropriation. All these design factors and choices provide valuable information for linking bombing cases to each other and for linking devices to one person based on the designs, materials, and parts that are used.

Explosive compounds can also be improvised and made from surprisingly simple ingredients. Recall that combustion and explosions require fuel and an oxidant; these ingredients can be combined or, in the case of **molecular explosives**, found within a single molecule. Nitroglycerin (NG) is one example (Figure 13.14). Notice that this molecule contains only nitrogen (N, blue), hydrogen (not shown), and oxygen (O, red). Nitroglycerin is a sensitive material that will burn when touched with a flame or detonate when hit with a hammer. The reaction forms hot expanding gas products such as CO_2, CO, N_2, and H_2O, and each molecule contains more than enough

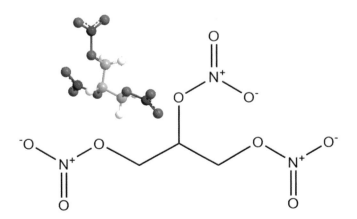

FIGURE 13.14 Structure of nitroglycerin, two- and three-dimensional. This is an example of a molecular explosive.

FIGURE 13.15 Aftermath of the Oklahoma City bombing where ANFO was used.

oxygen for the reaction to occur; therefore, NG is a molecular explosive. As an example of how fast a detonation can occur, the detonation velocity of NG when confined is ~25,000 ft/s; this is how fast the detonation wave moves through the NG.

Other improved explosives include ammonium nitrate and fuel oil (ANFO), which was used in the Oklahoma City bombing in 1995 (killing 168 people). The ammonium nitrate (NH_4NO_3) is the oxidant in this explosive mixture. Many terrorist bombings use peroxide-based explosives, such as triacetone triperoxidase (TATP), which is not difficult to make from easily available ingredients. The 2005 bombings in the London subway system were committed with TATP; 52 people died in that attack. Even sugar can be used as a fuel in an improvised explosive (Figure 13.15).

13.4 FORENSIC ASPECTS OF EXPLOSIVES

Bombings produce a wide range of physical evidence types that are often scattered over significant distances. Even a small device can eject shrapnel and other debris hundreds of feet from the site of the blast. If people are injured or killed, the scenes will be chaotic with first responders on the scene. In addition to the blast crime scene, there is almost always a secondary scene—the location where the bomb was created. These locations are particularly dangerous due to the materials present and the presence of booby traps that may be in place. Searching these scenes requires specialized expertise in explosives and hazardous materials as well as crime scene investigation.

Once collected, laboratory analysis focuses on identifying the explosive used as well as other ingredients used in the device. The types of chemical analyses used are like those discussed in Chapter 6 (forensic toxicology). Surprisingly, there are usually significant amounts of residue of the explosives left on debris, and thus the analysis for the main ingredients is typically straightforward. However, determining the chemical composition of the explosives is only one type of testing applied in IED cases. In fact, types of evidence that you might think would be obliterated in an explosion often can be recovered at the scene. Figure 13.16 shows an example in which a latent fingerprint (which we will discuss in Chapter 14) was found on the inside of a fragment

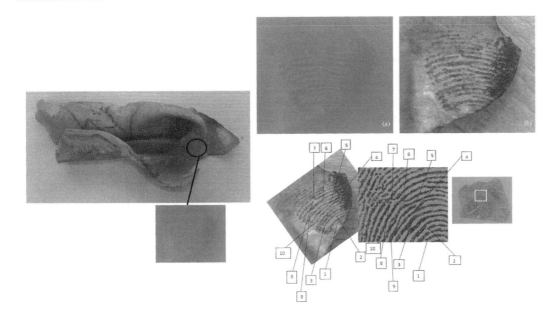

FIGURE 13.16 Recovery of latent fingerprints from a pipe bomb. (Reproduced from Bond, J. W., and Brady, T. F., Physical Characterization and Recovery of Corroded Fingerprint Impressions from Postblast Copper Pipe Bomb Fragments, *Journal of Forensic Sciences* 58, no. 3 [May 2013]: 776–781. With permission from Wiley.)

recovered from a copper pipe bomb. The prints were revealed using specialized techniques applied to digital images taken of the fragments. The quality of the print shown was sufficient for examination and comparison purposes.

DNA evidence can also be critical in bombing cases. Figure 13.17 shows the results of experiments conducted on backpacks used to contain pipe bombs. On the left are the intact backpacks that were first sterilized to remove any DNA and then provided to volunteers who used them as regular backpacks for several days. The labels refer to locations (e.g., z = zipper) that were swabbed after the blast (if recovered). Pipe bombs in the packs were then set off (right of figure) under controlled conditions and fragments recovered and correlated to the locations, as shown in the figure. The fragments were swabbed and analyzed for DNA (Chapter 10) targeting nine STR loci. The results are shown in Figure 13.18. Backpacks 1 and 2 were sampled after sterilization and not used again; no DNA was found on these two samples. Another backpack (11) was handled but not detonated and served as a positive control. The plot shows that most of the swabs (~85%) were found to contain the DNA of the person that handled it prior to the blast. For example, 10 fragments corresponding to points shown in Figure 13.17 were recovered, and all 10 samples had the same

FIGURE 13.17 Study of backpacks used as part of bombings. (Reproduced from Hoffmann, S. G., et al., Investigative Studies into the Recovery of DNA from Improvised Explosive Device Containers, *Journal of Forensic Sciences* 57, no. 3 [May 2012]: 602–609. With permission from Wiley.)

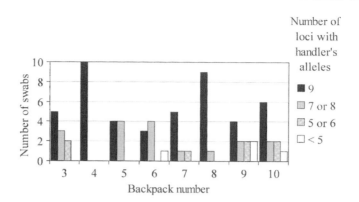

FIGURE 13.18 Study of backpacks used as part of bombings. (Reproduced from Hoffmann, S. G., et al., Investigative Studies into the Recovery of DNA from Improvised Explosive Device Containers, *Journal of Forensic Sciences* 57, no. 3 [May 2012]: 602–609. With permission from Wiley.)

FIGURE 13.19 A diffused pipe bomb. (Photo from the U.S. Air Force/Master Sgt. C. J. Reeves.)

alleles (at nine loci) as the person who handled the backpack. This study showed that it is possible to recover DNA from the bags or backpacks despite the damage done during a blast.

Often the most challenging part of the forensic investigation of a bombing is recreating the bomb from the fragments to learn how it was assembled, what parts were used, and what level of sophistication was needed. A bomb made with a cell phone timer and initiation system suggests more sophistication than a pipe bomb, as shown in Figure 13.19, or a pressure cooker bomb. If a serial bomber is suspected, the design of the device is vital to link bombings together as likely being perpetrated by the same person. Finally, many bombings involve several types of digital evidence (Chapter 20). For example, cell phone and security camera images were critical in helping to catch the Boston Marathon bombers. In that case, investigators collected images and asked the public to send in videos and images, and as a result, the suspects were identified early in the investigation.

13.5 REVIEW MATERIAL: KEY CONCEPTS AND QUESTIONS

13.5.1 KEY TERMS AND CONCEPTS

Blasting cap
Blast wave
Deflagration
Detonation
Detonation cord
Detonation velocity
Explosion

Flame front
Linear burn rate
Molecular explosive
Overpressure
Pipe bomb
Pressure wave
Propagation reaction
Shock wave
Shrapnel

13.5.2 REVIEW QUESTIONS

1. Dust explosions can be destructive and deadly. How can this occur? How would you classify this reaction?
2. Can gasoline detonate? Why or why not?
3. How does a pressure wave like those shown in Figure 13.7 differ from a detonation wave such as that shown in Figure 13.11?
4. Assume that a nail is ejected from a pipe bomb at a speed of sound (~1100 ft/s). Assuming it doesn't slow down appreciably, how far will it travel in 4 seconds?
5. In the top frame of Figure 13.13, you can see a pipe bomb being set off. Do you think the reaction was a deflagration or a detonation? Why?
6. Recently (2018), a young man set off a series of bombs in Austin, Texas, that killed two people. Video footage from a FedEx drop-off was useful in identifying him. He killed himself by setting off a device in his car when cornered by police. What types of physical evidence would be used to link the series of bombs to this one individual?
7. In Figure 13.14, sealant was used around the fuse. What was the reason for doing this?
8. What is the difference between a blast wave and a shock wave?
9. What is the difference between detonation and deflagration?

13.5.3 ADVANCED QUESTIONS AND EXERCISES

1. Research the Boston Marathon bombing from 2013. What type of bomb was used? What was the fuel? How was it detonated? What caused most of the injuries? What steps could be taken to reduce the risk of someone doing the same thing?
2. You are tasked as the lead investigator at a bombing scene where a large pipe bomb with ball bearings inside was used. How far from the point of the blast should your evidence recovery team go? What are you basing this distance on?
3. What is a "dirty bomb"?

BIBLIOGRAPHY AND FURTHER READING

BOOKS

Meyer, R., Kohler, J., and Homburg, A. *Explosives*. 7th ed. Weinheim: Wiley VCH, 2016.
Thurman, J. *Practical Bomb Scene Investigation*. 3rd ed. Boca Raton, FL: CRC Press, 2017.

WEBSITES

Link	Description
https://www.fbi.gov/history/famous-cases/	Links to several famous bombing cases
https://www.youtube.com/watch?v=uFQdcKJUijQ	Wonderful video regarding the science of explosives
https://explosives.org/	Home page of the International Society of Explosives Engineers

Fingerprints

14.1 PATTERN MATCHING DISCIPLINES IN FORENSIC SCIENCE

The next five chapters describe forensic disciplines that are, or incorporate elements of, **pattern matching**. At their heart, these disciplines focus on comparisons to make decisions regarding similarity. DNA shares elements of pattern matching disciplines, such as fingerprints, in that samples are analyzed and the patterns of the short tandem repeats (STRs) of one sample are compared with those of another. However, the term *match* in the context of DNA is expressed as a probability, not as an absolute certainty.

Because of the way DNA testing evolved in forensic science and because it has proven to be so invaluable, other disciplines are reexamining the methods used and how the results are expressed. DNA results are always presented as probabilities, such as "the DNA type from the evidence matched the suspect *and* there is a 1 in 3 billion chance that someone selected at random would have the same DNA profile." The part of the statement that comes after the word *and* is critical; it conveys to investigators, lawyers, judges, and juries a way to assign a weight to that information. A 1 in 3 billion chance is a much different than a 1 in 100 chance or a 1 in 5 chance. When so much is at stake, it is vital to have such information to put the weight of the evidence in the proper context.

In the past, pattern-based disciplines reported matches without providing additional information, qualifications, or probabilities. This practice has been challenged, particularly since publication of the National Academy of Sciences report (Chapter 1), and forensic researchers and practitioners are responding. For example, recent testimony regarding fingerprints is being presented along with **random match probabilities**, such as used in DNA. A random match probability is simply the probability that if someone were selected at random, their DNA type would be the same as that found in the case. Suppose a forensic examiner compares a latent print from a crime scene and determines that it is most similar to a fingerprint collected from a suspect. In the past, the examiner might call it a match and say no more; now increasingly this can be qualified with the probability that someone selected at random will have the same characteristics as that from the case. Developing these probabilities is not a trivial project for any type of pattern matching discipline, but progress is being made on all fronts.

Pattern matching usually involves classification procedures. Consider the comparison of two fibers—one from a crime scene and one from the carpet in a suspect's car. The composition of the fibers (cotton, nylon, polyester) is a **class characteristic**, as would be the color, diameter, and cross section under the microscope. If both fibers fall into the same categories, such as both being blue nylon with a 1.0 mm diameter and an oval shape, then they could have come from the same source. This is an example of inclusive evidence. However, if one has an oval cross section and the other has a trilobal cross section, they could not have come from the same source (exclusion). In this type of classification analysis (**successive classification**), it is not possible to conclusively prove that both fibers came from the same source. Why? Because there are a lot of blue nylon fibers in the world that have ~1 mm diameter and an oval cross section. Therefore, the term *match* is misleading and greatly overstates the value of the fiber evidence. Therefore, the terms *identification* and *individualization* are rarely used in forensic comparisons anymore. *Individualization* was a term that used to be used to describe evidence that could have come from one and only one source; however, DNA typing and the research it drove has proven that to be a nearly impossible task. Those making critical decisions in cases (investigators, judges, lawyers, and juries) are much better served when the results, along with qualifying information such as probabilities and known error rates, are reported.

A common way to generate a known error rate is to present a large number of practitioners with samples and ask them to analyze them and report results. The samples are blind in the sense that the person receiving them does not know the correct answer. The results are returned and compiled by the study organizers, and the number of incorrect results compared with the correct results constitutes an error rate. Although greatly simplified, this is how **black box studies** are conducted. Typical error rates are small, but never zero, and knowing and reporting error rates provides the context needed to weigh the value of the evidence. The errors may have several explanations, ranging from an honest mistake to improper training or a procedural error. Thus, such studies also help drive continuing improvement in pattern matching disciplines such as fingerprints.

14.2 FINGERPRINTS AS A MEANS OF IDENTIFICATION

Fingerprints is one of the oldest and most important forms of evidence in all forensic science. The use of these curious, highly individual **friction ridge skin** patterns on the end joint of the fingers as a means of personal identification dates back many centuries. In forensic science, we think of fingerprints as being used primarily to help locate, identify, and eliminate suspects in criminal cases. Fingerprints are also important in making unequivocal identifications of human remains when more conventional methods of postmortem identification cannot be used. Fingerprints may also be thought of as one member of a class of **biometric identifiers** that would also include retina or iris patterns and face thermography. Today you can purchase phones and computer accessories that lock out devices until approved by your fingerprints. There are two characteristics of fingerprints most important for their use as a means of personal identification: (1) every fingerprint is unique (to an individual), and (2) fingerprints do not change during a lifetime (unless there is damage to the dermal skin layer). For example, John Dillinger, a notorious gangster of the 20th century, used acid to alter his fingerprints. He was killed in a shoot-out on July 22, 1934 (Figure 14.1).

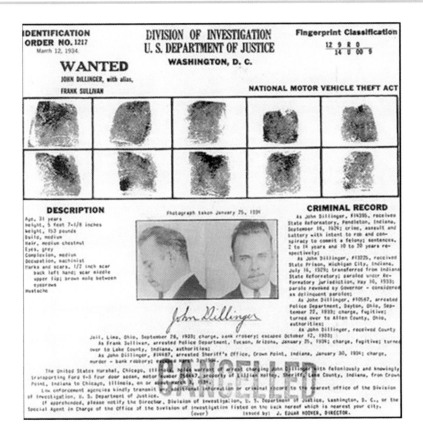

FIGURE 14.1 The 10-print card of John Dillinger. The "wanted" information was canceled after his death. (Image courtesy of the FBI.)

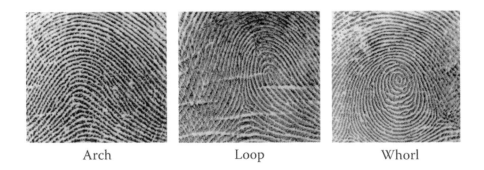

Arch   Loop   Whorl

FIGURE 14.2 Basic fingerprint patterns.

14.3 FINGERPRINT PATTERNS

There are three basic fingerprint patterns: **arch**, **loop**, and **whorl** (Figure 14.2). Within these major classes, fingerprint examiners commonly recognize other categories. Arches, for example, can be plain or **tented**. Loops can be **radial** or **ulnar**, depending on whether the slope of the print pattern is in the direction of the inner arm bone (radius) or outer arm bone (ulna). Whorls are the most complex of fingerprint patterns, and there are several whorl categories, such as **central pocket**, **double loop**, and **accidental**. Loop and whorl patterns contain definable features called the **core** and the **delta**. These are important in 10-print fingerprint classification and in comparisons. Figure 14.3 shows some of the additional fingerprint patterns. The core and delta of a fingerprint pattern are also indicated in one of the frames.

Within fingerprint patterns, there are features called **minutiae**. Once evidentiary and reference (inked) fingerprints have been oriented and found to have the same general ridge flow or

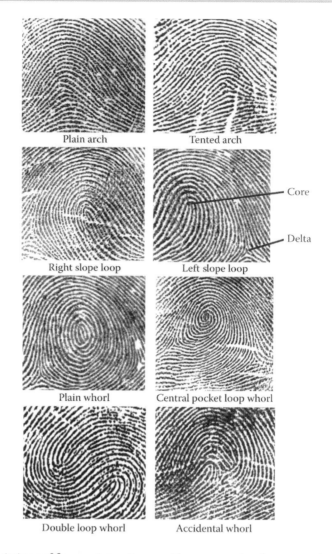

Plain arch

Tented arch

Right slope loop

Left slope loop

Core

Delta

Plain whorl

Central pocket loop whorl

Double loop whorl

Accidental whorl

FIGURE 14.3   Variations of fingerprint patterns with an example of a core and a delta (minutiae).

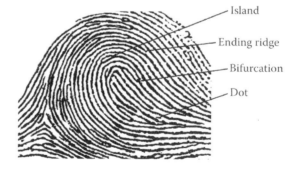

Island

Ending ridge

Bifurcation

Dot

FIGURE 14.4   Fingerprint minutiae.

pattern type, these features are used to compare the fingerprints and decide whether they are or are not from the same person. The ridges of the fingerprint form the minutiae by doing one of three things: ending abruptly (ending ridge), splitting into two ridges (**bifurcation**), or being short in length, like the punctuation mark at the end of a sentence (**dot**). Combinations of these minutiae also have names. For example, two bifurcations facing each other form what is called an **island**. Some minutiae are shown in Figure 14.4.

MYTHS OF FORENSIC SCIENCE

Fingerprints can always make an identification.

While undoubtedly one of the most powerful and useful forms of physical evidence, their use in not infallible. There are several reasons that mistakes occur. Latent fingerprints collected at crime scenes often only show a portion of a fingerprint, and the quality of that print may be poor. Also, not everyone's fingerprints are in a database, and thus there may be nothing to compare a crime scene print to. Finally, as in all operations performed by human beings, errors and mistakes inevitably occur. Procedures and protocols are designed to minimize such occurrences, but there is no such thing as an error rate of zero. This is one reason that the term *identification* when talking about DNA and fingerprints is falling out of favor; the preferred way to present this type of evidence is by a probability.

14.4 FINGERPRINT CLASSIFICATION

It was apparent to the early fingerprint examiners that a manageable, consistent classification system was essential if large sets of fingerprint files were to be useful for criminal identification. In the United Kingdom and the United States, the classification systems are variants of the one developed by Sir Edward Henry.

The modified **Henry system** as used in the United States is a scheme for the classification of 10-print sets, or a fingerprint card, for one individual. Figure 14.5 shows a typical fingerprint card of the kind long used by law enforcement and other agencies to maintain fingerprint records. Use of the classification system enabled efficient searching of large files. Keep in mind, though, that classification is based on having all 10 prints. Thus, a 10-print card from a fingerprinted person can be classified and filed, but you would need all 10 prints from a person to use his or her classification to search in the large file. The system therefore allowed organized maintenance of the large files maintained by many law enforcement agencies, but the files could not easily be searched manually for a single print. This is a significant problem because when a crime is involved, it is typical to recover single prints, or sometimes even partial single prints.

Until the development of computerized fingerprint search systems, it was impractical to search large files for a single print because it took so much time. Partial prints could obviously be compared with prints from cards on file, but it was necessary to have a possible suspect or suspects first to know which cards to retrieve for manual comparison. Not that long ago, all fingerprint examiners and identification personnel were extensively trained in fingerprint classification using the modified Henry system, as were many police officers. Figure 14.6 shows a fingerprint card image with an overlay of a print recovered from a crime scene. Notice that the quality of the crime scene print is much poorer than that of the 10-print card; image quality remains a significant challenge even for computerized systems. With the advancement of computer scanning, image analysis, searching, and database software, manual classification is no longer required, and it is possible to search the database for a single fingerprint or partial print instead of manually sorting through thousands of 10-print cards.

The acronym ACE-V describes methodology frequently used to classify fingerprint patterns. It stands for analysis, comparison, and evaluation that leads to verification. Analysis includes things such as the quality and orientation of the latent print. Comparison involves examination of the unknown print and potential sources in terms of class characteristics, such a loop, whorl, and arch, and location and appearance of minutiae. Verification follows based on the procedures of the laboratory and involves review of findings by at least one other examiner.

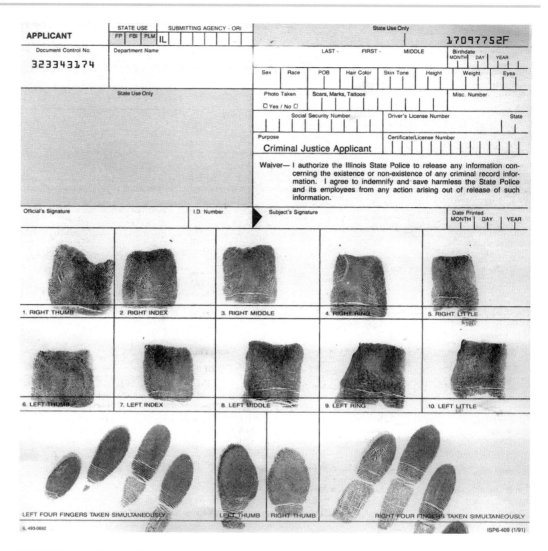

FIGURE 14.5 Typical 10-print card.

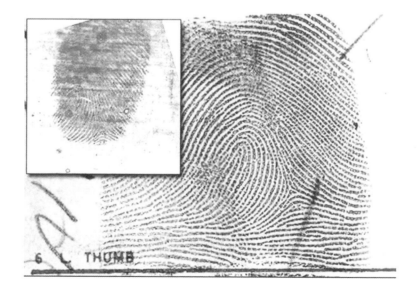

FIGURE 14.6 A fingerprint recovered from a crime overlaid on the same finger from a 10-print card. The tiny red arrows were created by AFIS. (Image courtesy of the FBI.)

THE MAYFIELD CASE

The process of comparing latent prints to knowns has occasionally resulted in an incorrect identification, even when experienced examiners conducted the comparison. In 2004, suspected terrorists set off bombs on several commuter trains in Madrid, Spain. Spanish police recovered some latent prints from evidence believed to be closely associated with the bombers. Images of the latent fingerprints were distributed internationally to law enforcement agencies to identify them. The Federal Bureau of Investigation (FBI) was one of the agencies that examined the latent fingerprints and searched its **Integrated Automated Fingerprint Identification System** (IAFIS) files for potential matches. The FBI eventually identified one of the latent fingerprints as belonging to a man named Brandon Mayfield. Mayfield, who lived in Portland, Oregon, was Muslim, although he was a U.S. citizen and had been in the Army. The fingerprint identification eventually led to Mayfield's arrest. Later, it became clear that the latent print was not Mayfield's but from an Algerian national named Daoud. The incident resulted in an investigation by the U.S. Department of Justice's Office of Inspector General, which issued a report in March 2006. The FBI has taken steps to change and improve its internal procedures to prevent such an occurrence from happening again.

14.5 COMPUTERIZED FINGERPRINT SYSTEMS

Computer storage and retrieval systems for fingerprints were originally developed for law enforcement applications. Efforts to develop these systems began in the early 1960s. In the United States, these efforts were the result of collaboration between the FBI, which maintains the largest (and the only national) fingerprint database, and scientists at the National Bureau of Standards (which later became the National Institute of Standards and Technology [NIST]). The law enforcement–based automated systems are commonly called the **Automated Fingerprint Identification System** (AFIS). There are two principal applications. The first is searching large files for the presence of a 10-print set of prints (taken from a person). The second is searching large files for single prints, usually developed latent fingerprints (see below) from crime scenes. In effect, AFIS is searchable databases of photographic and scanned images, but developing the appropriate scanning, storage, and searching programs was and continues to be challenging.

An interesting and sometimes frustrating problem with AFIS is that the systems were developed by several different companies, and as a result, there are many different fingerprint databases. Furthermore, it is not always easy for different systems to "talk" to each other. This problem, called lack of **interoperability**, means that it may be necessary to submit a recovered fingerprint to several different databases in different jurisdictions to do a complete search. For example, a large city may have one fingerprint database, while the state in which it is located may have another. Currently, NIST is working to develop procedures to enhance interoperability, but it is important to keep two key points in mind. First, there is no one single database in which all fingerprints exist; second, not everyone's fingerprints are in databases. If you have not been fingerprinted, your prints are not in any database and thus would never be evaluated as part of *any* search. The same would be true of a criminal who had never been fingerprinted. Thus, just because fingerprint impressions are recovered at a crime scene or off physical evidence, this does not mean that the person can easily be identified by database searches.

In part to address this, the FBI made their criminal database of known fingerprint cards available to other law enforcement agencies. This system is called the Integrated Automated Fingerprint Identification System. Using **IAFIS**, a latent print examiner can search unknown latent impressions in a neighboring state or several states. This is the national criminal database maintained by the FBI of all the 10-print cards received from all over the country. To date, this system has been very successful in developing leads and sometimes solving cold cases. An overview of AFIS and IAFIS is shown in Figure 14.7.

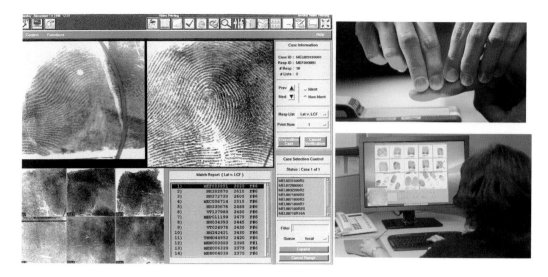

FIGURE 14.7 Left: A typical screen shot of an AFIS search. Top right: Obtaining a print on an automated reader. Lower right: An analyst working with IAFIS. (Images courtesy of the FBI and the Maine Crime Laboratory.)

IMAGE QUALITY AND AFIS

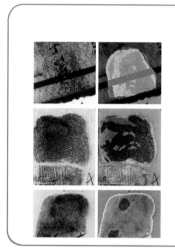

AFIS is based on databases of images, the quality of which is critical. The image at left, provided by NIST, shows how a computer program can be used to evaluate the quality of the image, which in turn can impact the ability to correctly identify it. Notice how colors are used to encode regions of inadequate quality or missing information. This is critical information given that latent prints recovered from scenes and evidence are usually only partial prints and, depending on age and surface, may not be high quality.

14.6 FINGERPRINTS AS PHYSICAL EVIDENCE

At crime scenes and/or on items of evidence, there are essentially three types of fingerprints that may be encountered: **patent**, **plastic**, and **latent**. A patent (or visible) print is one that needs no enhancement or "development" to be clearly recognizable as a fingerprint. Fingerprints on a 10-print card are patent prints. Such a print is often made from grease, dark oil, dirt, or even blood, rendering it visible and recognizable, and possibly even suitable for comparison without additional processing.

A plastic print is a recognizable fingerprint indentation in a soft receiving surface, such as butter, silly putty, and tar. Such prints have a distinct three-dimensional character, are immediately recognizable, and often require no further processing.

A latent print is one that requires additional processing to be rendered visible and suitable for comparison. The processing of latent prints to render them visible, and hopefully suitable for comparison, is called development, enhancement, or visualization. An enormous amount of literature has grown on this subject in the past 30 years or so. Many strides have been made in the area due to clever applications of chemistry and physics principles coupled with better understandings of the composition of latent residues.

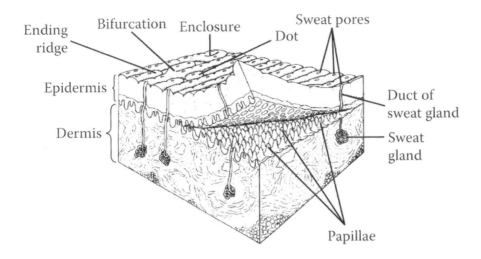

FIGURE 14.8 Friction ridge skin structure. (Image courtesy of the FBI.)

14.6.1 COMPOSITION OF LATENT FINGERPRINT RESIDUES

The starting point for understanding latent print development techniques is the fingerprint residue itself. Friction ridge skin (Figure 14.8), as found on the surfaces of the fingers, hands, and bottoms of feet, has pores through which small sweat glands can empty their contents onto the skin surface. These sweat glands, called **eccrine glands**, produce the watery-type sweat composition that forms the basis for latent fingerprint residue. Another type of sweat gland, called **apocrine**, located in other parts of the body, produces sweat that contains organic molecules, such as fats and proteins. This material can become part of the latent print residue from a person touching those areas of the body. In addition, an almost unlimited variety of substances from the environment can get onto the friction ridge skin and can then be deposited into latent residue when the person touches a surface. Thus, although there are several common constituents of most latent print residues based on the composition of sweat, the proportions may vary, and there are many other compounds and materials from the environment that can be present as well.

Most methods for the development of latent prints were developed based on knowledge of the latent print residue composition. Usually, a method must be capable of detecting or visualizing one of the compounds or elements present in latent residue and applied to target that compound or element. For the application to be successful, it must be possible to apply the method to evidentiary fingerprints on the variety of surfaces where they are found, and without destroying the integrity of the impression pattern. The methods commonly employed can be broadly divided into three groups: physical, chemical, and special illumination, or a combination of these methods, many of which involve laser or narrow-band-pass illumination.

It should be noted here (and will be reiterated below) that fingerprints developed at a scene must be photographed prior to any lifting or other collection effort. Prints developed in the laboratory or identification unit must also be photographed to document the results as well as the location of the print on the evidence item. Investigators often face tough decisions about whether to employ fingerprint enhancement procedures at a scene or seize the evidentiary item and submit it to the lab or identification unit. Many factors will be involved in resolving the matter, including the training and experience of the investigator, how well equipped he or she is at the scene, and whether the evidentiary item can be readily seized and transported. As we have seen many times before, the role of the crime scene investigator and documentation becomes critical in these situations.

14.6.2 PHYSICAL METHODS

Classically, physical methods are those that do not involve any chemicals or chemical reactions. They work by applying fine particles to the fingerprint residue, where they adhere preferentially, thus creating contrast between the ridges and the background. The most well-known one is

FIGURE 14.9 A magnetic brush with black powder is used to "dust" a latent fingerprint to make it visible.

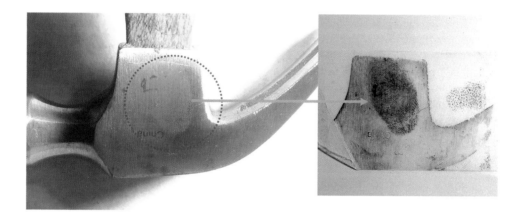

FIGURE 14.10 A latent (nearly invisible) fingerprint on a hammer was developed with black magnetic powder and lifted from the hammer using tape.

powder dusting—a mainstay of latent fingerprint detection for a century or more. Most common powders are inorganic and come in several colors. There are also a large variety of fingerprint brushes available. The principle of powder dusting is simply that the powder particles adhere to the latent residue. Careful use of the proper brush and powder often results in the development of excellent prints. Black powders are generally superior to the other colors. Black powders are produced in a way that yields more uniform particle size and generally produces better results. The technique is illustrated in Figure 14.9.

Also illustrated is the lifting of a developed latent impression using transparent lifting tape (Figure 14.10). The tape lift is mounted on a backing card with a color maximally contrasting to that of the powder (e.g., white backing for black powder). One-piece lifters, also known as **hinge lifters**, are commercially available for this purpose. A variant of the simple brush and powder combination is the magnetic brush (Figure 14.10), the original trademark version of which is called the **Magna Brush**. The brush is a small retractable magnet that is used with magnetic powders available in several colors. The principle of magnetic powder enhancement is the same as that for conventional powder, namely, adherence of the fine particles to fatty components of the residue.

The magnetic brush technique is more useful on some surfaces than conventional powder dusting, mainly because the magnetic wand can be used to remove any excess powder from the substratum. It also has the potential to be a gentler technique, in the sense that there is no brush, and thus no bristles, so it is less likely to damage the latent print ridges in the brushing process.

Another physical latent print developing procedure involves a **small particle reagent** (SPR). Typically applied by spraying or immersion, the most common formulation of SPR is a fine suspension of molybdenum disulfide in a detergent solution. The particles adhere to the lipid components of the residue. SPR is most commonly used on evidence that has been formerly wet.

14.6.3 CHEMICAL METHODS

The greatest progress in latent fingerprint visualization techniques in the past quarter century has involved chemical and instrumental/special illumination techniques. Classically, the chemical techniques were silver nitrate, **iodine fuming**, and **ninhydrin**. Silver nitrate is not used any longer because there are much better techniques available. The principle of its use lay in the reaction of the silver with the chloride in the fingerprint. The silver chloride was then photoreducible to silver, which contrasted with the background. A somewhat related method that still involves silver is called **physical developer** (PD), and it is discussed below.

Elemental iodine (I_2) is one of the compounds in nature that sublimes; that is, it can pass from a solid to a vapor without becoming liquid. Iodine sublimes (evaporates directly from a solid to a gas) easily with only moderate warming. The vapor is directed toward the fingerprint residue with a so-called iodine fuming gun (essentially a plastic blowpipe containing some iodine), or an object to be fumed can be placed in a closed cabinet, which is then filled with iodine fumes. Even though it has long been placed under the "chemical methods" category of latent print development, the iodine probably does not react with any of the components in the residue. It probably interacts with the fatty components in such a way that it is trapped in the residue, giving the ridge features a dirty-brown-colored appearance. The iodine-developed color is not stable in the latent print, however, and iodine prints must be quickly photographed. There are chemical methods for rendering iodine prints a permanent color that will not fade.

Since around 1910, ninhydrin has been known to react with amino acids, forming an adduct called **Ruhemann's purple** (named for one of the original observers). In the mid-1950s, Swedish scientists noted that ninhydrin reacted with the amino acid components of latent fingerprint residue and took out a patent on the process. Ninhydrin is applied by spraying, painting, or dipping. The ninhydrin reaction is slow unless accelerated by heat in the presence of humidity. Ninhydrin develops bluish-purple fingerprints (Figure 14.11) and is extremely useful on porous surfaces (such as paper), but occasionally the bluish-purple color may not show sufficient contrast with the background substratum. Ninhydrin is also useful as a preliminary treatment in a sequential process followed by other chemicals and viewing under laser or alternative light source illumination. Another significant factor in ninhydrin use for latent prints has been the design, synthesis, and evaluation of a series of ninhydrin analogs, perhaps the most important of which is 1,8-diazafluoren-9-one (**DFO**). Treatment of ninhydrin-developed prints with $ZnCl_2$ (zinc chloride) provides the basis for the use of laser and special illumination methods. The metal salt treatment converts Ruhemann's purple to another compound that can be excited by blue-green light using a laser or alternate light source to yield strong fluorescence. Many modifications of postninhydrin latent print treatments have been designed to maximize fluorescence under illumination by commercially available lasers or broad-band-pass-filtered alternate light sources. There are a host of other chemicals that react with amino acids yielding colored or fluorescent products. However, none have been as important in operational latent print work as ninhydrin (and posttreatments) and DFO.

The most important other chemical procedure is treatment with **cyanoacrylate** esters ("**superglue**"). Superglue "enhancement" was first observed by Japanese scientists in the National Police Agency. The method quickly caught the attention of latent print examiners all over the world. In the early 1980s, fingerprint examiners working in the U.S. Army Criminal Investigation Laboratory in Japan, and a little later in the U.S. Bureau of Alcohol, Tobacco and Firearms Laboratory, introduced alkyl-2-cyanoacrylate ester (superglue) fuming as a method for latent print development in the United States. Superglue can be induced to fume, and the fumes will interact with latent fingerprint residue by polymerizing *in situ*, yielding a stable, even robust,

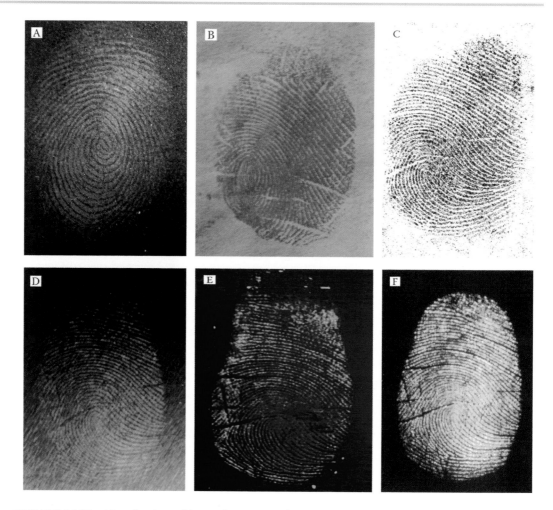

FIGURE 14.11 Visualization of latent fingerprints by various methods: (A) oblique lighting, (B) PD, (C) ninhydrin, (D) cyanoacrylate (superglue), (E) superglue with additional treatment and laser illumination, and (F) superglue with additional treatment under ultraviolet light.

friction ridge impression off-white in color (Figure 14.11). Items to be processed by glue fuming are usually placed into well-sealed cabinets where the glue is induced to fill the cabinet with vapors. The process requires a certain level of relative humidity, and a moisture source is always placed in fuming cabinets. Glue fuming can be accomplished in a closed cabinet or container without any "acceleration," but it is very slow. Typically, a strong alkali or heat is used to accelerate fuming. The progress of latent residue development in the fuming cabinet can be monitored by placing a test latent print onto a piece of aluminum foil, or similar surface, and placing the object into the cabinet, where it can be viewed during fuming. Latent fingerprints underdeveloped with superglue can simply be further fumed, but overfumed latent fingerprints may be ruined.

Like ninhydrin, glue fuming is an excellent method for developing latent fingerprints. But it may be most useful operationally as an initial step in protocols where the cyanoacrylate-developed prints are further treated and examined. The simplest enhancement of a cyanoacrylate-developed print is dusting it with powder. Other posttreatments of cyanoacrylate-developed prints include "dye stains," which induce luminescence or fluorescence in the residue when it is illuminated with a laser or alternate light source at the appropriate wavelength. Gentian violet, Coumarin 540 laser dye (Figure 14.11), Rhodamine-6G (Figure 14.11), and other treatments have been used for the enhancement of cyanoacrylate-developed print luminescence under alternate light or laser illumination.

The last chemical method we discuss is PD. Despite the name, PD is a chemical method. It has become common in many laboratories to use PD as a follow-up method to ninhydrin or DFO on porous surfaces, such as paper. While ninhydrin or DFO reacts with water-soluble components in the latent print residue, PD reacts with the lipid and other water-insoluble components (Figure 14.11). PD is a photographic-type process based on the deposition of silver onto latent fingerprint residue from a ferrous/ferric redox couple and silver salt mixture in solution.

14.6.4 COMBINATION/SPECIAL ILLUMINATION

Sometimes, a latent fingerprint can be visualized by simple oblique lighting (Figure 14.11). This technique might be considered physical, since it involves no chemicals or special light sources. A latent amenable to visualization by oblique lighting also illustrates that the definition of a latent fingerprint varies; at least some kind of pattern was visible to have prompted the observer to obliquely illuminate the surface, thus "enhancing" it.

Latent prints may also be viewed under alternate light (525 nm) or laser illumination. Alternate light is generally superior to incident white light in revealing ridge detail. Alternate light sources are exceptionally bright light sources. Alternate light sources are supplied with colored filters, which serve to filter the source light so that the developed latent print can be viewed with light of a narrow wavelength range. The filter is selected to induce illumination at the wavelength known to excite the chemical compounds used in the latent development procedure. Other filters may be used to view the luminescence resulting from the alternate light source excitation. The luminesced light will be of longer wavelength than the excitation, and the filter may provide sufficient contrast between the background and the luminescence of the latent print ridges to give a useful ridge detail image that can then be photographed and used for comparison. A laser is also a very high-intensity light source, but it emits light of a single wavelength (monochromatic light). There are several commercially available lasers. Methods have been developed over the years to take advantage of the excitation wavelengths afforded by these lasers.

14.7 LINKING A FINGERPRINT TO A PERSON

Everything in this chapter so far—indeed, the reason that there is a chapter on this subject at all—comes down to the use of fingerprints as a means of identification. As noted earlier, the uniqueness of fingerprints is a matter of common knowledge. Advertisements and commercials commonly use phrases like "as unique as a fingerprint." Even the term *DNA fingerprints*, as undesirable and sometimes misleading as it is, was coined to reflect the notion that a DNA profile might be as individual as a fingerprint. But most people have never given much thought to the process by which fingerprint identification is actually done.

In modern practice, once a fingerprint is collected, either by scanners or at a crime scene, the first step is to enter it into AFIS. Although the algorithms used depend on the system, a search of a database will yield a list of the most likely sources along with a numerical score. The search does not produce a match, but only a list of candidates for a latent fingerprint examiner to check and compare. In other words, a person always makes the final decision, not the computer. Fingerprint examiners are extensively trained and required to accumulate considerable experience before being entrusted with the responsibility of making identifications. Thus, in addition to the general principles and approaches used to make identifications, the knowledge, training, and experience of the examiner also come into play.

In the case of latent examinations, perhaps the first issue that comes into play is determining what is called the "suitability" of the latent fingerprint for identification. Here, the examiner must decide if sufficient quality and quantity of the ridge detail is present in the latent to make it possible to compare with a known print. As was shown in Figure 14.6, the quality of a print obtained from evidence is never as good as a 10-print card or the equivalent scanned image. Evidentiary prints are also often partial prints in that only a small area of a finger's image is available for comparison. This must be taken into consideration in any fingerprint examination.

In comparing a list of candidates to an evidentiary fingerprint, several factors are considered. Typically, the examiner must first analyze the latent, determine its proper orientation, decide if there are any color reversals or other unusual circumstances, decide suitability, and then proceed to the comparison. Comparison with the known takes place at several levels, such as the overall

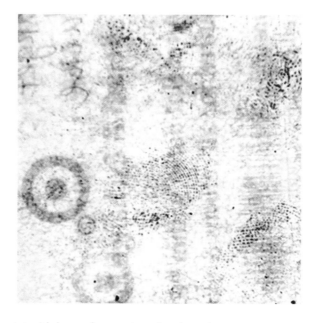

FIGURE 14.12   Receipt with latent fingerprints developed on paper.

pattern (arches, ridges, etc.) and directionality. Next, the individual characteristics (minutiae) are compared as to type of features and their locations, as well as pore shape, locations, numbers, and relationships. Any unexplained difference between the known and latent prints during this process would result in the conclusion that the known is excluded as a source of the latent. This is one possible outcome of the evaluation decision. If every compared feature is consistent with the known, and there are enough features sufficiently unique when considered as a whole, the examiner can confirm the source. Since this process, although rigorous, has subjective elements, the identification should always be double-checked by an independent examiner. Currently, the probability of an identification is not provided in the same sense that a DNA identification is, supported by a probability, but there are research efforts underway that are geared toward being able to do so.

14.8 RETURN TO THE SCENE OF THE CRIME

In our crime scene scenario (Figures 3.2 through 3.4 and 3.13 through 3.16), the crime scene investigators would be searching the house and truck for latent fingerprints. The challenge will be to decide which prints are critical and which are not. The home belongs to the couple (victim and suspect), so it will be full of latent fingerprints from both, and collecting them really does not help determine what happened and what did not happen during the commission of the crime. If the drugs are involved (Figure 3.14), then it might be useful to obtain prints from the paraphernalia. Also, finding prints that do not belong to the victim or suspect at the home might be useful, but again, it is difficult to know or see such evidence, let alone decide how much value it has. However, the truck would be a more promising place to find latent print evidence relative to the case. Recall that the body of the victim was wrapped in plastic and covered in a blue tarp. A receipt (Figure 14.12) was recovered in the truck bed that indicated the purchase of the same type and make of tarp the day of the murder. If the suspect's prints were found on that receipt and on the tarp (Figure 14.13), this would support the hypothesis that he purchased the tarp with the intent to bury his wife in it. The receipt is a good example of partial prints and shows how critical the quality of the print would be in trying to link it to a specific person. It is also an example of the value of development methods for visualizing latent prints.

Furthermore, if prints could be developed on the plastic that she was wrapped in (Figure 14.14), that would demonstrate that he at least touched that plastic and would have to be able to explain how that occurred if not during the crime. This evidence would be even stronger if his prints

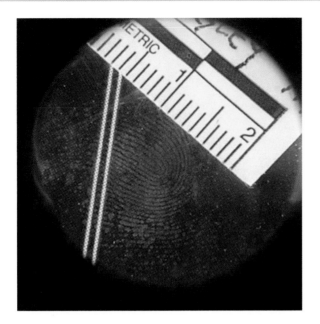

FIGURE 14.13   Latent fingerprint on the tarp.

FIGURE 14.14   Latent prints developed with superglue from the plastic sheeting used to wrap the victim's body.

were the only ones found on the plastic. On the other hand, if someone else's prints were found on the plastic, the tarp, or the receipt, this could lead to the identification of another suspect. In our scenario, it would be the suspect's fingerprints found on these items, which would support the prosecution's version of the story, in which he killed his wife.

14.9 REVIEW MATERIAL: KEY CONCEPTS AND QUESTIONS

14.9.1 KEY TERMS AND CONCEPTS

Accidental whorl
Apocrine gland
Arch
Bifurcation
Biometric identifiers

Black box studies
Central pocket whorl
Class characteristic
Core
Cyanoacrylate
Delta
Dot
Double-loop whorl
Eccrine gland
Friction ridge skin
Henry system
Hinge lifters
Individualization
Interoperability
Iodine fuming
Island
Latent print
Loop
Magna Brush
Minutiae
Ninhydrin
Patent print
Pattern matching
Physical developer
Plastic print
Powder dusting
Radial loop
Random match probability
Ruhemann's purple
Successive classification
Superglue
Tented arch
Ulnar loop
Whorl

14.9.2 REVIEW QUESTIONS

1. If a fingerprint was found in blood, would it be considered patent, latent, or plastic?
2. What is the primary value of fingerprints as evidence?
3. What is friction ridge skin? Where is it found on the human body?
4. What is AFIS? Indicate what the letters stand for and describe the system and how it helps in fingerprint identification.
5. What are the three main types of fingerprints that can be found at a scene?
6. What is meant by "development" or "enhancement" of a latent fingerprint?
7. What is an example of a physical method for enhancing latent fingerprints?
8. What is an example of a chemical method for enhancing latent fingerprints?

14.9.3 ADVANCED QUESTIONS AND EXERCISES

1. Suppose you work as a latent print examiner. You have partial prints from a case to characterize. What steps could you take to ensure that you are not influenced by confirmation bias?
2. Research the Will West case. What is interesting about it? What does it tell us about reporting zero error rates for any forensic technique?

3. Black box studies are very effective at quantifying an underlying error rate. What will such studies not tell you? How would you combine a black box study with another procedure to obtain the missing types of information?

BIBLIOGRAPHY AND FURTHER READING

BOOKS

Champod, C., Lennard, C. J., Margot, P., and Stoilovic, M. *Fingerprints and Other Ridge Skin Impressions.* 2nd ed. Boca Raton, FL: CRC Press, 2016. ISBN: 978-1498728935.

Daluz, H. M. *Fundamentals of Fingerprint Analysis.* Boca Raton, FL: CRC Press, 2014. ISBN: 978-1466597976.

Federal Bureau of Investigation. *The Science of Fingerprints: Classification and Uses.* Washington, DC: U.S. Government Printing Office, 1998. Stock number 027-001-00033-5.

Ramotowski, R., Ed. *Lee and Gaensslen's Advances in Fingerprint Technology.* 3rd ed. Boca Raton, FL: CRC Press, 2014. ISBN: 978-1420088342.

ARTICLES

Abraham, J., Champod, C., Lennard, C., and Roux, C. Modern Statistical Models for Forensic Fingerprint Examinations: A Critical Review. *Forensic Science International* 232, no. 1–3 (Oct 2013): 131–150.

Cadd, S., Islam, M., Manson, P., and Bleay, S. Fingerprint Composition and Aging: A Literature Review. *Science and Justice* 55, no. 4 (Jul 2015): 219–238.

Girod, A., Ramotowski, R., Lambrechts, S., Misrielal, P., Aalders, M., and Weyermann, C. Fingermark Age Determinations: Legal Considerations, Review of the Literature and Practical Propositions. *Forensic Science International* 262 (May 2016): 212–226.

Girod, A., Ramotowski, R., and Weyermann, C. Composition of Fingermark Residue: A Qualitative and Quantitative Review. *Forensic Science International* 223, no. 1–3 (Nov 2012): 10–24.

Page, M., Taylor, J., and Blenkin, M. Uniqueness in the Forensic Identification Sciences—Fact or Fiction? *Forensic Science International* 206, no. 1–3 (Mar 2011): 12–18.

WEBSITES

Link	Description
https://www.nist.gov/programs-projects/latent-print-afis-interoperability-working-group	NIST group working on the interoperability of fingerprint databases
https://www.fbi.gov/services/cjis/fingerprints-and-other-biometrics#Fingerprints	FBI fingerprints and biometrics
www.theiai.org	The International Association for Identification
https://www.nist.gov/news-events/news/2017/05/do-you-have-what-it-takes-be-forensic-fingerprint-examiner	NIST site—see if you can be a fingerprint examiner!

Firearms and Tool Marks

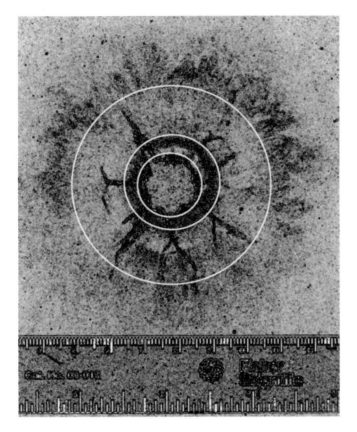

15.1 MARKINGS ON METAL

Both firearm and tool mark examiners use microscopic examinations of markings found on metal surfaces. In the case of firearms examinations, the marks made by a firearm on a fired bullet or cartridge casing are useful for determining how likely it is that a given weapon was used to fire that ammunition. In tool mark examinations, the microscopic features of a tool mark are used to determine if it is possible that a specific tool made a mark. What these two areas have in common is the creation of markings due to metal-on-metal contact. When marks are made on metal surfaces, they are not subject to change in the same way as marks on softer surfaces and as such can be a useful source of physical evidence. These marks are yet another example of Locard's principle in that the contact between two metal surfaces leaves markings on both surfaces; these markings are what firearm and tool mark analysts study. Depending on the laboratory, forensic scientists in this section can also be involved in related tasks, such as evaluating the safety and operability of a gun, measuring its overall length and that of the barrel, estimating distances from weapons to targets, reconstructing shooting events, and restoring serial numbers that have been scratched or scraped off metal surfaces.

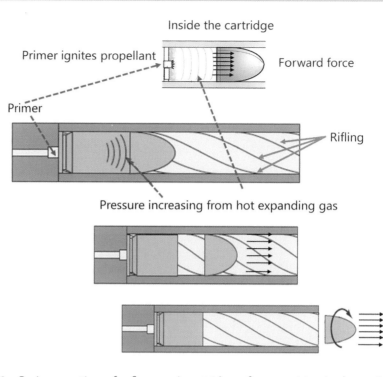

FIGURE 15.1   Basic operation of a firearm. A cartridge of ammunition is shown in the barrel of a gun.

15.2 HOW FIREARMS WORK

There are many types of firearms available, and thus many variations that are seen in forensic laboratories. Pistols, rifles, and shotguns are the most commonly seen as evidence. In all firearms, hot expanding gas generated by the combustion of burning **propellant** (gunpowder) drives the projectile (the bullet) forward and out of the barrel at high speed. The process is shown in Figure 15.1, which would apply to a rifled barrel such as in a handgun or rifle. When the primer bursts into flames (top of the figure), it ignites the propellant in the cartridge. As the propellant burns, large volumes of hot gas are produced and the pressure insider the cartridge rises rapidly. When the pressure becomes high enough, it pushes the bullet forward, down the barrel of the weapon, and out the muzzle. How the ammunition is produced is critical; if too much gas is produced too quickly, the cartridge and barrel of the weapon can burst open; too little pressure and the bullet becomes stuck in the barrel. The bullet also spins; we will discuss why shortly.

15.2.1 TYPES OF FIREARMS

We break the discussion of the common type of firearms seen as evidence into handguns and long guns. **Long guns** are rifles and shotguns, while handguns are as the name implies, a weapon meant to be handheld. There are two types of handguns. In a **revolver** (Figure 15.2), the cartridges are held in firing chambers in a rotating cylinder. **Single-action** revolvers are fired by manually cocking the **hammer** and then pulling the trigger; **double-action** weapons rotate the cylinder and bring the hammer down on the cartridge with a single trigger pull. Pulling the trigger causes the **firing pin** to strike the primer, which ignites it and starts the process shown in Figure 15.1. Cocking the hammer both rotates the cylinder to place one of the chambers under the hammer and cocks the firing mechanism.

The other type of handgun is a semiautomatic pistol (Figure 15.3). These weapons capture some of the hot gases produced by the firing and use it to eject the cartridge case and put a new cartridge into the chamber. Pulling the trigger results in the firing pin striking the primer as in the revolver. Some of the energy produced by the fired cartridge is used to drive the **slide**

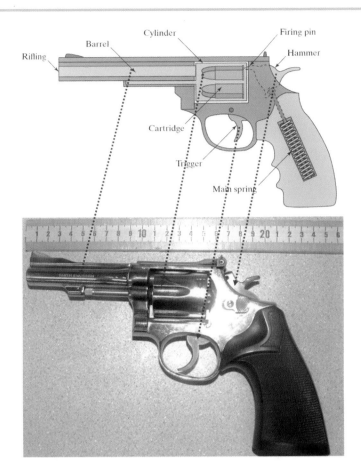

FIGURE 15.2   Revolver and parts.

backwards, extracting the spent cartridge from the chamber, ejecting it from the weapon, and cocking the firing mechanism. A spring is compressed when the slide moves back. At the end of the backward motion, the spring uncoils, pulling up an unfired cartridge from the magazine and loading it into the firing chamber. At the end of this cycle, the pistol is left loaded and cocked; it requires only the pull of the trigger to fire another shot.

Unlike handguns, long guns (rifles and shotguns) are designed to be held in two hands when being fired from the shoulder. There are several types of rifles seen in forensic laboratories, such as semiautomatics, lever actions, and bolt actions, where the term *action* refers to the mechanism of cycling or cocking the weapon. Shotguns can be semiautomatic, lever action, or pump action. Shotgun barrels are not rifled, and they fire many small pellets or balls rather than a single bullet.

15.2.2 RIFLING, CALIBER, AND GAUGE

In most guns, the barrel has **rifling**, which imparts spin on the bullet, as shown in Figure 15.1. Rifling consists of a series of grooves (**lands and grooves**) cut into the barrel at an angle; the hot bullet is soft enough to conform to the grooves and thus acquires spin. The twist angle and land/ groove patterns of weapons vary and create markings found on bullets. The number of lands and grooves and the direction and angle of twist are class characteristics. Giving the bullet spin is important for accuracy and helps to keep the bullet on the trajectory that it has when leaving the barrel. Rifling in handguns is shown in Figure 15.4.

The **caliber** of a gun originally referred to the diameter of the barrel of a rifled pistol or rifle; however, the term can also refer to the size of cartridges used in firearms. Caliber is measured from the tops of the lands and is given in hundredths or thousandths of an inch or in millimeters. Common calibers include 9 mm and .22, .38, .40, and .45 for pistols. The last four measurements are in inches. Examples of rifle calibers are .22 and .30-06 (inches). The caliber of a gun is now

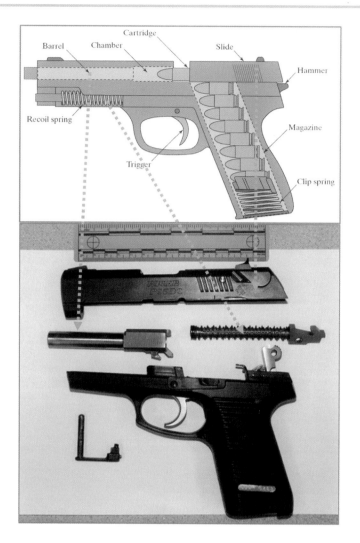

FIGURE 15.3   A semiautomatic pistol and parts.

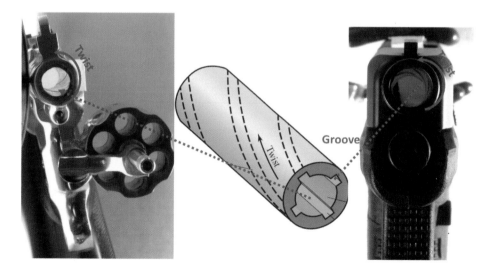

FIGURE 15.4   Rifling in a revolver (left) and a semiautomatic pistol (Right). Notice the difference between the patterns; bullets fired from the two weapons would be marked differently as a result. The lands are the elevated regions between grooves.

descriptive, and the actual barrel diameter may vary slightly from the barrel diameter. With shotguns, the term *gauge* is used to describe the size. Originally, the gauge of a pellet referred to how many pellets of a given size (the same as the barrel diameter) were needed to reach a weight of 1 pound. Twelve-gauge pellets weighed approximately 1/12 of a pound each and would fit in the barrel of a 12-gauge shotgun. A higher gauge means smaller pellets and thus smaller barrels, so a 12-gauge shotgun has a barrel of larger diameter than does a 16-gauge shotgun, just as 12-gauge shot is larger than 16-gauge shot.

GUNSHOT WOUNDS

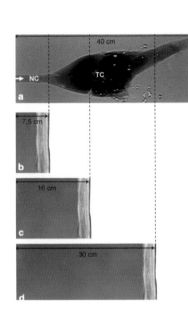

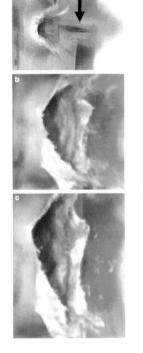

High-powered weapons such as rifles fire projectiles that travel very fast and thus carry a large amount of energy. As a result, the bullets can do enormous damage to the body. In a recent study, bullets from a high-powered rifle were fired into a specialized gelatin material to visualize what happens internally. In the figure at left, the top shows the block as the bullet travels through it. Notice the large cavity of gas formed in the middle of the block about 40 cm (almost 16 inches) into the gelatin. The lower frames show the height of the cavity at different distances into the gelatin. These experiments demonstrate why exit wounds are often larger than entrance wounds made by these types of weapons. How big the exit wound is depends on where the cavity forms. This was illustrated by additional experiments using a modified material meant to represent human tissue. In the top frame of the figure at right, the bullet is shown leaving the material. The size of the exit wound is much larger when the cavity reaches maximum size just at it reaches the edge of the tissue.

Source: Thierauf, A., et al., The Varying Size of Exit Wounds from Center-Fire Rifles as a Consequence of the Temporary Cavity, *International Journal of Legal Medicine* 127, no. 5 (Sep 2013): 931–936. Figures reproduced with permission. Copyright Springer.

15.2.3 AMMUNITION AND PROPELLANTS

Modern ammunition is typically supplied as a self-contained **cartridge** (Figure 15.5). The projectile or bullet is placed atop a cartridge case made of brass or other metal and crimped tightly into place. The seal must be tight enough to allow pressure to build up behind the bullet when the propellant burns, but not so tight that the cartridge explodes. The cartridge contains the powered propellant. The **primer** is mounted in the base of the cartridge. The primer contains a small amount of a shock-sensitive explosive. When the firing pin strikes the primer, it generates a flash that travels through small openings to ignite the propellant. The case is rarely packed full of propellant but rather is usually half full or less.

Bullets may be lead, lead alloy, semijacketed, or full metal jacket; the jacket is typically brass or copper. The bullet shown in Figure 15.5 is a copper-jacketed lead bullet. Because lead is a soft

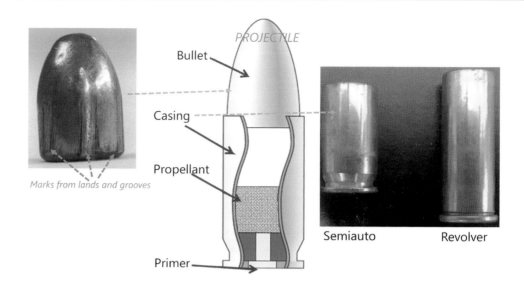

FIGURE 15.5 Ammunition.

metal, lead bullets are marked by the rifling as shown in Figure 15.5. Semijacketed bullets commonly consist of a lead core covered with a thin jacket of brass. The brass typically covers the sides of the bullet, leaving the lead core exposed at the nose (as in hollow-point bullets). The jackets of hollow-point and soft-point bullets may separate from the lead cores when the bullets enter the body of a shooting victim. Treating medical personnel or forensic pathologists must make every effort to recover the jacket because only the jacket bears markings made by the weapon's rifling.

Cartridges may be either **rimfire** or **centerfire**. Rimfire cartridges have the primer composition in the rolled rim of the cartridge. Small-caliber cartridges such as 0.22 caliber are rimfire. These cartridges are fired by the weapon's firing pin striking the cartridge rim. A centerfire cartridge has a primer cap placed in the center of the cartridge base. Centerfire cartridges are produced in a wide range of shapes. In Figure 15.5, the cartridge shown on the right is for a revolver and is cylindrical without any necking. This is because the empty cartridges are not automatically ejected from the weapon but are manually removed. The case on the left is designed for a semiautomatic pistol and has a necked down area at the bottom. This is where the ejection mechanism (**ejector**) grabs the casing for ejection out of the chamber.

The base of a cartridge bears information such as the vendor of the ammunition (usually abbreviated, as F for Federal, Rem for Remington, R-P for Remington–Peters, or WIN for Winchester) and the gauge or caliber. These markings are referred to as the **headstamp**. The headstamp may also consist of a logo (e.g., a diamond on .22-caliber Winchester rimfire cartridges). In Figure 15.6, a fired cartridge (9 mm) headstamp is shown. The crater mark in the middle was made by the firing pin impacting the primer (inner circle) of this centerfire-type cartridge. The Bureau of Alcohol, Tobacco, Firearms and Explosives (ATF) maintains a database of headstamp images that is listed at the end of the chapter.

15.2.3.1 PROPELLANTS

The most common type of modern propellants (also called smokeless powder) seen in forensic casework is called **double-base** powder, which uses **nitrocellulose** and **nitroglycerin** to produce hot expanding gas to drive the bullet forward. Nitrocellulose used in smokeless powders is produced by treating the cotton (cellulose) with nitric and sulfuric acids, resulting in nitration of the molecules. This material is also called **guncotton**. Nitroglycerin is manufactured by nitration of natural or synthetic glycerin, an oil. Nitroglycerin is considered an energetic plasticizer. It softens the propellant granules, raises their energy content, and reduces their absorption of moisture. Smokeless powders used in small-caliber ammunition contain a variety of other ingredients, such as stabilizers and plasticizers. **Stabilizers** react with the acidic breakdown products

FIGURE 15.6   Example of a headstamp on a fired cartridge.

FIGURE 15.7   Magnified image of handgun propellant.

of nitrocellulose and nitroglycerin. Typical stabilizer concentrations range from 0.5% to 1.5%. Plasticizers soften the propellant granules and reduce the absorption of moisture.

The manufacturer of smokeless powder controls many characteristics of the powder. The energy of the powder is determined by the formulation of the powder (i.e., whether the powder is single base or double base); it is primarily a function of the flame temperature and the average molecular weight of the muzzle gases. The burning rate depends on the composition and the porosity of the powder grains. The surface area of the powder grain is determined by its geometry. Powder grains may be spheres, cylinders, disks, or flakes; the grains may also be perforated. The surface area of the powder grains is the major control factor that the manufacturer uses to adjust the performance of a smokeless powder product.

Smokeless powders are produced by one of two processes: the **extruded powder** process and the **ball powder** process. In the extruded powder process, the nitrocellulose and other ingredients are kneaded together with an organic solvent to form a doughlike mass. The dough is extruded through small openings in a steel die. A rotating blade cuts off lengths of extruded dough. Examples of this are shown in Figure 15.7. The disks are cut short and the rods are cut

FIGURE 15.8 A comparison microscope and example images.

longer; smaller ball shapes are also shown. The dimensions of the extruded grains are controlled by the size of the openings in the die, by the rate of extrusion, and by the speed of the cutting blade. The extruded powder grains are then coated and glazed with graphite. In the ball powder process, the nitrocellulose and other ingredients are mixed with solvent and dropped into hot water to form spherical grains. The solvent is removed by the hot water, and the grains are allowed to harden. The spherical grains are sorted by size, and particles in the desired size range are subjected to further processing. The grains are coated and may be passed between rollers to flatten them into disks or flakes. In both processes, the final product is a mixture of several batches of smokeless powder. The batches are blended to create a powder with a specified burning rate.

15.2.4 TYPES OF MARKINGS AND FORENSIC EXAMINATION

The marks that are studied by firearms examiners are tool marks that arise from the many opportunities for metal-to-metal contact. The bullet in Figure 15.5 shows marks made on the bullet as it was hot and traveling down the barrel in contact with the lands and grooves. As noted above, lands, grooves, and twists are class characteristics; the shape of the firing pin is also a class characteristic. In Figure 15.6, you can see where the firing pin hit the primer. The patterns of these marks are used to determine if a given gun could have fired a given bullet, or if it could not. The instrument used to study firearms is called a **comparison microscope** (Figure 15.8). The device consists of two microscopes connected by an optical bridge that allows the examiner to adjust and align the separate images for direct comparison. In the figure, the images of two cartridge cases are aligned, studied, and compared. Digital images are critical to the examination and subject to image analysis using software and databases. In Figure 15.8, the crater features are firing pin impressions, and if you look closely, you can see a ridged pattern to the right of the firing pin impression on both primers. Another example is shown in Figure 15.9 at a higher magnification. This is also a comparison microscope image, but the split between the two cartridges is very difficult to see given how similar the patterns are. This example shows the importance of the fine structure called **striations**, which are patterns of lines that run roughly parallel to each other. These can be made by dragging one metal surface over another or when a striated metal surface, such as a firing pin, strikes another surface, such as a primer. Striations are also seen as a fine structure within the markings that result from the bullet moving across the lands and grooves in a barrel (Figure 15.10). As seen in this figure, the patterns can be quite detailed, and the greater the magnification, the greater the detail available to study. These striations are loosely analogous to the minutia of fingerprints and are the basis upon which examiners determine how likely it was that a given bullet was fired from a given gun.

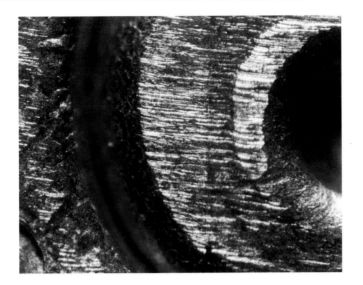

FIGURE 15.9 Higher magnification of firing pin impressions with striations. The line between the two images is about 65% of the way across to the right.

FIGURE 15.10 Striations on bullets.

The methods of analysis and comparison of firearms markings is evolving and is conceptually like what we discussed in Chapter 14 regarding fingerprints. As you can see with the images shown in this chapter, it is possible for the analyst to line up two images from two samples and subjectively compare them. If the markings are not at all similar, it can be relatively simple to determine that the two bullets were likely not fired from the same weapon. However, as we saw with fingerprints (and is common with all types of pattern evidence), it is becoming critical that some form of probability or statistical measure of the "goodness" of the match be developed and stated by examiners. This is a challenging proposition and the subject of debate, discussion, and current research in the forensic community.

When firearms evidence such as a bullet or cartridge case is submitted without a firearm, the analysis begins with determining what type of weapon was likely used to discharge the ammunition. With a bullet, for example, the markings and dimensions can reveal the caliber of the weapon, the number of lands and grooves in the barrel, and the direction of the twist of the lands and grooves. The width of the lands and grooves can further reduce the number of weapons that could have been used to fire the gun. For example, some guns have barrels that are made with a right-hand twist in the land and groove pattern, while other manufacturers used a left-hand twist. With

FIGURE 15.11   Examples of ejector marks on cartridge cases.

cartridge casings, the appearance of the firing pin impression can help limit the potential sources; for example, the square impression shown in Figure 15.6 is characteristic of Glock guns. A discharged cartridge will also have marks from where the back of the barrel contacts the back of the chamber (**breech face marks**) and **ejector marks** if the weapon was a semiautomatic (Figure 15.11).

If a firearm is submitted as evidence, the question is often if that weapon was used to fire a given bullet or was the source of a cartridge case. For example, a bullet or bullet fragments may be recovered during an autopsy and a gun might be found in the possession of a suspect. The initial examination is straightforward and involves determining the caliber and other factors that might eliminate a gun as a potential source. To determine if that gun could have been used to fire the bullet, the firearms examiner conducts test firings into a bullet trap. Most firearms laboratories use some type of water trap to catch fired bullets. Using the traps allows for the recovery of intact bullets that have not passed through any other materials, such as clothing, walls, flesh, or bones. Once recovered, the test bullet or cartridge is then mounted along with the case sample on a comparison microscope for examination and documentation through digital imaging. There are also instrument systems such as the **Integrated Ballistic Identification System** (IBIS) that collect images and access databases for comparison in the same way that the Integrated Automated Fingerprint Identification System catalogs and compares fingerprints.

Comparing a test-fired bullet to a case sample involves comparing and aligning features using the comparison microscope. Most examples shown in photos in this chapter illustrate cases in which the markings on the test and case sample appear to be quite similar; however, the practice of saying that a bullet was definitely fired from a specific gun is no longer considered to be as simple as a yes/no decision. This is the same evolution we described for latent fingerprints in Chapter 14. Increasingly the expectation is that the degree of similarity be expressed as a number or probability, which is a more complex challenge, and there continues to be debate about how to make such comparisons quantitative. As an example, look again at Figure 15.10, which shows the alignment of striations on two bullets. They appear similar, but how similar? They are not identical, but are they sufficiently alike to conclusively state that both were fired from the same weapon? More realistically, what is the likelihood or odds that both were fired from the same gun? The firearms community continues to study methods that can be used to state the degree of similarity of markings in a quantitative way.

15.3   TOOL MARKS

With firearms, we are concerned with markings, including those made by the barrel rifling on bullets, firing pin impressions, and breech face markings. Tool mark analysis is similar in that we are interested in the marks made on or in a material by some type of tool.

FIGURE 15.12   Tool mark made by pushing a screwdriver across clay. Notice that striations are created like those seen on bullets.

An example would be a mark left by a screwdriver on a metal window frame. An example of such a mark is shown in Figure 15.12. In firearms examination, the markings of interest are produced by metal-to-metal contact, but tool marks may be generated on many surfaces, such as wood. However, caution is required when attempting to interpret marks on soft surfaces because they can change or deform. Interpretation of this type of mark can be problematic if the characteristics of the surface are not considered.

In general, soft metals such as lead, copper, and brass are excellent recipients of tool marks. Many plastics are also good surfaces for the retention of tool marks, as are painted surfaces. Other surfaces, such as raw wood and hard metal, are poor substrates for tool marks. Raw wood is a poor substrate for sliding tool marks because its grain structure has the same dimensions as the striations in the typical tool mark. Hard metal surfaces are poor recipients of tool marks because their hardness prevents them from being marked by tools.

The three categories of tool marks are compression (or indented) tool marks, sliding tool marks, and cutting tool marks. Compression tool marks result when a tool is pressed into a softer material. Such marks often show the outline of the working surface of the tool, so class characteristics of the tool (such as dimensions) can be determined. The characteristics of the tool may be more difficult to discern in compression tool marks. Sliding tool marks are created when a tool slides along a surface; such marks usually consist of a pattern of parallel striations. Cutting tool marks are a combination of compression and sliding tool marks. One of the important variables with tool marks is the angle of the tool to the surface. For example, pulling a screwdriver along a surface at a 90° angle and dragging it along the same surface at a 45° angle will change the characteristics of the mark.

The characterization of tool marks is like that of markings associated with firearms. Test marks can be compared with case marks, and castings can be made from tool marks, as shown in Figure 15.12. As with firearms, debate and research continues in regards to how to characterize findings. It is often easier to tell that a mark was not made by a given tool, but beyond that, there is no consensus yet on how to state the degree of similarity of marks or to determine how common a type of marking is across large populations. For example, screwdrivers are mass-produced items purposely designed to meet specific criteria. Two screwdrivers made at the same time by the same process should leave the factory with very similar surfaces and thus would be expected to leave similar markings on similar surfaces. How much like each other? How many other screwdrivers could create similar marks? Recall the discussion at the beginning of Chapter 14 regarding fingerprints and pattern evidence in general. Calling something a match is best presented along with data that allows laypeople to fairly interpret the value of evidence, even if only a measurement of the known error rate.

To take it a step further, suppose two screwdrivers are made at the same time under the same conditions and leave the factory as nearly identical as possible. One of these is purchased by a mechanic who uses the tool several times a day and thus experiences wear. The other is purchased by someone who puts it in a drawer at home and uses it once a month. The acquisition of **wear characteristics** by the first screwdriver means that the markings it creates could be much different than those of the other, which is rarely used. How different are these marks? Of course, that is difficult to answer, but in a case, such information might be vital. These questions are those being debated and studied now, and the goal is like that with fingerprints and other types of comparisons—how can these be made more quantitative and statistically based? It is not an easy or simple question, but the goal is to make the analysis more objective than it has been in the past.

MYTHS OF FORENSIC SCIENCE

You can tell the caliber of a gun from this size of the wound it makes.

The reason this is not possible is that many factors determine the size of a gunshot wound. Entrance wounds are usually smaller than exit wounds, and the distance between the muzzle and the skin can make a significant difference. The bullet might also be tumbling, which can alter the size and shape of a wound. The figure below was from a study that examined the relative size of entrance wounds of three different sized projectiles. The darkened area is a ring of scraped skin around the wound. In some cases, the entrance wound is no more than a tiny slit or puncture that is much smaller than the bullet itself.

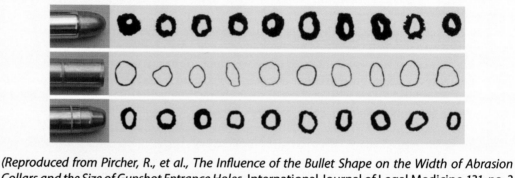

(Reproduced from Pircher, R., et al., *The Influence of the Bullet Shape on the Width of Abrasion Collars and the Size of Gunshot Entrance Holes*, International Journal of Legal Medicine *131, no. 2* [Mar 2017]: 441–445. With permission from Springer.)

15.4 ADVANCES IN FIREARM AND TOOL MARK CHARACTERIZATION

Because images are used in firearm and tool mark analysis, they can be studied and analyzed in ways similar to the way fingerprint images are studied and stored in databases. The development of image analysis tools has opened the way for mathematical and statistical analysis of markings and methods of comparison that allow for quantitative descriptions of similarity. An example of one such method is shown in Figure 15.13. Here, two cartridge cases known to be fired from the same weapon were imaged, processed, and combined. The left side is a coded grayscale image, and the right, a color-coded version. The gray portions of the color scale are locations on the two cases that are not as important in establishing a match. Red areas correspond to elevated patterns ("peaks") on the two breech faces and firing pins that are similar, green areas represent similar "valleys," and yellow areas show where peaks and valleys overlap. The color coding is determined by mathematical and statistical calculations.

Another example, this time related to tool marks, is shown in Figures 15.14 and 15.15. In this project, image processing techniques were applied to tool marks from a screwdriver to help compensate for differences in the angle between the screwdriver and the surface. As noted above, that angle can have a significant impact on the appearance of the tool mark and striations. Thus, it is important for the examiner to have a way to account for any such differences when comparing test marks to marks associated with a case. Figure 15.14 shows a screwdriver and a close-up of the blade surface that would make a pattern when applied to a surface. Since the angle of the blade to the surface is a factor, imaging processing can assist in accounting for this, as shown in Figure 15.15. The initial image is processed to reveal critical features for a comparison between a test sample and a case sample, including dimensions of markings.

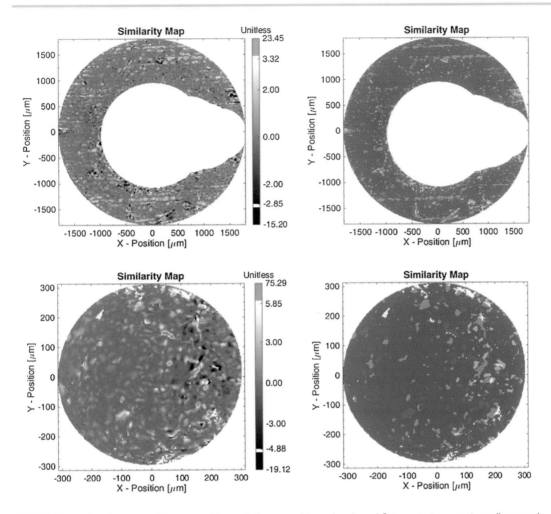

FIGURE 15.13 Processed images of breech face markings (top) and firing pin impressions (bottom) of two cartridge cases that were known to be fired from the same weapon. (Reproduced from Ott, D., et al., Applying 3D Measurements and Computer Matching Algorithms to Two Firearm Examination Proficiency Tests, *Forensic Science International* 271 (Feb 2017): 98–106. Article is available via open access.)

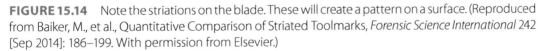

FIGURE 15.14 Note the striations on the blade. These will create a pattern on a surface. (Reproduced from Baiker, M., et al., Quantitative Comparison of Striated Toolmarks, *Forensic Science International* 242 [Sep 2014]: 186–199. With permission from Elsevier.)

15.5 RELATED ANALYSES

Firearms examiners are frequently responsible for other types of tests associated with weapons and marks on metal surfaces. One of these is estimating the distance of a gun from the target surface, which is referred to as **distance estimation** and is a critical element in **shooting reconstructions**. It is not possible to exactly determine how far a gun barrel was from a target when

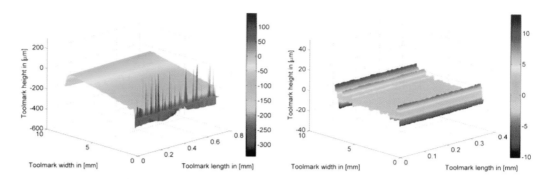

FIGURE 15.15   Raw and processed three-dimensional images of a tool mark made with the screwdriver. (Reproduced from Baiker, M., et al., Quantitative Comparison of Striated Toolmarks, *Forensic Science International* 242 [Sep 2014]: 186–199. With permission from Elsevier.)

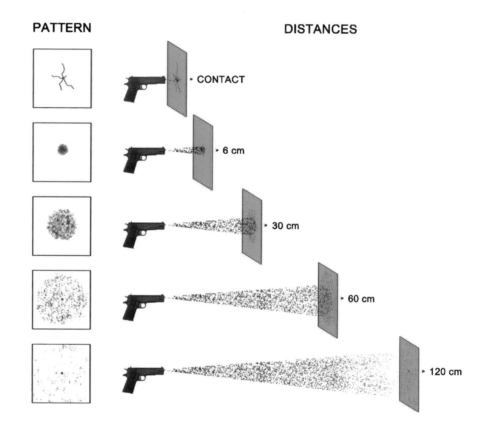

FIGURE 15.16   How distance from the target affects residues on the target. (Reproduced from Lopez-Lopez, M., and Garcia-Ruiz, C., Recent Non-Chemical Approaches to Estimate the Shooting Distance, *Forensic Science International* 239 [Jun 2014]: 79–85. With permission from Elsevier.)

it was fired, but it is possible to establish likely ranges of distance. This is due to the number of variables involved in shootings, such as type of weapon, ammunition, movements at the scene, weather, and air movement. Distance determinations are established by a series of experiments in which the gun in question is loaded with ammunition that is as similar to that in the case as possible. A series of shots are fired at known distances into targets the mimic case samples (i.e., types of clothing). The process is illustrated in Figure 15.16; the farther the muzzle from the target, the less residue deposited.

Depending on the target material and the distance, the residue pattern may be easy to see. As shown in Figure 15.17, when the target is white paper, the variation in residues deposited is easy to see. However, some form of development is usually undertaken to enhance the pattern and

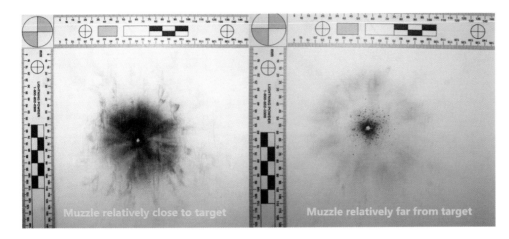

FIGURE 15.17   Undeveloped residues on white target material.

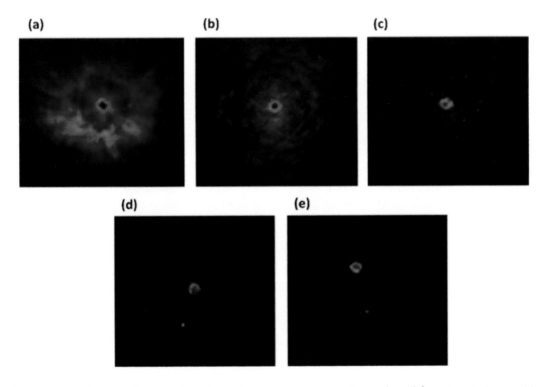

FIGURE 15.18   Visualization of lead for distance estimation. (Reproduced from Lopez-Lopez, M., and Garcia-Ruiz, C., Recent Non-Chemical Approaches to Estimate the Shooting Distance, *Forensic Science International* 239 [Jun 2014]: 79–85. With permission from Elsevier.)

make it easier to see. This becomes particularly important when the distance is greater than a few inches. The treatments used can be a chemical treatment that results in a color change, specialized lighting, or an instrumental technique. An example is shown in Figure 15.18, in which a specialized x-ray instrument (which we will discuss in more detail in Chapter 17) is used to map the presence of lead on targets that ranged from 5 cm away from the muzzle (~2 inches, frame a) to 100 cm (~3.2 feet, frame e). In casework, the case sample, such as clothing from a victim, would be treated in the same way and the pattern compared with the test fires to estimate the range of the weapon to the victim when the shot was fired.

Finally, firearms analysts are sometimes tasked with trying to recover serial numbers that have been scratched or scraped off metal surfaces. This process of **serial number restoration** utilizes chemical and electrochemical methods to recover all or part of a serial number that has been

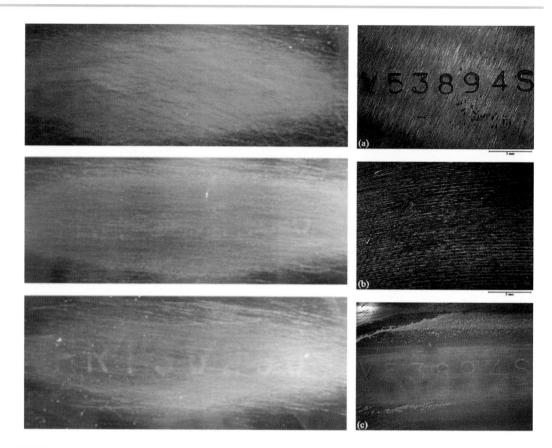

FIGURE 15.19 Restored serial numbers on glass (left) and steel (right). (Left image reproduced from Miller, R. J., The Restoration of Serial Numbers on Vehicle Glass Using Hydrofluoric Acid, *Forensic Science International* 228, no. 1–3 [May 2013]: 28–31. With permission from Elsevier. Right image reproduced from Fortini, A., et al., Restoration of Obliterated Numbers on 40NiCrMo4 Steel by Etching Method: Metallurgical and Statistical Approaches, *Journal of Forensic* Sciences 61, no. 1 [Jan 2016]: 160–169. With permission from Wiley.)

stamped into a metal surface. The process of stamping involves striking down on the metal surface with another metal surface, and the force of that impact damages the structure of the metal directly underneath the stamp. When someone scrapes a number off the surface, they typically stop when the number is obliterated. However, the damaged zone goes deeper than the visible markings. The goal of treatment of the surface is to use methods that will dissolve the damaged area more quickly than the surrounding undamaged area of the metal. If this process is not carefully monitored, the undamaged part of the surface will dissolve away as well, so it is important to stop the process at the appropriate time. Examples are shown in Figure 15.19. Interestingly, the process works even on glass (left side of figure). Serial numbers are obviously not stamped into glass, but the same principle applies; the process of marking the glass causes damage that extends beyond and below the area where the serial number is easily seen.

15.6 REVIEW MATERIAL: KEY CONCEPTS AND QUESTIONS

15.6.1 KEY TERMS AND CONCEPTS

Ball powder process
Breech face marks
Caliber
Cartridge
Centerfire

Comparison microscope
Distance determination/estimation
Double action
Double base
Ejector
Ejector marks
Extruded powder process
Firing pin
Gauge
Grooves
Hammer
Headstamp
Lands
Lands and grooves
Long guns
Nitrocellulose
Nitroglycerin
Primer
Propellant
Revolver
Rifling
Rimfire cartridge
Serial number restoration
Shooting reconstruction
Single action
Slide
Striations
Wear characteristics

15.6.2 REVIEW QUESTIONS

1. In what types of marking evidence do striations appear? How are they similar and how are they different?
2. Serial numbers cannot be stamped in glass. How do you suppose they are created?
3. How can a firearms examiner determine if a cartridge case came from a revolver or a semiautomatic handgun?
4. Explain why the angle between a tool and a surface matters for evaluation of tool mark evidence.
5. Explain how a gun works.
6. Why do bullets spin when leaving the barrel? Why is this important?
7. How is the distance from a target estimated?
8. A burglar uses a pry bar to open a door. What type of tool mark(s) would be left on the wooden door frame?
9. What is the difference between a rimfire and centerfire cartridge?

15.6.3 ADVANCED QUESTIONS AND EXERCISES

1. What would be some examples of class characteristics created by tools such as a hammer or a screwdriver?
2. Revisit the crime scene scenario described in Chapter 3 and review Figure 3.9. Based on the description of witnesses and finding the cartridge casing, what might a firearms examiner be asked to do? What else would have to be found?
3. Assume that the autopsy showed that the woman died of blunt force trauma. What does this suggest about the cartridge casing found at the primary scene?

4. Distance determinations can provide critical information in cases where a shooting has taken place and the person who fired the weapon claims self-defense. What can distance tell the investigators?

BIBLIOGRAPHY AND FURTHER READING

BOOKS

Baldwin, D., Birkett, J., Facey, O., and Rabey, G. *The Forensic Examination and Interpretation of Tool Marks.* Hoboken, NJ: John Wiley & Sons, 2013.

Heard, B. *Firearms and Ballistics: Examining and Interpreting Forensic Evidence.* 2nd ed. Hoboken, NJ: John Wiley & Sons, 2008.

Huske, E. E. *Practical Analysis and Reconstruction of Shooting Incidents.* 2nd ed. Boca Raton, FL: CRC Press, 2015.

Petraco, N. *Color Atlas of Forensic Toolmark Identification.* Boca Raton, FL: CRC Press, 2010.

Shina, J. K. *Forensic Investigation of Unusual Firearms: Ballistic and Medico-Legal Evidence.* Boca Raton, FL: CRC Press: 2014.

Walker, R. E. *Cartridges and Firearm Identification.* Boca Raton, FL: CRC Press, 2012.

Wallace, J. S. *Chemical Analysis of Firearms, Ammunition, and Gunshot Residue.* Boca Raton, FL: CRC Press, 2008.

Warlow, T. *Firearms, the Law, and Forensic Ballistics.* 3rd ed. Boca Raton, FL: CRC Press, 2012.

ARTICLES

Baiker, M., Petraco, N. D. K., Gambino, C., Pieterman, R., Shenkin, P., and Zoon, P. Virtual and Simulated Striated Toolmarks for Forensic Applications. *Forensic Science International* 261 (Apr 2016): 43–52.

Bolton-King, R. S. Preventing Miscarriages of Justice: A Review of Forensic Firearm Identification. *Science and Justice* 56, no. 2 (Mar 2016): 129–142.

Bunch, S., and Wevers, G. Application of Likelihood Ratios for Firearm and Toolmark Analysis. *Science and Justice* 53, no. 2 (Jun 2013): 223–229.

Gerules, G., Bhatia, S. K., and Jackson, D. E. A Survey of Image Processing Techniques and Statistics for Ballistic Specimens in Forensic Science. *Science and Justice* 53, no. 2 (Jun 2013): 236–250.

Lock, A. B., and Morris, M. D. Significance of Angle in the Statistical Comparison of Forensic Tool Marks. *Technometrics* 55, no. 4 (Nov 2013): 548–561.

Maitre, M., Kirkbride, K. P., Horder, M., Roux, C., and Beavis, A. Current Perspectives in the Interpretation of Gunshot Residues in Forensic Science: A Review. *Forensic Science International* 270 (Jan 2017): 1–11.

Petraco, N. D. K., Shenkin, P., Speir, J., Diaczuk, P., Pizzola, P. A., Gambino, C., and Petraco, N. Addressing the National Academy of Sciences' Challenge: A Method for Statistical Pattern Comparison of Striated Tool Marks. *Journal of Forensic Sciences* 57, no. 4 (Jul 2012): 900–911.

Spotts, R., Chumbley, L. S., Ekstrand, L., Zhang, S., and Kreiser, J. Optimization of a Statistical Algorithm for Objective Comparison of Toolmarks. *Journal of Forensic Sciences* 60, no. 2 (Mar 2015): 303–314.

WEBSITES

Link	Description
https://www.atf.gov/firearms/national-integrated-ballistic-information-network-nibin	National Integrated Ballistic Information Network website
https://www.atf.gov/firearms/automated-firearms-ballistics-technology	IBIS overview
https://www.theiai.org/disciplines/firearms_toolmark/	The International Association for Identification page for firearms and tool marks
https://afte.org/	Association of Firearm and Tool Mark Examiners
https://nij.gov/journals/274/Pages/firearm-toolmark-examination.aspx	The science behind firearms from the National Institute of Justice
https://afte.org/resources/headstamp-guide	AFTE Headstamp Guide

Tread Impressions

16.1 ANOTHER FORM OF IMPRESSION EVIDENCE

Fingers, guns, and tools leave pattern evidence; treaded items, such as tires and shoes, do the same. Tires and shoes are mass-produced items, so at first it might seem that the best forensic scientists could do with a tire or shoe impression is link it back to the type of tire or a brand of shoe. However, often much more can be done with this type of evidence. Suppose you and a friend both purchase a pair of shoes from the same store on the same day. They are the identical type and brand of shoes and they came off the assembly line one right after the other. Initially, the treads will probably be nearly identical. You wear these shoes every day and use them to play basketball on outdoor concrete surfaces. Your friend wears the shoes occasionally and mainly indoors. Over time, the tread patterns will become worn, and these wear characteristics can be critical in differentiating your shoes from your friend's shoes. This is the same idea we discussed in Chapter 15 regarding screwdrivers. The way in which the item is used changes the features and patterns that it leaves behind.

Tread impressions are also another example of pattern matching in which successive classification is used. Markings that arise from wear and tear are loosely analogous to minutiae in latent fingerprints and are the key to differentiating one shoe from another. As with all the pattern recognition disciplines, computer imaging, analysis, and databasing are used to assist in classification, differentiation, and providing statistical and probability-based methods of analysis and presentation.

Locard's principle is also at play. As you walk through day, your shoes contact many different surfaces, such as concrete, tile, grass, dirt, and carpet. The shoes then deposit these acquired materials back onto other surfaces that they subsequently track over. This is transfer evidence, and it also can become the material that creates a pattern. Muddy shoeprints on the garage floor are an example of this—the mud creates the impression and came from mud that the person

FIGURE 16.1   Use of oblique lighting to visualize a latent shoeprint. Lighting is coming from a source on the floor and to the left of the print.

walked through at some earlier time. Shoes can leave patent (visible) and latent (invisible) two-dimensional impressions, like an outline of a shoe in the dust on a floor. On softer surfaces, such as sand, soil, or snow, they often will cause permanent deformations of that surface in the form of **three-dimensional impressions**, such as tire tracks in mud. These types of impressions are conceptually like tool marks in that there is depth to the impression. The same ideas apply to tire marks and tire tread impressions—two and three-dimensional impressions can be created, and the tire collects and deposits transfer and trace evidence. As with other types of pattern evidence, digital imaging and databasing are becoming key tools in the analysis of tread impressions. This chapter focuses on footwear impressions, but the same general principles apply to tire impressions.

16.2 FOOTWEAR IMPRESSIONS

Patent shoeprints are those that are easily seen, such as bloody shoeprints on a light-colored surface. Locating most footwear impressions, however, requires additional effort. Making a slow visual search, followed by darkening the room and searching for impressions with a high-intensity oblique light source (Figure 16.1), often reveals many impressions that could not otherwise be seen. Sometimes a developing reagent or technique is used to help visualize impressions for digital imaging. In some cases, impressions found at a crime scene are taken back to the laboratory for imaging and analysis. With two-dimensional impressions, it is often possible to lift the impression after it has been properly documented.

There are many methods and materials for lifting. Electrostatic lifting (electrostatic detection apparatus [ESDA]; described in detail in Chapter 18) is a method that utilizes a high-voltage power source to create a static electrical charge that enables the transfer of a **dry origin impression** (made by dry material such as dust) from the surface to a special black lifting film. The shoeprint shown in Figure 16.1 could be lifted in this manner. An example showing how an impression is lifted is seen in Figure 16.2. Notice the difference between the impression on the floor and the lifted version; much more detail is visible.

If an impression will not lift with the electrostatic method, the impression either is a wet origin (made by something wet, like blood or mud) or is composed of other materials that have bonded to the surface. Different methods of preservation of the impression are needed in such cases.

Three-dimensional impressions are typically cast in **dental stone**. Dental stone is like plaster but sets harder and retains detail while being easy to use. Although dental stone can be mixed in a bucket, the more popular and common use for casting footwear impressions involves placing a 2-pound portion into several zip-lock bags to have on hand when needed at a crime scene. The proper amount of water for the 2-pound portion can then be added to the bag at the crime scene and combined within the bag. When mixed, the solution is about the consistency of thin pancake batter, but it hardens into a strong and durable material that can be cleaned of debris. The importance of that characteristic is shown in Figure 16.3.

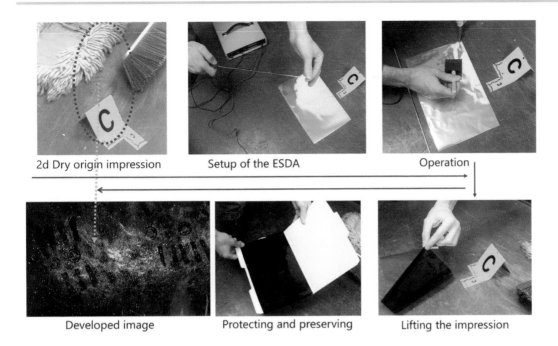

2d Dry origin impression   Setup of the ESDA   Operation

Developed image   Protecting and preserving   Lifting the impression

FIGURE 16.2   Using ESDA to lift a shoe impression. The initial print is shown in the circled region top left, between the broom and the mop. A static charge is created on the Mylar material that attracts and attaches the impression to that material, which preserves the impression for later analysis.

FIGURE 16.3   A cast just removed from dirt/mud (top) and a cleaned cast (bottom).

A particularly challenging type of impression to preserve is impressions made in snow. Dental stone cannot be directly poured into the impression because as it dries, it gives off heat that will melt and alter the impression. One way to prevent this is to spray the impression first with a material that will insulate the impression from the heat. The material is waxy, so it conforms to the impression while protecting it. An alternative is to use a sulfur mixture that can be melted and

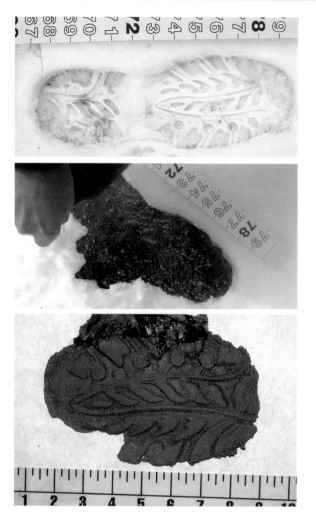

FIGURE 16.4 Process of casting an impression in snow using sulfur. The sulfur mixture is cooled before pouring it into the impression.

allowed to cool before pouring it into the impression. This process is shown in Figure 16.4. The sulfur cools very quickly.

Finally, elimination and comparison images are collected using a compressible foam. For example, suppose a three-dimensional impression is found at a scene, cast, and returned to the lab. If a suspect is identified and shoes are collected, the examiner needs to make an impression for comparison. This can be done by casting as well, or an impression can be made by pressing the shoe into the foam, which is designed to retain features. An example is shown in Figure 16.5.

16.3 EXAMINATION OF FOOTWEAR EVIDENCE

The forensic examination of footwear is like that of other types of pattern evidence that we have described previously, such as fingerprints and tool marks. Case samples are compared with impressions from potential sources and with elimination samples. As with other types of pattern evidence, it is usually easy to eliminate a shoe as a possible source of an impression but challenging to express the degree of similarity between a questioned impression and possible sources.

Footwear impressions recovered from the crime scene area are usually assumed to be those impressions made by the shoes of the perpetrator or perpetrators; however, impressions left by other persons at the crime scene, or additional impressions of unknown origin, may also be among those recovered. Thus, elimination samples must be collected from footwear worn by medical personnel, police officers, or other persons, who, in addition to the suspect, could possibly have

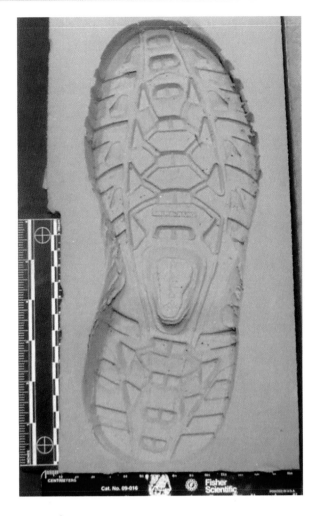

FIGURE 16.5   Impression in foam.

left the recovered impressions. By documenting the shoes of others who walked through the crime scene area, their shoe designs can be eliminated as those of the perpetrator.

Once a suspect is identified, shoes that might have been worn at the time in question would be collected and evaluated. Footwear examinations involve comparisons of both full and partial crime scene impressions with known shoes. The comparison process utilizes side-by-side and superimposition methods, assisted by low magnification, specialized lighting, and known impressions. The most obvious characteristic of any impression is its design or pattern, and so this is normally the first area to be compared. Look back at all the pictures of tread impressions in this chapter and you can easily identify different patterns. There are many thousands of shoe designs, with new ones arriving on the market as older designs are discontinued. There are commercial databases of shoeprints available to forensic practitioners, such as SoleMate®, which contains thousands of example patterns and impressions.

Depending on how the shoe has been made, the specific features of that design can even vary slightly among different shoes of that general design. Any design features that are evident in the questioned impression must correspond sufficiently with the suspect shoe for the examiner to determine the brand and model of a shoe. If it does not, the suspect shoe can easily be eliminated. If the design of the impression corresponds with the respective area of the suspect shoe, then the examination process continues.

Each shoe design manufactured comes in many sizes. The soles of these different sized shoes have different dimensions or proportions of their design throughout the size range. Consequently, the physical size and shape feature of soles are of comparative value during examination of the crime scene impressions. During the examination, the shoe design and the physical size and shape features of that design are evaluated together. When a crime scene impression corresponds

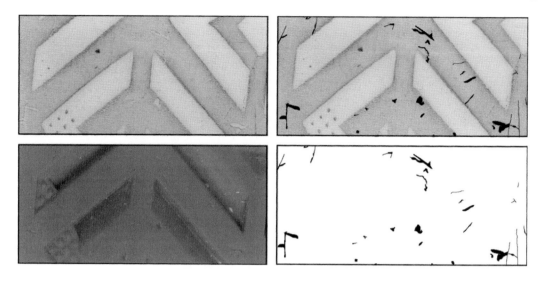

FIGURE 16.6   Examples of wear/accidental features. (Reproduced from Speir, J. A., et al., Quantifying Randomly Acquired Characteristics on Outsoles in Terms of Shape and Position, *Forensic Science International* 266 (Sep 2016): 399–411. With permission from Elsevier.)

in both design and physical size of that design with the suspect's shoe, the association can have high evidentiary value.

Beyond size and design, the next level of comparison is often of wear characteristics or other types of marks or features. Examples of wear or **accidental marks** are shown in Figure 16.6. These are also referred to as wear characteristics. The shoe sole is shown on the lower left, and you can see several small nicks and gouges in the plastic. In this example, the study authors obtained an image of the sole (upper left) and used software to mark the small features (upper right). The lower right frame shows just those marks. Using image processing in this way is often a key step to developing methods to characterize these marks by general shape and frequency of occurrence.

Another example of computer-based processing and modeling is shown in Figure 16.7. Here, the authors of the study developed a model to estimate what a sole impression would look like as time passes and the shoe undergoes wear and aging. This could be of value in a case in which a footwear impression is found and processed at a scene, but considerable time passes (months or years) before a suspect is identified and sample shoes are collected. Even if one of the collected shoes did make the impression at the scene, accumulated wear could make comparisons more challenging. As the figure shows, the other application of a model like this would be to estimate how much time (as measured by accumulated wear) has passed from when a shoe was new (A in the figure) to when the impression was collected (B).

16.4 TIRE IMPRESSIONS

Tire impression evidence shares many characteristics with footwear impression evidence. Tire impressions can be two- or three-dimensional. Tire tread impressions reflect the tread design and dimensional features of the individual tires on a vehicle. A tire tread impression is depicted in Figure 16.8. These tread impressions can be compared directly with the tread design and dimension of the tires from a suspect vehicle. **Tire tracks** are the relative distance between two or more tires of a vehicle. Tire tracks reflect general information about the vehicle that left the impressions. By measuring the dimensions of tire tracks at the crime scene, it may be possible to determine or approximate the track width, wheelbase, or turning diameter of the vehicle that created those impressions.

Thousands of tire designs are made for a variety of vehicles and in a variety of sizes. Since the 1960s, the annually published *Tread Design Guide* by Tire Guides, Inc., has provided photographs

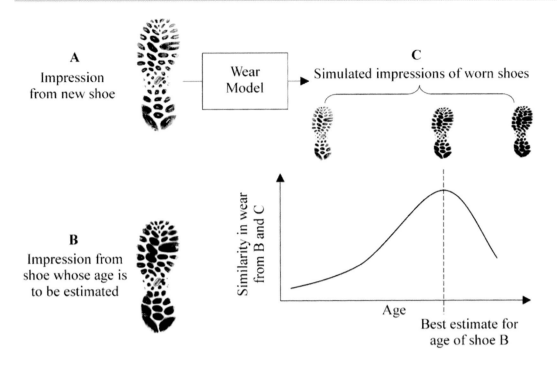

FIGURE 16.7 A proposed method to model wear on a shoe. (Reproduced from Skerrett, J., et al., A Bayesian Approach for Interpreting Shoemark Evidence in Forensic Casework: Accounting for Wear Features, *Forensic Science International* 210, no. 1–3 (Jul 2011): 26–30. With permission from Elsevier.)

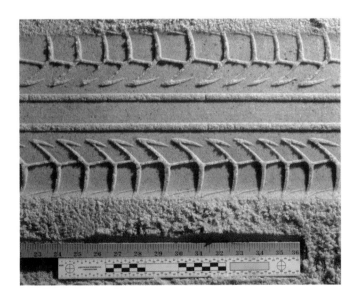

FIGURE 16.8 Example tire impression.

of most tires' designs. The publication *Who Makes It and Where* lists where tires are manufactured. These guides allow classifications based on tread designs and other features.

One factor that adds to the value of tire impression evidence is the additional information that can be obtained about the vehicle the tires are mounted on. Tire track width, wheelbase, turning diameter, and the relative positions of multiple turning tracks are collectively referred to as tire track evidence. In some cases, only one tire impression may be found at a scene, but in others where all four tires create usable impressions or patterns, significantly more information is available.

Track width (**stance**) is the measurement made from the center of one wheel or impression to the opposite wheel or impression (Figure 16.9). Interestingly, a vehicle's front and rear track widths are engineered to be slightly different. As a vehicle travels forward in a straight line, the rear tire tracks will track over the top of all or most of the tracks left by the front tires. As a result, the front wheel measurement may not be clear. Regardless of traveling straight or in a turn, the rear track width will always record accurately, so the most reliable crime scene track width will always be obtained from the rear tire tracks.

The **wheelbase** of a vehicle, the measurement between the centers of the hubs of the front wheels to the centers of the hubs of the rear wheels, is hard to obtain from impression evidence if made by moving tires. If a vehicle parked in snow, or briefly during a light rain, it may be possible to see four patches that mark the bottoms of the four tires, but that is the best case for obtaining a wheelbase measurement. The **turning diameter** of a vehicle is the diameter of the circle a vehicle makes when its steering wheel is fully turned; sometimes this can be determined from impression evidence at the scene.

Finally, when a vehicle travels in a straight line, the rear tire tracks run almost directly over the tracks of the front tires. For that reason, there are normally only two tracks to measure in a straight-traveling vehicle—those of the rear tires. When a vehicle is turning, the front and rear tires track separately, and the rear tires will track to the inside of the path of the front tires (Figure 16.10). This can be valuable information, particularly if the vehicle had different tires of more than one design or tires that were in different conditions of wear.

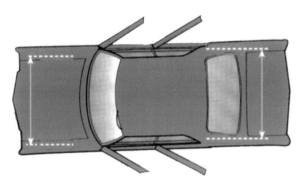

FIGURE 16.9   Track width. Note the difference between the back and front distances.

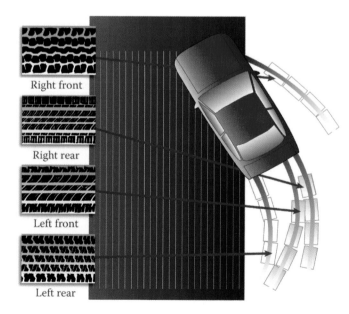

FIGURE 16.10   An example of impressions left by a vehicle with different wear on all four tires.

16.5 REVIEW MATERIAL: KEY CONCEPTS AND QUESTIONS

16.5.1 KEY TERMS AND CONCEPTS

Accidental marks
Dental stone
Dry origin impression
Stance
Three-dimensional impressions
Tire tracks
Turning diameter
Wheelbase

16.5.2 REVIEW QUESTIONS

1. Look closely at the top frame of Figure 16.4. Can you identify trace evidence present? How does this illustrate Locard's exchange principle?
2. Is a bloody shoeprint on a dry floor a wet origin or dry origin impression? Two- or three-dimensional impression?
3. Which of the following types of tread impressions would likely show small accidental marks (as shown in Figure 16.6)? In which cases would it be difficult to see such marks? Explain why.
   - Dry origin impression, dusty tile floor
   - Dry origin impression, dusty concrete floor
   - Bloody impression on carpet
   - Bloody impression on hardwood floor
4. Why does oblique lighting make latent tread impressions easier to see?
5. Why is it critical to have high-resolution images of shoe wear and tire impressions?
6. Figure 16.10 shows impressions from four tires. Could this happen if the car was not turning? Why?

16.5.3 ADVANCED QUESTIONS AND EXERCISES

1. Review the crime scene scenario presented in Chapter 3, including all the relevant figures and images. Identify evidence that would be considered as tread wear and classify each as two- or three-dimensional.
2. Suppose that casts are made of the shoe and tire tread impressions documented at the secondary crime scene. Using class characteristics alone, the forensic examiner is able to state that the shoe impressions came from two different shoes and that neither shared class characteristics with any of the shoes belonging to the suspect. The tire tread impressions share class characteristics with those on the suspect's truck, but the soil recovered from the tire treads on that truck do not share any of the characteristics of the soil at the burial site. Based on all the information provided in the scenario, offer several possible occurrences that could have led to this finding. Be mindful to avoid confirmation bias. For each hypothesis developed, explain what additional evidence will have to be collected to confirm or refute your ideas.

BIBLIOGRAPHY AND FURTHER READING

BOOKS

Bodziak, W. J. *Forensic Footwear Evidence*. Boca Raton, FL: CRC Press, 2016.
Bodziak, W. J. *Tire Tread and Tire Track Evidence: Recovery and Examination*. Boca Raton, FL: CRC Press, 2008.

ARTICLES

Basu, N., and Bandyopadhyay, S. K. Crime Scene Reconstruction—Sex Prediction from Blood Stained Foot Sole Impressions. *Forensic Science International* 278 (Sep 2017): 156–172.

Battiest, T., Clutter, S. W., and McGill, D. A Comparison of Various Fixatives for Casting Footwear Impressions in Sand at Crime Scenes. *Journal of Forensic Sciences* 61, no. 3 (May 2016): 782–786.

Richetelli, N., Lee, M. C., Lasky, C. A., Gump, M. E., and Speir, J. A. Classification of Footwear Outsole Patterns Using Fourier Transform and Local Interest Points. *Forensic Science International* 275 (Jun 2017): 102–109.

Richetelli, N., Nobel, M., Bodziak, W. J., and Speir, J. A. Quantitative Assessment of Similarity between Randomly Acquired Characteristics on High Quality Exemplars and Crime Scene Impressions via Analysis of Feature Size and Shape. *Forensic Science International* 270 (Jan 2017): 211–222.

Speir, J. A., Richetelli, N., Fagert, M., Hite, M., and Bodziak, W. J. Quantifying Randomly Acquired Characteristics on Outsoles in Terms of Shape and Position. *Forensic Science International* 266 (Sep 2016): 399–411.

Stoney, D. A., Bowen, A. M., and Stoney, P. L. Loss and Replacement of Small Particles on the Contact Surfaces of Footwear during Successive Exposures. *Forensic Science International* 269 (Dec 2016): 78–88.

WEBSITES

Link	Description
https://www.swgtread.org/footwear/make-model-determination	Footwear database description
https://archives.fbi.gov/archives/about-us/lab/forensic-science-communications/fsc/july2009/review/2009_07_review02.htm	Federal Bureau of Investigation article on footwear with examples
www.tireguides.com	Source of information on tires and treads, current and past
http://www.forensicsciencesimplified.org/fwtt/how.html	Concise overview of tread impressions

Trace Evidence and Microscopy

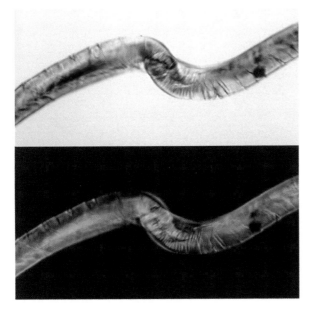

17.1 WHAT IS TRACE EVIDENCE?

Perhaps in no other type of forensic discipline is Locard's exchange principle (Chapter 2) more evident or important. Trace evidence can literally be anything. Transfers can be **primary** (soil from a grave site embedded into the shoe treads of the killer) or **secondary** (you sit in my car and dog hair from my dog is transferred to your clothing). Quite often, the evidence that is transferred is microscopic in size, adding to the challenge of analysis. Not surprisingly, the most useful tool of the forensic trace analyst is the microscope, and the analysis conducted is referred to generically as microanalysis.

Microanalysis is the application of a microscope and microscopical techniques to the observation, collection, and analysis of microevidence that cannot be clearly observed or analyzed without such devices. Microanalysis today generally deals with samples in the milligram or microgram size ranges. Analysis with a microscope may be limited to observations of morphology or involve the collection of more sophisticated analytical data, such as optical properties, molecular spectra, or elemental analysis. Regardless of what tool or instrument is used in the analysis, trace evidence relies heavily on successive classification based on class characteristics, as was described in the beginning of Chapter 15. The purpose of most of these analyses is to determine whether a link of persons, places, and things can be established, and the strength of that association.

MYTHS OF FORENSIC SCIENCE

This soil matches the soil where the body was found!

There are many variants of this type of statement regarding trace evidence, such as "the glass fragments in her clothes match the broken window" or "the fiber from the body matches the shirt worn by the suspect." As was discussed at the beginning of Chapter 15, the term *match* in this context is misleading, is incomplete, and overstates the value of the evidence. A better phrasing is that the questioned sample (soil, glass, fiber, etc., sometimes referred to as **Q**) shares characteristics of the standard sample (known, **K**). This type of examination is an example of successive classification, but it cannot lead to a definitive conclusion that the Q and K samples came from the same source. Successive classification leads to results of exclusion, inclusion, or inconclusive.

As an example, suppose a victim of a homicide is wrapped in a blue wool blanket before being buried in a shallow grave deep in a forest. The person who wrapped the body in the blanket and buried the victim will, according to Locard's exchange principle, likely get fibers from the blanket on their clothing and in the car trunk in which the body was transported. Soil from the grave site will be trapped in the treads of their shoes and can then be transferred to any surface those shoes cross. If a suspect is arrested, trace evidence such as fibers and soil can be used to establish an association between the suspect and the victim or to indicate that the evidence does not show any association. If, for example, the suspect's car is searched and blue wool fibers are found in the trunk along with soil like that at the grave, this is an association since both samples share similar class characteristics. This is inclusive evidence, but it does not mean the person is guilty; the evidence does not support this by itself. Rather, the data becomes part of the investigation and as always, the relevant evidence must be internally consistent to support one version of events over another.

Consider another version of this example. Suppose the suspect's car trunk has several blue fibers, but after microscopic examination, these are shown to be nylon and not wool. Does that mean the suspect is innocent? Again, this evidence alone does not support a conclusion either way. It must be integrated into the larger context of the investigation. What the evidence does suggest is that the body, when wrapped in the blanket, was never in that car trunk. As we discussed in Chapter 3, the context is critical to any finding, and with trace evidence this is particularly clear.

The analysis of trace evidence is also one of the most obvious examples of comparative analysis and successive classification in forensic science. In our blue wool blanket example, the goal of microscopic analysis is to compare fibers from the blanket used in the crime (the known sample) to fibers recovered as part of the investigation (the questioned). Often these two samples are referred to as the Q and K. The goal of trace evidence analysis is to determine if the Q and K samples could have had a common source (i.e., both could have come from the blanket used in the burial; inclusive evidence) or to show that they could not. The blue nylon fibers could not have come from a blanket made of only blue wool. This latter case is an example of **exclusionary evidence**—the Q and K did not come from the same source (here, the same blanket). Inclusionary evidence is the opposite but with an important consideration. If a blue wool fiber is found in the trunk of the suspect's car, then Q and K could have come from the same source, but with current methods, it is not possible to say that Q and K *definitely* came from the same blanket.

Now suppose that the suspect's car is found to contain many blue wool fibers with dozens in the trunk, and soil similar to that at the grave site is found in his shoes and on the floormat in the car. This is all inclusionary evidence that strengthens the apparent association of the suspect with transporting the body and the scene where it was buried. If only a few fibers are found, the association is not as strong. However, it is not possible to assign a value or probability to the association in the same way that it can be applied in DNA analysis. Even so, the results of the analysis of trace evidence are often critically important in linking people, places, and things through detailed analysis and the exchange principle.

17.2 MICROSCOPY

17.2.1 BASIC MICROSCOPY

A variety of microscopes are available for use in a forensic laboratory. Although many tools are available to the trace analyst, we touch upon only a few: visible and infrared spectrophotometry via a microscope, and basic scanning electron microscopy (SEM) with energy-dispersive x-ray spectroscopy (EDS). Before delving into these instruments, we begin with a discussion of the microscope itself and an extension called **polarizing light microscopy** (PLM). The microscope most likely to be employed first in the examination of evidence is the **stereo binocular micro-scope** (Figure 17.1). It is often employed in the preliminary evaluation of submissions, and for the location and recovery of microscopic particles and materials from their substrates, such as blue wool fibers from the carpet in a car trunk. This microscope is a compound type. **Total magni-fication** (TM) is computed by the power of the **objective lens** (OBJ), or first lens, multiplied by that of the **eyepiece lens** (EP), or finial lens.

A lens is an optical component that may be composed of one or multiple elements. The stereo microscope is constructed from two similar but separate optical microscopes for observation by each eye simultaneously. The views are separated by a small angle, usually about 15°, so that each eye sees the subject from a slightly different perspective. This generates a three-dimensional image in the same way that our eyes do. Most observations performed with stereo microscopes are carried out with reflected light analogous to how we normally see objects.

Many significant preliminary and other analytically important observations are made with this microscope. The layer structure of a recovered paint chip, including the color of each layer, and an estimate of the curliness of a human hair are only two examples. The stereo microscope is also frequently employed for viewing an object while it is being prepared for further analyses.

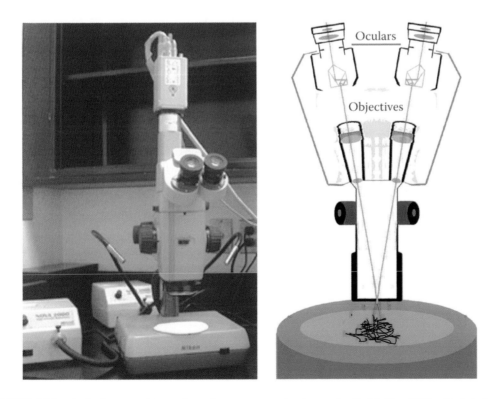

FIGURE 17.1   Left: A stereo binocular microscope and schematic. Right: The OBJ, which is in the eyepiece, is shown.

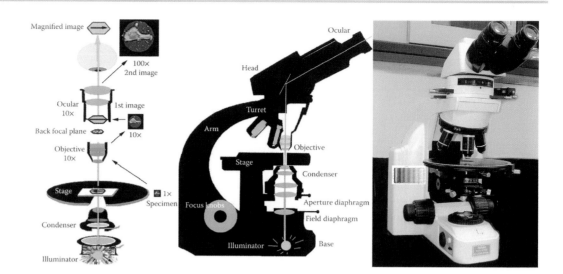

FIGURE 17.2 Schematic and photo of a typical forensic microscope.

The second most common type of microscope encountered in the laboratory is the compound binocular microscope (Figure 17.2). Most people are familiar with such microscopes because they are routinely found in schools and medical laboratories. Although this microscope employs two eyepieces, both eyes see the same image because each eyepiece magnifies an image formed by a single common objective. Most often this microscope employs transmitted, bright-field illumination for viewing. In transmitted light, the sample is transparent or mostly transparent. Most of the illumination passes through the subject and some passes around it.

This microscope is capable of TMs in the range of 25–1200 times (×) greater than the object, with 40–400× magnification commonly encountered in forensic laboratories. The TM is the product of the OBJ magnification multiplied by EP magnification: OBJ× multiplied by EP× = TM×; for example, (10× OBJ) × (10× EP) = TM of 100.

Of similar importance is analytical information about a sample that can be obtained without the use of sophisticated techniques or the addition of complex accessories to a microscope. Analytical information is obtained by measuring a characteristic that is observed. Determining the color and layer structure (number of layers and their order) of a paint chip sample obtained by simple viewing of the object is an example. Valuable additional information can be added to the basic data, if the information is further qualified by comparison to a standard—for example, detailing of the color of a soil by comparing it to a collection of color standards.

Using a specialized scale (a micrometer) allows for measurements to be made, such as length, width, and diameter. This method is known as **micrometry**. The thicknesses of the layers of a paint chip and the modification ratio of a synthetic fiber determined by viewing a cross section are examples of quantitative information that can be obtained. In some cases, microchemical tests that aid in the identification of a material can be performed on micro- and ultra-micro-sized samples. For example, if a grain of white powder is thought to contain calcium carbonate, adding a tiny drop of acid can help make that determination; if fizzing is observed, that is consistent with carbonate.

17.2.2 POLARIZING LIGHT MICROSCOPY

When a compound microscope is fitted with certain accessories, it is converted to a polarized light microscope, one of the most powerful analytical tools available to forensic science. The basic requirements are that two polarizing elements are positioned in the optical path of the microscope. The first, called the **polarizer**, is placed prior to the sample, normally in the condenser mount just prior to the lenses. The second, called the **analyzer**, is positioned in the body of the microscope, usually in an intermediate accessory tube between the objectives and the viewing head that holds the EPs.

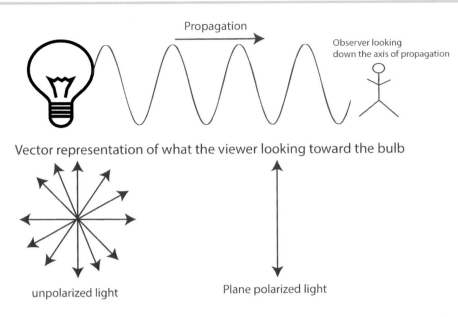

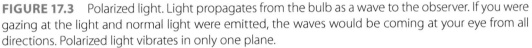

FIGURE 17.3   Polarized light. Light propagates from the bulb as a wave to the observer. If you were gazing at the light and normal light were emitted, the waves would be coming at your eye from all directions. Polarized light vibrates in only one plane.

Light is a wave phenomenon. Its characteristics are velocity (c; the speed of light), wavelength (λ), and frequency (υ; how many waves pass in a second) related to color; the amplitude (a; height of the wave) related to brightness; and vibration direction, which is always perpendicular to the direction of propagation (travel). Wavelength and frequency are inversely related: the longer the wavelength, the lower the frequency of vibration. Shorter-wavelength light, violet light, and ultraviolet light have higher energy than longer-wavelength red light and infrared light. Normal light is randomly polarized (lower frame of Figure 17.3). If the vibration is restricted to only one direction, it is referred to as **plane polarized light**. This is how polarized sunglasses cut glare—only the light traveling in one plane is allowed through the lenses to your eyes, and the others that cause the glare are filtered out. Polarization does not have anything to do with color, only the angle of the direction of the approaching light wave.

Light can become partially or totally polarized by reflection, adsorption, and propagation through an **anisotropic** material (a material that is not the same in all directions). The earliest devices employed to generate plane polarized light were obtained by cutting and polishing anisotropic materials along certain directions within a crystal and cementing them together in a certain orientation so that light transmitted along the optic axis of the microscope was plane polarized.

Today, plane polarized light is obtained using polymer films in which the molecules are very highly oriented and have been treated with a dye so that they almost totally absorb light vibrating in all but one direction. This single direction is called the **privileged direction**. When two polarizers are placed in such a way that light passes through one and then the second and the privileged directions of each are perpendicular, no light will emerge from the second. This condition is referred to as **crossed polars** and results in complete extinction of transmitted illumination. If there is no sample placed on the microscope stage and crossed polars are set, the field of view is completely dark; no light reaches the eye (Figure 17.4). The same is true if the light moves through a (same in all directions) material.

When an object is placed in the illumination path of a polarizing light microscope and between the two polarizers, it may affect the vibration direction of the plane polarized light reaching it from the first polarizing element. If this is the case, the material is anisotropic, and it will resolve the original vibration's intensity into two perpendicular vibration directions. Each of these resulting rays, except in certain special directions of propagation, will have a different **refractive index** (RI), and the difference of these indices is referred to as the **birefringence** (ΔRI). The maximum birefringence is an analytical and identifying characteristic of a material. Both rays travel through the material at different velocities, and this results in a phase shift of the rays when they

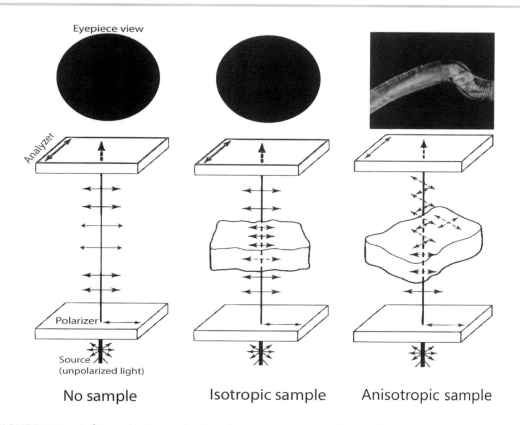

Eyepiece view

No sample   Isotropic sample   Anisotropic sample

FIGURE 17.4 Left to right: Crossed polars, light propagating in the privileged direction, no sample. Middle: Isotropic sample, anisotropic sample. Notice that the background is black; only where the sample interacts with the light is there color.

emerge from the material. When the thickness is in micrometers and the resultant difference is multiplied by 1000, the retardation (R) of the sample measured in nanometers is obtained:

$$R\left(nm\right) = \Delta RI \times Thickness\left(\mu m\right) \times 1000$$

When the rays with the two separate vibration directions pass through the second polarizer (analyzer), they are both resolved into rays vibrating in the same direction and are then able to interfere with each other and can produce **interference colors** (Figure 17.5). The amount of interference and the resultant intensity depend on the phase difference between the rays. If the illumination is white light, the various wavelengths interfere in different amounts; certain colors are intensified, and others are decreased or even eliminated due to destructive interference. The result is an interference color associated with sample retardation (Figure 17.5). In Figure 17.5, the image of the cotton fiber at the right side of the figure shows interference colors. Under normal lighting conditions, this is a simple white cotton fiber with no color; under crossed polars, interference colors are seen, but the fiber itself has not become colored. Another familiar example of interference colors is colors that you see when light reflects off a thin layer of oil or the surface of a DVD. The colors are the result of the way light waves interact, not from the material itself.

An analytical working tool referred to as the **Michel–Lévy chart** (Figure 17.6) relates the birefringence, thickness, and retardation properties. If the microscopist directly measures any two components, the third can be easily determined. PLM is particularly valuable for the characterization of synthetic fibers, such as cotton, rayon, and nylon. The polarizing light microscope is the instrument of choice to characterize many forms of microscopic materials, especially because analytical measurements can usually be made nondestructively. These measurements can lead to unambiguous identifications and can aid significantly in the goals of association and individualization; however, PLM is not the only microscopic tool available for trace evidence analysis, and often it is the combination of techniques that is critical for the characterization of evidence. A few of these other techniques are described below.

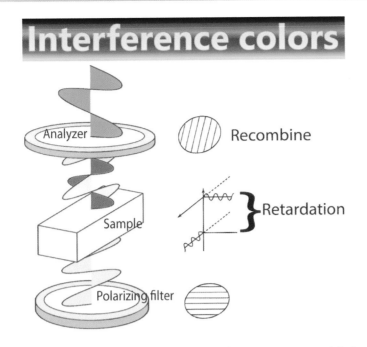

FIGURE 17.5 When plane polarized light passes through anisotropic materials, interference colors can be seen.

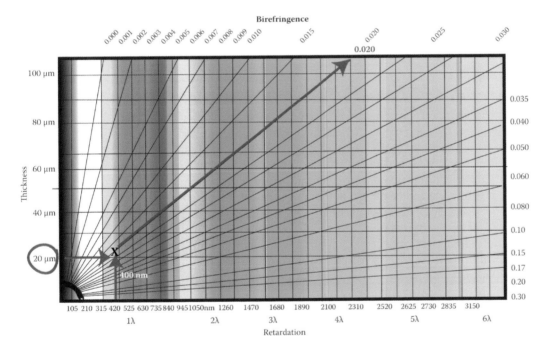

FIGURE 17.6 A Michel–Lévy chart. In the example, the sample is 20 μm thick with yellow-orange interference colors observed, resulting in a birefringence of 0.020.

17.2.3 COMPARISON MICROSCOPY

Comparison microscopes, previously described in Chapter 15, vary in their design and application to the analysis of evidence, but are similar in design principle. They are two microscopes linked by an **optical bridge** so that the observer can simultaneously view two independent images in one field, each from a separate objective (Figure 17.7). The optical bridge often has a mechanical screen to provide a split field of view with a variable point of demarcation. These

FIGURE 17.7 Left: A comparison microscope with two devices connected with an optical bridge. Right: The image that would appear on the screen.

bridges also allow superimposition of the two images. These are like dual stereo microscopes, but they lack the three-dimensional imaging common with dual stereo devices. Large tool marks and fabrics are often examined with this instrument using reflected light. Firearms and tool mark examiners (Chapter 15) make extensive use of these types of comparison microscopes. The classical, transmission illumination, bright-field microscopes, and even polarizing light microscopes, are often linked with a bridge so that very small samples, such as hairs and fibers, can be critically examined side by side in the field of view.

17.3 MICROSCOPES COMBINED WITH SPECTROSCOPY

Microspectrophotometry (MSP) is an area of microscopy that over the past 25 years has become widely used in trace evidence analysis. The principal types are visible and infrared microspectrophotometers, and each requires a different instrument design because of the nature of the radiation employed to characterize the exhibit. Visible microspectrophotometers (colorimeters) lend themselves well to the accurate measurement of color by eliminating the subjectivity that is inevitable when a human observes and describes color. This person-to-person variation arises from individual variations in color perception. These instruments generate transmission, reflection, or absorption spectra from various translucent and opaque samples, and the spectrum is an objective quantitative characterization of color that avoids human subjectivity. Examples of the most common applications are spectra obtained from colored fibers and paint surfaces. An example of microspectrometry applied to two different colored fibers is shown in Figure 17.8. The left side of the scale is the blue end of the range, while the right side is the red end. The red fiber reflects red and absorbs blue, while the blue fiber does the opposite.

Another example of microspectrometry that is becoming more common in trace evidence analysis is Raman microspectrometry. Raman spectroscopy is based on how light from a source is scattered by the electron cloud of a molecule. The photons that are scattered will have either less energy than the original source photons or more energy than the original. These shifts (Raman shifts) are directly related to the vibrational energy levels of the bonds of the molecules, and therefore the spectra obtained contain similar, if not the same, information as the infrared spectra. Modern Raman instruments can be handheld or attached to a microscope in the laboratory and used the same way as the example shown in Figure 17.8.

Figure 17.9 shows an interesting application of a portable Raman system to differentiate hair dyes. A hair dyed with different blue dyes might look the same to our eyes and even have similar visible spectra (Figure 17.8), but here the Raman spectra are noticeably different.

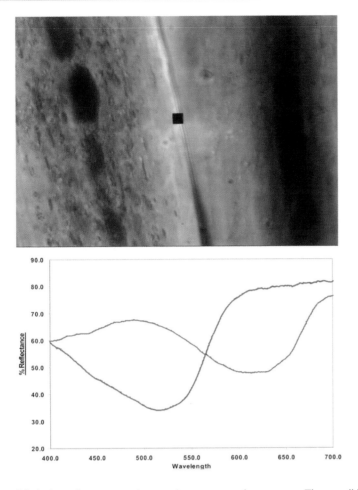

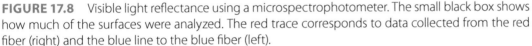

FIGURE 17.8   Visible light reflectance using a microspectrophotometer. The small black box shows how much of the surfaces were analyzed. The red trace corresponds to data collected from the red fiber (right) and the blue line to the blue fiber (left).

17.4 SCANNING ELECTRON MICROSCOPY

The **scanning electron microscope** is a powerful addition to a forensic laboratory that permits the viewing of samples at much greater magnification and resolution than is possible by light microscopes. The image of the hair shown in Figure 17.9 was collected using a scanning electron microscope. Magnification is possible in the range of 10–100,000 times. In forensic labs, the lower magnifications are of more importance with few samples requiring more than 5000× magnification. Very rarely is magnification above 25,000× needed. When the scanning electron microscope is combined with an **energy-dispersive x-ray spectrometer**, the technique becomes even more powerful. The SEM/EDS combination can readily resolve a particle or structure smaller than 1 μm in size, while generating spectra revealing the elemental composition of the object.

The principle of operation is that an electron beam is accelerated by a high potential difference, usually 10,000–30,000 electron volts. This beam is then focused using electromagnetic lenses to a small beam spot and swept over the sample (Figure 17.10). The beam causes many interactions slightly below and at the surface of the sample. **Backscattered electrons** (BSEs) and **secondary electrons** (SEs) are emitted from the surface and converted to an electrical signal by an appropriate detector. The position of the sweeping beam is coordinated with the sweep of a cathode ray tube observation screen, and the intensity of the signal from the detector is converted to brightness. SEM images are not colored. The screen size is fixed, and the analyst uses the controls to vary the size of the portion of the sample scanned. The relationship of this scanned area to the viewing screen is the magnification of the microscope.

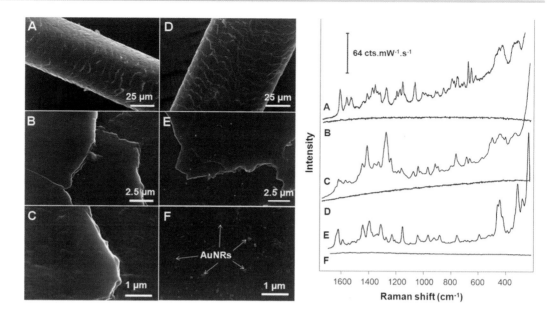

FIGURE 17.9   The use of a small Raman device to differentiate hair dyes. On the left are microscopic images of hair: (A–C) undyed and (D–F) dyed. Tiny nanoparticles used in the analysis are shown in F. To the right are spectra of three different blue hair dyes. (Reproduced from Kurouski, D., and Van Duyne, R. P., In Situ Detection and Identification of Hair Dyes Using Surface-Enhanced Raman Spectroscopy (SERS), *Analytical Chemistry* 87, no. 5 [Mar 2015]: 2901–2906. With permission from the American Chemical Society 2015.)

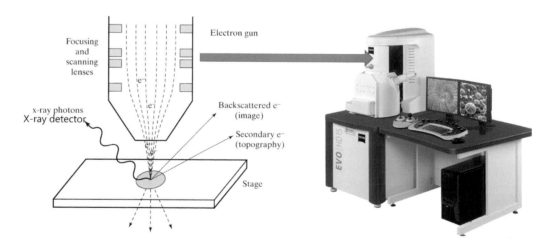

FIGURE 17.10   SEM interactions and instrumentation. (Image courtesy of Zeiss from Flickr, https://www.flickr.com/photos/zeissmicro/6908552901.)

This electron beam causes many other interactions with the sample, two of which are described here. The first is the emission of x-rays that are characteristic of the element that the electron collides with in the sample. A typical type of x-ray detector is an energy-dispersive design, or energy-dispersive x-ray spectroscope. SEM/EDS is used for paints and other types of trace evidence; perhaps the most common forensic use is for the detection of **gunshot residue** (GSR), which was introduced in Chapter 15 and will be described in more detail in a later section.

17.5 EXAMPLES OF TYPES OF TRACE EVIDENCE

17.5.1 GLASS

Glass, a common type of microscopic evidence, is a reasonably hard, transparent, or translucent material composed of fused inorganic materials. Upon cooling, it is **amorphous** in nature and, as noted previously, isotropic. Glass is found in many shapes, sizes, colors, and types. Its uses range from containers to optical devices. It has a wide variety of chemical compositions, by both design and happenstance. Variation of its elemental formulas can significantly alter its characteristics, and therefore often its ultimate uses. For example, glass with a high boron content is resistant to thermal shock and is employed in laboratory glassware and cookware. Inexpensive soda lime glass is usually high in sodium and calcium content and is found as containers, windows, and many other products. The addition of high-atomic-number elements increases the RI of glass, causing it to sparkle and serve decorative and aesthetic purposes. Because glass has so many uses, possesses different qualities, breaks easily, and ejects very small fragments in different directions that are retained by garments, it is frequently encountered as transfer evidence. Whether flat, container, decorative, optical, or other glass, varying its composition allows it to be discriminated by physical, optical, and elemental characteristics.

A first examination of glass should be directed to physical properties that can be evaluated macroscopically or with a stereo microscope. Examination of a broken window can reveal whether the impact that caused the fracture was a low-velocity blunt trauma or a high-velocity point trauma. Figure 17.11 shows breaks in sheet glass caused by multiple impacts. One can observe **radial cracks**, originating from the impact point and propagating away, and **concentric cracks**, which seem to make a circle around the impact point. By noting that some cracks terminate at their intersections with others, one can conclude that terminated cracks were caused by a later impact. Figure 17.12 displays a cross section of a flat glass impacted by a high-velocity projectile. If enough of the impact point is intact, one will note evidence of the core ejected from the far side of the glass upon impact.

Another fact that can often be ascertained from examination of such fractures is the side from which the force that caused the fracture was applied. When glass fractures, the edges often show characteristics referred to as **conchoidal lines**. When the fracture examined is located prior to any concentric crack and is not too far from the point of impact, then the surface opposite that part of the mark that appears to contact the surface at or near a right angle is the side from which the impact force originated. The acute marks point back toward the propagation point of the crack.

Even if tiny bits of glass are recovered, the optical properties of small fragments can be successfully evaluated, and these characteristics can be reasonably discriminating. Small fragments of glass can be removed from larger pieces and the optical properties measured. The most significant property is the RI. The RI values of small glass chips cannot be measured directly but may be measured indirectly by one of the immersion methods based on submersion of a sample in a liquid, usually referred to as oil, even if it is not organic in origin. Employing one of the **immersion methods** allows an analyst to determine when the RI of the sample matches that of the liquid

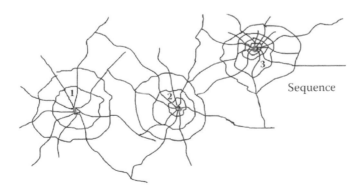

Sequence

FIGURE 17.11   Glass showing how cracks can be used to determine the order of impact. The numbers correspond to the order in which the impacts occurred. All the cracks radiating out from "1" are uninterrupted while those from impact 2 and 3 stop where they intersect cracks already created.

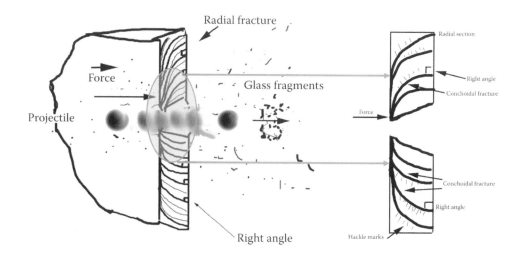

FIGURE 17.12 Coring effect fractures and close-up. Features such as these can be used to determine the direction of impact.

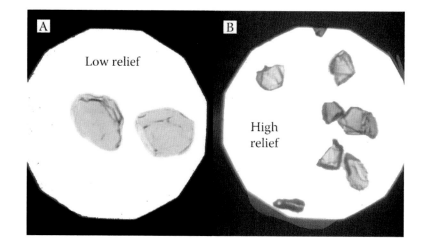

FIGURE 17.13 Contrast and media. (A) RI of media close to that of the sample, low contrast. (B) High contrast means the RI of the media is not close to that of the sample.

medium at a given λ, and then he or she measures the liquid or reports the predetermined RI of that liquid. When the RI of the sample is near or matches the RI on the medium, the contrast of the sample will be low and it will be difficult to see in the liquid.

When the RI differs by an appreciable amount, the contrast will be significant and the sample will be easy to observe. See Figure 17.13 for examples of high and low contrast. At first impression, one might not want to attempt this measurement as it seems to require good luck or extended effort to find the matching liquid by trial and error. However, a number of methods can determine whether the liquid medium or the solid sample has the higher RI, thereby giving direction for the next choice of liquid. Oblique illumination, dispersion staining color, and the movement of the **Becke line** are the most common techniques. To use the Becke line method, a microscope is focused on the sample and the focus is then raised. The distance between the sample and the microscope objective is increased. When this is done, a halo or brightness near the edge of the sample, the Becke line, will move into the material of greater RI, whether it is the sample or the mounting medium. Figure 17.14 shows examples of the appearance of a Becke line in and out of a sample.

An automated method for determining the match point of a glass chip employing commercial instrumentation has been available for more than a decade and is called the **glass refractive index measurement** (GRIM). A GRIM instrument employs computer control of the heating stage and a video detector to determine the match point (i.e., the point at which the glass is no longer visible). Along with automating the calculation, the main advantage is removal of the operator's subjectivity in determining the match point.

PACKING TAPE

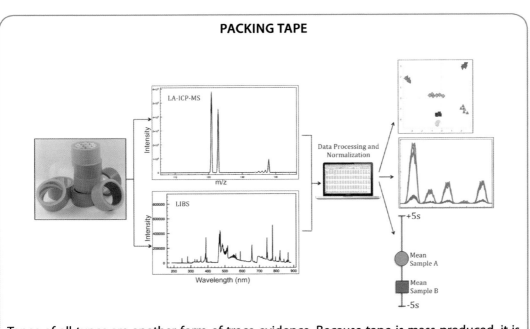

Tapes of all types are another form of trace evidence. Because tape is mass produced, it is challenging to differentiate tapes from one another and successive classification using physical methods (color, width, thickness, etc.) is of limited value. To add to the evidentiary value of items such as tape, chemical instrumentation is increasingly being used to provide better classification capabilities that can increase the value of tape evidence. In this study, researchers utilized a method in which a small tape sample is bombarded with a laser, which causes elements within to be ejected into a gas phase that is swept into a specialized mass spectrometer (collectively referred to as LA-ICP-MS). This instrument allows for the identification of elements and the estimation of relative amounts of each present in the sample.

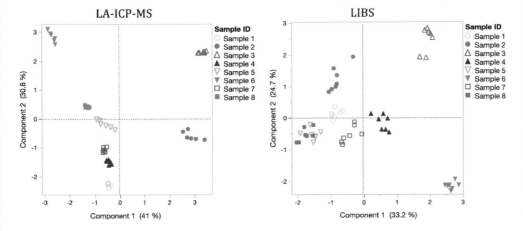

Using replicates and statistical analysis, more categories can be discerned than a simple analysis provides. The figure above shows a plot of different samples based on statistical analysis and combinations of variables; groups become evident. In this case, the clusters of samples are grouped based on similarity of chemical composition. The instrumental methods of analysis are joining microscopy as the workhorse methods of trace evidence analysis.

Source: Martinez-Lopez, C., et al., Elemental Analysis of Packaging Tapes by LA-ICP-MS and Libs, *Forensic Chemistry* 8 (May 2018): 40–48. Figures reproduced with permission. Copyright Elsevier 2018.

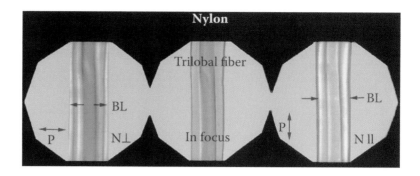

FIGURE 17.14   Example of Becke line (BL) method.

17.5.2 FIBERS

Fibers constitute another common class of microscopic transfer evidence. Natural fibers are those that have not been greatly altered in physical composition or characteristics by processing. Coloring and treatments that improve merchandising or performance do not change the classification. Manufactured fibers are produced from fiber-forming substances that can be synthetic polymers, transformed or modified natural materials, and glass. Processing is necessary to form the fiber. Synthetic fibers are manufactured from synthesized chemical compounds, such as nylon, that are then formed into fibers.

Although fiber evidence can be examined by instrumental techniques, the use of microscopy for initial examination and collection of the first analytical data is the accepted forensic procedure. Several of the microscopes mentioned in the first portion of this chapter are routinely applied to fiber analysis. PLM techniques are of particular value.

Examination of the fiber is carried out under crossed polars and also with plane polarized light to first determine if the fiber is isotropic or anisotropic. Almost all isotropic fibers are made from glass (i.e., fiber glass). A few synthetic fibers appear to be isotropic because their birefringence (ΔRI) is so small and the fibers are not very thick. For anisotropic fibers, the retardation is also determined; depending on the sample, there are many other characteristics that can be determined using PLM.

An analyst who is most interested in the fine surface structure and requires a higher-resolution image of a fiber may decide that the advantages of SEM observation are warranted. Along with the advantages mentioned earlier, SEM observation also allows a sample to be viewed with a greatly increased depth of field. This can be very useful when examination of a piece of textile is the task or a more in-depth study of fiber morphology is desired.

The abilities of SEM/EDS that allow the determination of elemental composition can also assist the criminalist in identification and comparison. As examples, the presence of chlorine in a preliminarily identified acrylic fiber indicates that the fiber is a modified acrylic. The detection of appreciable amounts of titanium dioxide in a fiber is indicative of the mineral being added to the fiber to act as a **delusterant**. When tin and bromine are found in a fiber in which they are not part of the expected chemical formulation, this indicates that a fire retardant may be present. When fibers are colored by incorporation of inorganic pigments in a polymer prior to extrusion, the elemental profile determined by x-ray spectra can aid in identification. Finally, MSP of fibers, particularly in the infrared region, is often performed to identify fibers and dyes.

17.5.3 PAINT

Paint samples are a major portion of microsamples submitted to crime laboratories. This class of transfer evidence can play a key role in investigations and possible prosecutions. Paint submissions usually involve vehicular accidents where contact between two objects is sought to be established or investigative information, such as make, model, and color of a vehicle involved in a hit-and-run, is desired. Less often, paint from an architectural source is submitted. These exhibits are usually related to investigations of crimes against property. On some occasions, samples of an artistic nature are submitted.

Paint includes a range of materials, from thin, translucent stains to heavy, opaque films. The **vehicle** in paint is the binder that holds all the components together and is usually of polymeric nature, consisting of natural or synthetic resins. The binder can form a surface film in many ways. When the film forms by the simple evaporation of the solvent system of the liquid, the paint is normally classified as **lacquer**. A characteristic of lacquers is the fact that they dissolve when subjected to many organic solvents. Pigments supply paint with color, hue, and saturation. Pigments may be organic or inorganic. Blues and greens are predominantly organic, whereas whites, yellows, and reds are inorganic. This is not a strict rule, and crossovers and mixtures are common in modern formulations.

On occasion, it may be possible to obtain a physical match between a paint chip and the source it was dislodged from. This is the strongest association that can be established. A physical "jigsaw" fit of edges or a match of surface markings on the questioned and known samples is convincing evidence of a common source; we discussed these methods briefly in Chapter 15 in regard to tool marks. These examinations are generally conducted macroscopically, using an illuminated desk magnifier, a stereo microscope at its lower range of magnification, and reflected light illumination at various incident angles.

If a physical match is not attained, the layer structure order, color, thickness, and other details are studied. Angle cuts and thin sectioning with a clean (new) scalpel blade can clearly reveal the layer structure. It may be necessary to embed the sample in a resin to obtain high-quality thin sections. Embedded samples make it possible to grind and polish a sample so that fine physical details, such as pigment size and distribution, can be evaluated by higher-resolution microscopes.

Infrared MSP is routinely employed for paint analysis. Transmission measurements can be obtained on thinly sliced or rolled samples and by compressing the paint in a diamond cell. Attenuated total reflection objectives allow for the collection of spectra from the surface of a paint sample without the need to prepare a thin specimen. The recent introduction of commercially available Raman microspectrometers added to the information that can be obtained from a paint sample. The analysis of paint by SEM/EDS is reasonably straightforward after sample preparation is complete.

17.5.4 SOILS

Soils are complex mixtures of materials of mineral, animal, and vegetable origin at various levels of change and decay. Many of the components are common. Some have been deposited by natural forces, while others have been delivered through the intervention of humans. The great variation of these combinations leads some to believe that soil has a unique composition in any given area and changes detectably every few feet. Light microscopes, particularly polarizing light microscopes, lend themselves well to the investigation of forensic soil samples. Many other techniques, some instrumental, are also applicable but will not be covered in this section. Physical characteristics, such as color, pH, and particle size, can be relevant. As a very basic introduction to this topic, one could consider the pollen content and mineral assemblages present in an exhibit. **Pollens** can be readily identified by their morphology using light microscopy or SEM. Although pollens are small and can be windblown, any reasonable concentration of a certain type can be a strong indicator of a location that becomes more specific when a number of pollens are identified.

The identification and quantitative estimation of the mineral content of soils have long been accepted as indicators of location. In these analyses, the more common minerals referred to as the light fraction are considered much less important than those of greater density. Separations are first conducted, and then the minerals are identified by colors, shapes, and optical properties, such as RI values and birefringence. When all the microscopically obtained data and those developed by other methods are considered, it is possible for a trained examiner to supply valuable investigative and probative information concerning soils.

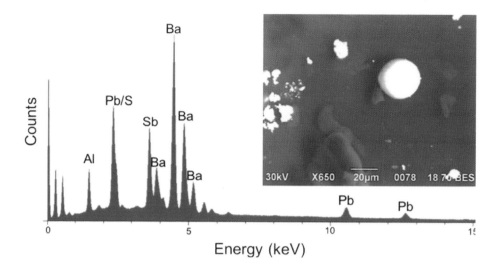

FIGURE 17.15   Examples of a GSR particle and the x-ray spectrum obtained. Note the presence of barium (Ba), antimony (Sb), and lead (Pb). (Reproduced from Trejos, T., et al., Fast Identification of Inorganic and Organic Gunshot Residues by LIBS and Electrochemical Methods, *Forensic Chemistry* 8 (May 2018): 146–156. With permission from Elsevier.)

17.5.5 GUNSHOT RESIDUE

We discussed guns, firearms, and how GSR is created in Chapter 15, but here we focus on how it is analyzed as trace evidence. GSR is currently defined as the inorganic residue produced from the primer of the ammunition, and there are standard methods of analysis using SEM/EDS that have been published and are widely used.

Whatever surface is to be tested is sampled using a small metal stub with an adhesive coating on the top. The surface can be someone's hand, clothing, or other substrates. GSR particulates range from about 0.5 to 5 μm, with some as large as 10 μm. The particles are located by viewing in a scanning electron microscope; the brighter the spot, the higher the mass of the atoms. This allows the operator to either manually ignore duller particles or set a brightness threshold for further analysis by the computer automated instruments.

The brighter particles are analyzed for their elemental content by EDS, and those with compositions, especially if spherical, are analyzed using the energy-dispersive x-ray spectroscope. Particles containing barium, antimony, and lead or those containing barium and antimony are considered characteristic of GSR. Other combinations, such as lead and antimony, are indicative of GSR. Figure 17.15 shows several GSR particles and an example of an x-ray spectrum of GSR.

17.6 RETURN TO THE SCENE OF THE CRIME

The crime scene scenario presented in Chapter 3 provides numerous opportunities for trace evidence collection and evaluation. The scenario also illustrates one of the challenges of trace evidence—sorting out the relevant from the irrelevant. Locard's principle applies to all types of contact, not just criminal. The evidence of past innocent deposition can easily be confused with evidence deposited because of a criminal act. Collecting every fiber from the body or from the truck would create a mountain of evidence, of which most will be irrelevant. Thus, the crime scene investigators call on their training and judgment to assist in sorting out what is critical and what is not. That is never an easy task, and this is one reason that thorough documentation of scenes is so critical. Examples of evidence that appear to be crucial in this scenario include the soil in the tire tread of the truck, the hair on the upholstery, and materials recovered from the victim's clothing and the tarp. There is likely limited value in trace evidence from the house given that the home was shared by the two people involved in the crime. However, if the wife rarely rode in or drove the husband's truck, finding evidence of her being in the truck could be useful information in the larger investigation.

17.7 REVIEW MATERIAL: KEY CONCEPTS AND QUESTIONS

17.7.1 KEY TERMS AND CONCEPTS

Amorphous
Analyzer
Anisotropic
Backscattered electrons
Becke line
Birefringence
Concentric cracks
Conchoidal lines
Delusterant
Exclusionary evidence
Gunshot residue
Immersion method
Inclusionary evidence
Interference colors
Lacquer
Michel–Lévy chart
Microanalysis
Micrometry
Microspectrophotometry
Optical bridge
Plane polarized light
Polarizer
Primary transfer
Privileged direction
Radial cracks
Refractive index
Retardation
Secondary electrons
Secondary transfer
Stereo binocular microscope
Total magnification
Vehicle

17.7.2 REVIEW QUESTIONS

1. What characteristic separates microscopic evidence from other evidence?
2. What instrument is employed for the collection and first evaluation of small evidence?
3. How is the TM of a microscope determined?
4. Explain plane polarized light and retardation.
5. What determinations about a glass fracture can be made by macroscopic examination?
6. What is the value of visible MSP for fiber comparisons?
7. What information about a paint sample can be obtained by use of infrared MSP?
8. What data is obtained from a paint sample by use of SEM/EDS?
9. What fraction or type of mineral is of most value for soil comparison?
10. What elements, when found in a spherical particle, are considered necessary to conclude that the particle is characteristic for GSR?

1. Review the material in Chapter 15 regarding how firearms work. Based on that, can you explain why GSR particulates are always roughly spherical?
2. Revisit the crime scene scenario from Chapter 3. Suppose investigators learn that the wife rarely rode in the truck. Fifteen hairs linked to the victim through DNA analysis are found in the bed of the truck and one is found in the cab on the driver's seat. Latent prints of the husband are found all over the truck, but none of the wife. What is your hypothesis to explain these findings? Does your hypothesis support one version of events over the other? What data or information would help prove or disprove your idea?

BIBLIOGRAPHY AND FURTHER READING

BOOKS

Bell, S., and Morris, K. B. *An Introduction to Microscopy*. Boca Raton, FL: CRC Press, 2009.

Petraco, N., and Kubic, T. *Microscopy for Criminalists, Chemists, and Conservators*. Boca Raton, FL: CRC Press, 2003.

Wheeler, B., and Wilson, L. J. *Practical Forensic Microscopy*. Hoboken, NJ: John Wiley & Sons, 2008.

ARTICLES

Chang, K. H., Jayaprakash, P. T., Yew, C. H., and Abdullah, A. F. L. Gunshot Residue Analysis and Its Evidential Values: A Review. *Australian Journal of Forensic Sciences* 45, no. 1 (Mar 2013): 3–23.

Kurouski, D., and Van Duyne, R. P. In Situ Detection and Identification of Hair Dyes Using Surface-Enhanced Raman Spectroscopy (SERS). *Analytical Chemistry* 87, no. 5 (Mar 2015): 2901–2906.

Maitre, M., Kirkbride, K. P., Horder, M., Roux, C., and Beavis, A. Current Perspectives in the Interpretation of Gunshot Residues in Forensic Science: A Review. *Forensic Science International* 270 (Jan 2017): 1–11.

Muehlethaler, C., Massonnet, G., and Hicks, T. Evaluation of Infrared Spectra Analyses Using a Likelihood Ratio Approach: A Practical Example of Spray Paint Examination. *Science and Justice* 56, no. 2 (Mar 2016): 61–72.

Stoney, D. A., and Stoney, P. L. Critical Review of Forensic Trace Evidence Analysis and the Need for a New Approach. *Forensic Science International* 251 (Jun 2015): 159–170.

WEBSITES

Link	Description
www.swggsr.org	Working group for GSR
http://www.forensicsciencesimplified.org/trace/	Overview of trace evidence

Questioned Documents

18

18.1 DOCUMENTS IN AN ELECTRONIC SOCIETY

The field of questioned documents is one of the older disciplines in the forensic sciences, dating back to ancient times. Perhaps the most famous case that utilized forensic document examination was the Charles Lindbergh baby kidnapping trial in 1937. In this case, handwriting analysis was used to study the ransom note. Today, questioned document analysis encompasses many other areas in addition to handwriting, to include scanners, copiers, chemical analysis of inks and printer toner cartridges, and even electronic signatures.

Handwriting analysis is a type of pattern evidence, and databases have been created to store example patterns of writing. Counterfeiting of currency and credit cards requires physical and chemical analysis of inks, paper, and plastics. In this way, questioned document analysis shares characteristics of trace evidence in that many different skills and tools are needed depending on the type of physical evidence involved in each case. We are a long way from being a paperless society, but even with increasing volumes of electronic documents and signatures, there is still a need for questioned document analysis, even if a document is not piece of paper. As with other pattern analysis disciplines, handwriting analysis has come under increasing scrutiny to become more quantitative and objective.

Figures 18.1 and 18.2 show an example of how computers and sophisticated algorithms are being used to encode features of writing, and then that data is used to determine the likelihood that handwriting supports a given hypothesis. In this example study, the goal was to estimate the likelihood that a document was written by a person native to Vietnam, as opposed to a person raised in Australia. Features of the writing were evaluated using two different computer algorithms and a large set of baseline data. The algorithms identified key features that were useful in making the differentiation in the writing based on nationality. The models were tested with other samples and were successful in more than 90% of the cases of correctly identifying the writer's nationality.

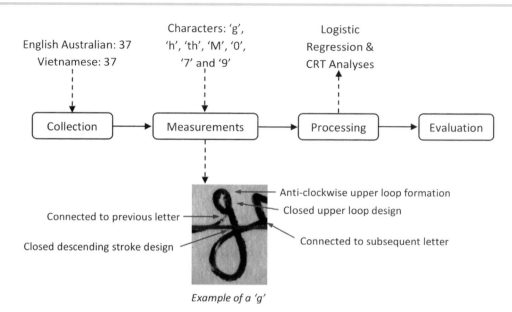

English Australian: 37
Vietnamese: 37

Characters: 'g',
'h', 'th', 'M', '0',
'7' and '9'

Logistic
Regression &
CRT Analyses

Collection → Measurements → Processing → Evaluation

Anti-clockwise upper loop formation
Closed upper loop design
Connected to previous letter
Connected to subsequent letter
Closed descending stroke design

Example of a 'g'

FIGURE 18.1   An example of how features of handwriting can be automatically encoded for computer-based methods of analysis. (Reproduced from Agius, A., et al., Using Handwriting to Infer a Writer's Country of Origin for Forensic Intelligence Purposes, *Forensic Science International* 282 [Jan 2018]: 144–156. With permission from Elsevier.)

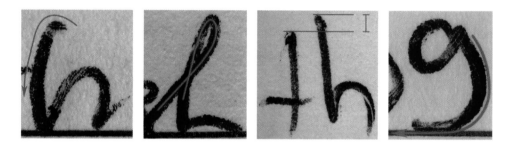

FIGURE 18.2   Critical features of the writing identified by the computer code. (Reproduced from Agius, A., et al., Using Handwriting to Infer a Writer's Country of Origin for Forensic Intelligence Purposes, *Forensic Science International* 282 [Jan 2018]: 144–156. With permission from Elsevier.)

MYTHS OF FORENSIC SCIENCE

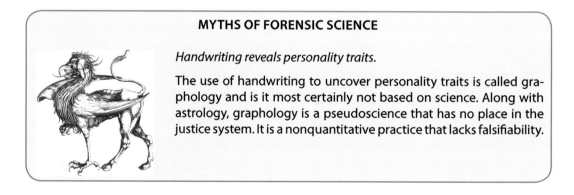

Handwriting reveals personality traits.

The use of handwriting to uncover personality traits is called graphology and is it most certainly not based on science. Along with astrology, graphology is a pseudoscience that has no place in the justice system. It is a nonquantitative practice that lacks falsifiability.

18.2 HANDWRITING COMPARISON

We still sign many documents by hand. When you purchase something with a credit card, you insert the card in a reader and, depending on the type and amount of the purchase, you may "sign" for it using a touchscreen. If a package is delivered to you at your home, you may do the same.

Legal documents still require handwritten signatures in most cases. Thus, whether physical or electronic, handwriting is still an important type of evidence in cases such as fraud or forgery. Handwriting is an acquired skill that becomes ingrained; it is habitual and assumed to be unique to each person. This uniqueness is a basic principle in document examinations.

18.2.1 COLLECTION OF WRITING STANDARDS

Before an examination can be made, known standard writing must be submitted for comparison purpose, and the known writing must be of the same form as the questioned writing. For example, if the questioned material is printed, the known standards must be printed; if the writing is in pencil, the standards should be written in pencil. Two classes of writing standards are utilized for comparison purposes. These are **non-request writing**, also known as spontaneous or undictated writing, and **requested writing**, or dictated **exemplars**. Non-requested writing examples would be old grocery lists, credit card slips, or signed documents that were done in the past and not related to the case at hand. This type of writing is likely to reveal the normal writing habits of the individual.

Requested exemplars are standards written at the request, and usually in the presence, of the investigator or examiner. Their advantage is that they provide writing that is comparable to the questioned material, and authentication is easily accomplished. The disadvantage with requested standards is that given the situation, the person may not write as they would on a typical day, or they may try to conceal or disguise their writing style. Ideally, the examiner has both, but that is not always the case.

18.2.2 PROCESS OF COMPARISON

A document examiner compares questioned handwriting or signatures side by side with the known standards. Handwriting attributes are examined both visually and microscopically. Everyone who looks at writing and signatures notices the most conspicuous features first, such as the slant of the writing and how the letters are formed. An examiner will look beyond the obvious features and study the subtle, inconspicuous aspects of the questioned signature or writing.

To account for variation in a person's writing, an examiner needs an adequate number of writing or signature standards to compare. Writing variation represents the alternative forms of a single handwritten characteristic found in a person's writing. One principle in document examination is that no two individuals write exactly alike, and another principle is that no one person writes the same way twice. Look at two examples of your signature; they will appear similar, but close examination will reveal differences, even if subtle. Thus, when comparing two written samples, the examiner must keep in mind the range and inherent variation in a person's writing style.

For example, notice the way a person writes *th* in the word *the*. Is the *t* higher than the *h*, lower than the *h*, or the same height? Look at your own handwriting and see which you do. An examiner would study all the words that a person writes that contain a *th*. Let us suppose that one individual habitually writes the *t* higher than the *h*. At times, the *t* may be much higher than the *h*, and sometimes just a little bit higher. Any slight change with respect to the height of the *t* would be considered variation within his or her writing, if the basic habit of making the beginning *t* higher than the *h* remains. These kinds of patterns and their detection is an important part of comparing two handwriting samples to try to determine if the same person did or did not write both.

18.2.3 TYPES OF FRAUDULENT WRITING

Types of fraudulent writing include **freehand simulations**, tracings, and normal hand forgeries. A freehand simulation is an attempt to draw the signature or writing of another person (Figures 18.3), usually when working with a model signature. This type of forgery requires that the individual maintain the same speed as the original writing and imitate the correct letter formations, height ratio, and pen pressure at the same time. This is not easy to complete. If his

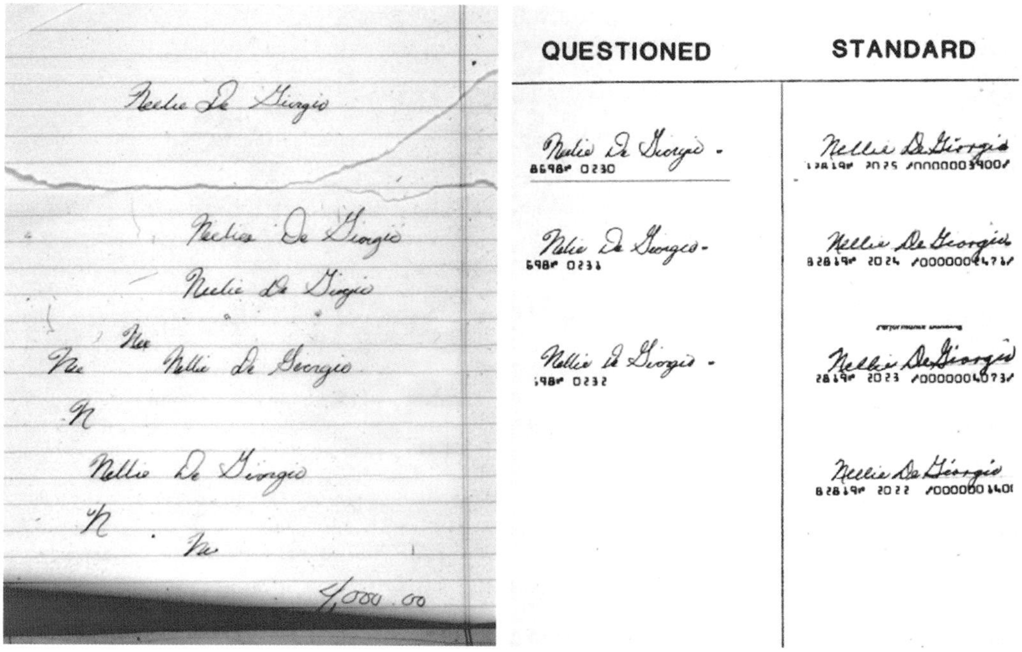

FIGURE 18.3   Writing samples. Left: A practice sheet by the forger. Right: Questioned writing compared with known writing.

or her attention is directed at maintaining the speed of the writing, as in a stylized signature, then less concentration can be focused on replication of the correct letter formations and **connecting strokes**, and the chances are greater that these characteristics will be wrong. If the forger attempts to make the letters and connecting strokes correct, then he or she will pay less attention to the speed of the signature. The line quality will suffer—the signature will display an awkward uneven line that appears hesitant and drawn. Adding to this challenge, the forger must continuously suppress his or her own writing habits so that they will not be revealed in the simulation. Upon closer microscopic examination, evidence may be observed of the forger's patching or retouching of the written line to make it appear closer to the genuine signature. Additional characteristic **pen lifts** may also be revealed.

Tracing, another common type of simulation, involves using an original signature or writing as a guide to produce a fraudulent document. Indications of a tracing include the presence of guidelines around the questioned signature, such as graphite from a pencil or remnants of a line from carbon paper. Indented impressions around the questioned writing may signify that an impression of the original signature was made with a pointed object and then a pen was used to go over the impression. A more direct approach is to take the original signature or writing, place the fraudulent document over it, and trace it. A tracing will normally reveal poor line quality. Instead of the written line appearing smooth and free flowing, it will be uneven and wavy and appear to have been drawn slowly. The final product will not reveal the shading differences in a freely and naturally executed signature that is written with a fair amount of speed.

In a **normal hand forgery**, the individual does not attempt to copy the victim's signature or writing. The forger either writes the name in his or her own writing style or tries to distort it. **Disguised writing** is another form of fraudulent writing. The individual attempts to alter his or her writing to be able to later deny that he or she was the author. Methods of disguise vary from altering the slant of writing to changing the size of writing. Some individuals alternate between uppercase and lowercase letters or printed and cursive forms of letters. Others add additional strokes to letters. The more a person writes, the more difficult it is to suppress habitual writing characteristics. Because the act of writing is an acquired skill that has become habitual, a person will find it challenging to maintain an effective disguise. The smaller, inconspicuous writing characteristics of the individual are revealed the more a person writes. It is difficult to write effectively and concentrate at the same time on altering learned handwriting habits.

Authentic conventional (handwritten) signatures

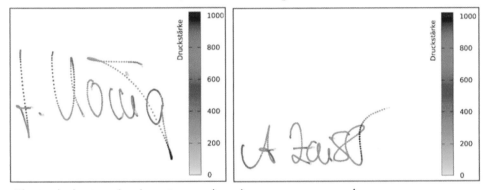

Plotted electronic signatures showing pressure used

FIGURE 18.4 Paper and electronic signatures compared. The pressure used on the screen is shown by the color scale. (Reproduced from Heckeroth, J., and Boywitt, C. D., Examining Authenticity: An Initial Exploration of the Suitability of Handwritten Electronic Signatures, *Forensic Science International* 275 [Jun 2017]: 144–154. With permission from Elsevier.)

The principles applied to manual handwriting are applicable to electronic signatures using pressure-sensitive touchscreens. Using computer-based signature devices such as pads to sign for credit card purchases has another advantage in that additional data can be recorded and utilized. Figure 18.4 shows data from a recent study related to electronic signatures. The top frame shows handwritten signatures and the bottom frame shows the electronic versions. The pressure used by the writer to create the signatures is shown by colors; the lighter the color, the less pressure that was used. Notice also how individual points on the electronic signature are shown. This type of data can be useful, as shown in Figure 18.5, where the top signature is authentic and the bottom is simulated. Look closely at the first blue letter and you can see a red + cursor showing critical points that are plotted and measured via software. The location of these critical points is useful for comparison and can provide quantitative data for establishing differences or similarities between writings.

18.2.4 FACTORS THAT CAN AFFECT HANDWRITING

Many factors can affect handwriting. For example, the health of the writer is a consideration. Various disorders can produce muscular weakness that prevents proper writing control or reduces the ability to hold a writing instrument correctly. Anything that causes loss of motor control or tremors in the hand will impact the handwriting. Age can also be a factor; generally, as we get older, our writing changes. This can complicate comparisons with known exemplars collected years earlier than the questioned writing in a case. Intoxication with drugs or alcohol also can change the appearance of a person's handwriting, with the effects varying from person to person.

USING SOFTWARE TO IDENTIFY FORGERIES

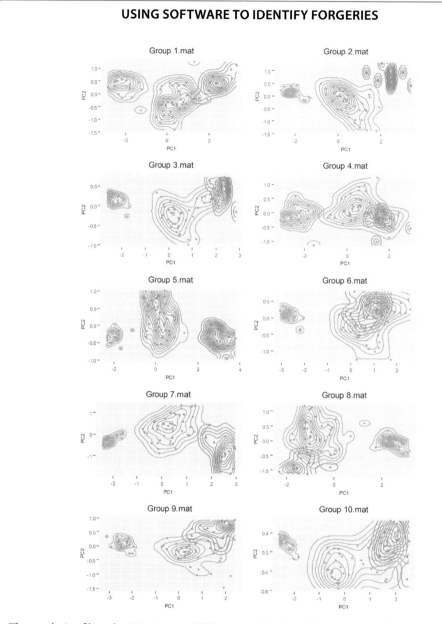

The analysis of handwriting to establish authorship (and thus uncover forgeries) is still a commonly requested forensic analysis. Fundamentally, the analysis involves comparisons of known authentic writing and writing in which authorship is in question. In this study, the authors utilized automatic feature extraction and an advanced computer procedure to see how feasible it was to identify forged signatures from authentic ones. Volunteers were asked to create forgeries of signatures using several methods, including tracing and freehand (called imitations), as well as simply signing a name without an attempt to make a copy (random). The program trained using sets of known authentic and forged samples and was then tested with different signatures using blind testing. The authors extracted features and compared how they were distributed between genuine signatures (red), random forgeries (green), and imitation forgeries (blue). In most cases, there was little or no overlap of features between genuine and forged signatures; however, there was significant overlap within the forged signatures based on the method used.

Source: Chen, X. H., et al., Assessment of Signature Handwriting Evidence via Score-Based Likelihood Ratio Based on Comparative Measurement of Relevant Dynamic Features, *Forensic Science International* 282 (Jan 2018): 101–110. Figure portion reproduced with permission. Copyright Elsevier.

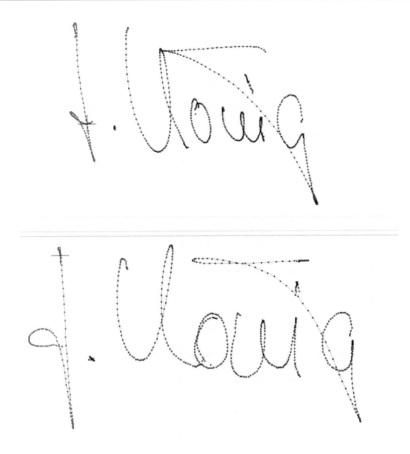

FIGURE 18.5 Signatures plotted in the software package described by the authors. (Reproduced from Heckeroth, J., and Boywitt, C. D., Examining Authenticity: An Initial Exploration of the Suitability of Handwritten Electronic Signatures, *Forensic Science International* 275 [Jun 2017]: 144–154. With permission from Elsevier.)

18.3 ALTERATIONS, OBLITERATIONS, AND INK DIFFERENTIATION

Often, a document examiner discovers and deciphers an alteration to or obliteration of a document or determines an ink differentiation. The obvious and discernible evidence of a second writing instrument may well prove that an addition has been made after the fact. In addition to a naked-eye or low-magnification visual examination of the questioned material, an examination may also incorporate the use of selective color filters or color filter combinations. This could be the method of choice, for instance, in the examination of obscured, bloodstained documents. Viewing by microscope may reveal subtle differences, such as a slight change in shading or hue, within the questioned material. Secondary lines, indicative of alteration when present, are usually discovered during this process. More in-depth examinations of the questioned document using selective portions of the electromagnetic spectrum can be accomplished either photographically or by a process generically known as **video spectral comparison** (VSC).

18.3.1 VIDEO SPECTRAL COMPARISON

As we discussed in Chapter 17, colors as perceived by human vision represent one small region across the electromagnetic spectrum. Trace analysts exploit interactions of evidence with other types of energy, most commonly infrared (IR) and, to a lesser extent, ultraviolet (UV). Questioned document examiners also exploit these regions during the analysis of paper and ink. Although we are not able to see the IR or UV regions of the light spectrum with the naked

FIGURE 18.6 The use of an ALS (lower right) and VSC (top right) to visualize erased writing on a whiteboard. (Reproduced from Toulomelis, W., The Decipherment of Latent White Board Entries, *Journal of Forensic Sciences* 61, no. 4 [Jul 2016]: 1146–1148. With permission from Wiley.)

eye, we can use instrumentation to convert those wavelengths into visible images when examining a questioned document. Different inks that appear alike may react quite differently when viewed under UV light or with the use of IR imaging techniques. This is the same principle that underlies the use of alternative light sources as we discussed in the context of the detection of biological evidence (Chapter 9; see Figure 18.2). Here, instead of looking for biological stains, the goal is to identify different inks and features that are not necessarily visible otherwise.

When examined using IR, an ink can be observed to luminesce or glow, be transparent, or appear unchanged, depending on its chemical properties. The IR examination may use specialized light filters and films for photographic imaging or equipment specifically designed for IR imaging. This process is referred to as video spectral comparison, and the equipment is usually referred to as a **video spectral comparator**. In addition to the comparator being used on ink examinations, this instrument is routinely employed to detect forgeries in passports, visas, identification cards, immigration documents, and so forth. Many security documents utilize luminescent features, such as holograms and watermarks. The comparator can easily check for authenticity. Additional features include an embedded information decoder. This is where personal information is invisible on passports.

VSC can also be useful for visualizing features that have been erased or obliterated. For example, if someone uses a thick permanent marker to obliterate writing in a ball point ink, the original writing often can be seen using VSC. An interesting example is shown in Figure 18.6, where VSC and an alternate light source (ALS) were used to see writing that was erased and not otherwise readable.

18.3.2 THIN-LAYER CHROMATOGRAPHY

Another useful analysis that is occasionally employed is thin-layer chromatography (TLC). To perform TLC, small "punches" of the ink are taken from a portion of the written line that is least likely to play a prominent part in any subsequent handwriting examination. Punches are made by using a blunted hypodermic needle, preferably one with a small diameter, such as those used for insulin injections. These work well and do minimal damage to the document. These punches (some three or four or more) containing portions of the suspect ink are placed into a small test tube, and the ink is separated from the paper portion of the punch by the introduction of a solvent. The resultant ink and solvent solution is spotted onto paper or glass TLC plates (Figure 18.7). The plate is placed in the solvent and allowed to develop. Portions of the ink spot separate into bands of color and migrate upward along with the solvent. Each of these (usually three or four) bands reaches a stopping place on the strip. A comparison of bands created by different suspect ink areas may at times allow for conclusive opinions of difference; however, even if the TLC bands created from different ink spottings appear in the same pattern, a conclusive opinion

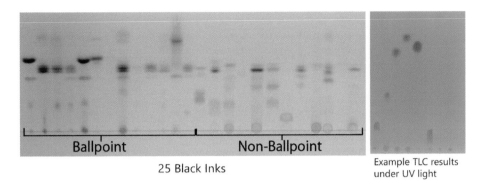

FIGURE 18.7 TLC examples. Left: Twenty-five black inks are visible. Right: An example of a TLC plate viewed under UV light. (Reproduced from Barker, J., et al., The Effect of Solvent Grade on Thin Layer Chromatographic Analysis of Writing Inks, *Forensic Science International* 266 [Sep 2016]: 139–141. With permission.)

that both the questioned inks are from a common source (or writing instrument) is not possible. Each ink manufacturer fills thousands of writing instruments, sometimes for different pen distributors, with ink of the same formulation. These formulations are changed infrequently over the years. The best that can be said is that the inks appear consistent and could have come from a common source.

18.4 INDENTED WRITING

Indented writing, or second-page writing, is the impression from the writing instrument captured on the second sheet of paper below the one that contains the original writing. This most often manifests itself on pads of paper. Indented writing can be a source of identification in anonymous note cases and is an invaluable investigative procedure when medical records are suspected of containing alterations. Often, a writing addition to a record or file can be revealed by an impression that has been transferred to the page below. Indented writing on subsequent pages may not agree with what appears on the surface of the document. Writing found to be out of position, missing, or added after the fact can often be demonstrated by recovering and preserving indented writing from other pages. Mystery novels and television and movie plots sometimes depict the recovery of indented writing as part of a clue. The method they usually show for reading indented writing from suspect pages is to rub a soft lead pencil over the surface of the document, causing the indentations to be highlighted in relief. Although entertaining, this technique is one way to destroy what might be valuable evidence and should serve as a warning against amateur examinations. Indented writing is normally recovered either photographically using oblique (glancing) light or by use of **electrostatic detection apparatus** (ESDA).

18.4.1 OBLIQUE LIGHTING

Until recently, the forensic document examiner applied **oblique lighting** (or glancing light) to the furrows of indented writing. An example of how oblique lighting works is shown in Figure 18.8. When a document with indented writing is illuminated this way, it is often possible to visualize and photograph the results. A combination of multiple exposures taken while moving the light source fills in the available indentations with shadow and effectively reproduces the indented writing. While such techniques are often acceptable, they lack the ability to recover invisible microscopic indentations (those occurring three or four pages down) and have an inherently lengthy processing time.

FIGURE 18.8   Oblique light. On the right, nothing is visible, but when illuminated from the side (an oblique angle), a footwear impression becomes visible.

18.4.2 ELECTROSTATIC DETECTION

The modern well-equipped forensic laboratory employs ESDA to recover indented writing. The technique was introduced in Chapter 16 as applied to shoe wear impressions. With ESDA, indented writing can be recovered three, four, or even more pages below the original writing. A preliminary examination eliminates documents or cases in which the material to be examined is unsuitable for the detection and recovery of indented writing by electrostatic detection. Documents that have been previously processed for latent fingerprints with ninhydrin or that have been saturated with fluids normally fall into this classification. Thick cardboard mediums are usually incompatible with ESDA. The document to be processed may need to be humidified slightly if it has been kept in, or had as its source, an arid environment, such as an interior page from a pad of paper. This helps the electrostatic charge to develop. In more humid climates, several hours of exposure to normal room air serves the same purpose for most documents.

The page suspected of bearing indentations is covered with a cellophane (Mylar®) material, which is then pulled into firm contact with the paper by a vacuum drawn through a porous bronze plate. This fastens the document and cellophane covering to the plate. The cellophane covering prevents damage to the original document. The examiner then subjects the document and cellophane to a repeated high-voltage static charge by waving an electrically charged wand over the surface of the document.

This results in a variably charged surface, with the heavier static charge remaining within any impressions, even those that are microscopic in depth. Using tiny glass beads as the carrier, black toner (like that used in dry-process photocopy machines) is then cascaded over the cellophane surface, or it can be applied by "misting," where the toner is sprayed over the paper within a chamber placed over the questioned document. The toner is strongly attracted to static electricity and is retained on the cellophane surface in accordance with the amount of residual static charge present at any given surface point. The areas of the document containing the higher static electric charge retain greater portions of the black toner, resulting in a deposit of toner aligned with the indentations in the paper.

The developed indentations are preserved by placing an adhesive-backed transparent plastic sheet over the cellophane while it is still being held in place by the vacuum of the ESDA. The advantages of electrostatic detection are twofold. The indentations are revealed on the protective cellophane surface and are fixed by applying pressure-sensitive adhesive plastic over the cellophane. The original document remains unharmed throughout the process. Second, ESDA is

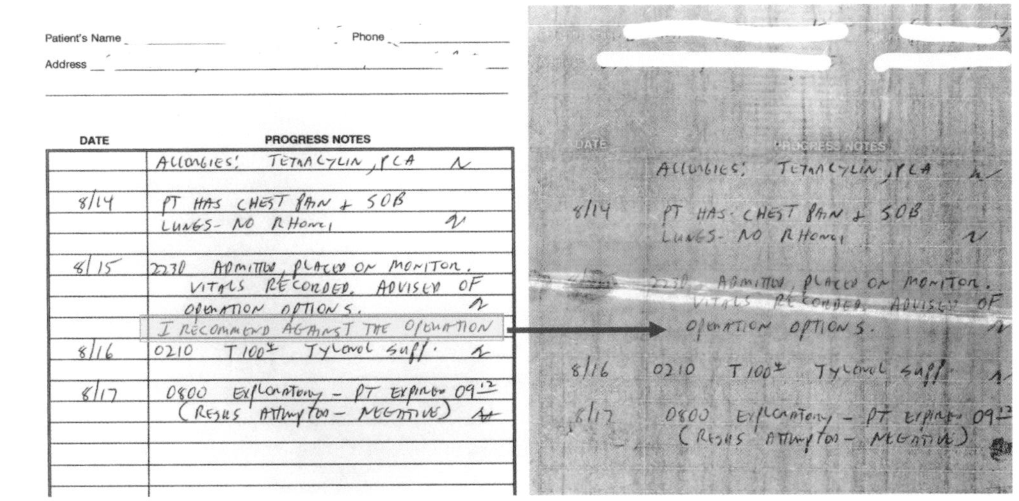

FIGURE 18.9 ESDA applied to medical notes. The original is on the left. On the right is an ESDA image collected from the page that was below the original, showing where the note page was altered; the phrase "operation options" was indented and thus what was originally written.

extremely sensitive, allowing indentations that are not revealed by any other method to be readily observed and recovered. If the recovered indented writing is of a high enough quality, the handwriting in the indentations, used in comparison with standard material, may even serve as a method to associate somebody with the questioned document. An example of ESDA being applied was presented in Chapter 16 (Figure 16.2) in the context of recovering footwear impressions in dust. The same concept applies here. Figure 18.9 illustrates an example in which ESDA was used in a case involving medical notes.

18.5 PHOTOCOPIERS AND LASER PRINTER EXAMINATIONS

18.5.1 HOW COPIERS WORK

Most photocopying machines and even laser printers operate in similar ways. The image of the document to be copied is captured by a camera lens and transferred to a cylindrical drum usually coated with a light-sensitive substance, such as selenium. This drum has been charged with static electricity, which dissipates when exposed to light. The image of the document transferred to the drum is made of light and dark areas that create similar or corresponding areas of more or less static electric charge on the drum. The drum is then bathed with toner, which is in either a dry or wet form depending on the specific machine. Toner has an inherent affinity for static electricity and clings to those areas of the drum in quantities proportional to the electrostatic charge. The toner in turn is transferred to a piece of paper that is then subjected to a fixing process, usually in the form of heat that fuses and attaches the toner to the paper. Paper is pulled through the photocopying machine by **grabbers**. Marks made by the grabbers are transferred to the paper. These may be small depressions at the edge of the paper, areas of toner, or toner-less spots on the finished photocopied document.

A machine may leave the photocopy characteristics that are specific to a particular make and model within a manufacturer's line. Such characteristics can include grabber marks, paper edge depressions, designs incorporated into specialized paper for specific machines, paper type, and toner type.

18.5.2 PHOTOCOPY EXAMINATION

Questioned photocopies can be examined visually for individual characteristic **trash marks** that may be made because of dirt, scratches, and other extraneous marks on the surfaces of the drum, cover, glass plate, or camera lens of a photocopy machine. A comparison of these marks on the

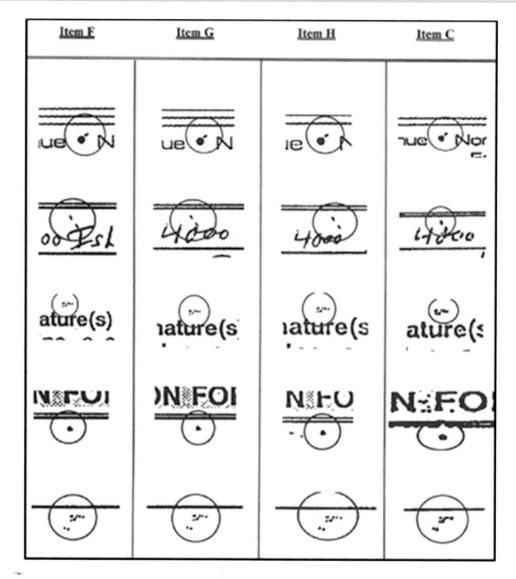

FIGURE 18.10   A demonstration chart showing the same photocopier trash marks (circled), demonstrating that these documents were copied on the same copy machine.

questioned document with those marks made by a specific machine can identify or eliminate that particular machine as the source of the document. Similarly, a side-by-side comparison of two or more questioned photocopies may reveal if they are the product of a common photocopy machine (Figure 18.10). As in photocopier identification, multiple generations of copies and the involvement of more than one photocopy machine severely limit the conclusiveness of the resulting opinions.

18.5.3 PHOTOCOPY FORGERY

When the document examiner examines a photocopy to determine the genuineness of the original signature as represented by the photocopy, the examination must take into account the possibility that a genuine signature was affixed to a fraudulent document and the composite, or paste-up, photocopied. This may result in what would appear to be a photocopy of an original document bearing a genuine signature. The same may be true of any other portion of a photocopy. Photocopies can be prepared from a composite of parts of two or more documents that, when copied, can appear to be a reproduction of a single document. The resultant copy, made from composites, may or may not display characteristics indicative of its production from two or more document sources.

Indications of spuriousness include misaligned typing; different fonts and font sizes; misaligned preprinted matter; incorrect vertical, horizontal, and margin spacing; "shadowing" in the joined areas; disproportionate area sizes; different preprinted material and ink densities; and missing portions of writing or printing (e.g., covered by the paste-up, too closely trimmed, masked by an opaque fluid). The trash marks surrounding the signature may be of greater or lesser quantity than those on the remainder of the document. This is especially true if either the model signature or document to be used in the paste-up was itself a photocopy. The best indication of a possibly fraudulent photocopy is a claim that the original document has "disappeared" or has been "misplaced."

18.6 REVIEW MATERIAL: KEY CONCEPTS AND QUESTIONS

18.6.1 KEY TERMS AND CONCEPTS

Connecting strokes
Disguised writing
Exemplars
Freehand simulations
Grabbers
Indented writing
Nonrequest writing
Normal hand forgery
Oblique lighting
Pen lift
Requested writing
Trash marks
Videospectral comparator

18.6.2 REVIEW QUESTIONS

1. In Figure 18.3, list at least three differences between the top and bottom signatures that could be measured and described numerically.
2. Can you determine the sex and age of the writer from his or her handwriting?
3. What is the difference between requested handwriting standards and nonrequested standards?
4. Explain the meaning of the term *class characteristics* in relation to handwriting.
5. Can a document examiner identify all types of writing?
6. Name one of the methods of ink differentiation for similar-appearing inks.
7. What is a photocopy trash mark and how does it occur?

18.6.3 ADVANCED QUESTIONS AND EXERCISES

1. Why is graphology not falsifiable? Be specific.
2. Studies are starting to appear in which sophisticated learning algorithms are being applied to see if it is possible to determine the sex and age of the writer. Will these be falsifiable?
3. Do some research and find out what forensic linguistics is and how it has been used in casework such as the investigation of the Unabomber.

BIBLIOGRAPHY AND FURTHER READING

BOOKS

Caligiuri, M. P., and Mohammed, L. A. *The Neuroscience of Handwriting*. Boca Raton, FL: CRC Press, 2012.

Kelly, J. S., and Lindblom, B. S., Eds. *Scientific Examination of Questioned Documents*. Boca Raton, FL: CRC Press, 2006.

Koppenhaver, K. M. *Forensic Document Examination: Principles and Practice*. Totowa, NJ: Humana Press, 2010.

Lewis, J. A. *Forensic Document Examination: Fundamentals and Current Trends*. San Diego: Academic Press, 2014.

ARTICLES

Barker, J., Ramotowski, R., and Nwokoye, J. The Effect of Solvent Grade on Thin Layer Chromatographic Analysis of Writing Inks. *Forensic Science International* 266 (Sep 2016): 139–147.

Braz, A., Lopez-Lopez, M., and Garcia-Ruiz, C. Raman Spectroscopy for Forensic Analysis of Inks in Questioned Documents. *Forensic Science International* 232, no. 1–3 (Oct 2013): 206–212.

Calcerrada, M., and Garcia-Ruiz, C. Analysis of Questioned Documents: A Review. *Analytica Chimica Acta* 853 (Jan 2015): 143–166.

De Alcaraz-Fossoul, J., and Roberts, K. A. Forensic Intelligence Applied to Questioned Document Analysis: A Model and Its Application against Organized Crime. *Science and Justice* 57, no. 4 (Jul 2017): 314–320.

Ezcurra, M., Gongora, J. M. G., Maguregui, I., and Alonso, R. Analytical Methods for Dating Modern Writing Instrument Inks on Paper. *Forensic Science International* 197, no. 1–3 (Apr 2010): 1–20.

Johnson, M. E., Vastrick, T. W., Boulanger, M., and Schuetzner, E. Measuring the Frequency Occurrence of Handwriting and Handprinting Characteristics. *Journal of Forensic Sciences* 62, no. 1 (Jan 2017): 142–163.

Li, B. A. An Examination of the Sequence of Intersecting Lines Using Microspectrophotometry. *Journal of Forensic Sciences* 61, no. 3 (May 2016): 809–814.

Muro, C. K., Doty, K. C., Bueno, J., Halamkova, L., and Lednev, I. K. Vibrational Spectroscopy: Recent Developments to Revolutionize Forensic Science. *Analytical Chemistry* 87, no. 1 (Jan 2015): 306–327.

Sun, Q. R., Luo, Y. W., Zhang, Q. H., Yang, X., and Xu, C. How Much Can a Forensic Laboratory Do to Discriminate Questioned Ink Entries? *Journal of Forensic Sciences* 61, no. 4 (Jul 2016): 1116–1121.

WEBSITES

Link	Description
http://www.asqde.org/	American Society of Questioned Document Examiners
http://www.forensicsciencesimplified.org/docs/how.html	A Simplified Guide to Forensic Document Examination

Forensic Engineering

19

19.1 PHYSICS AND FORENSIC SCIENCE

We had a taste of forensic engineering in Chapter 12 in relationship to fire investigation. A good deal of fire investigation, such as determining the cause and understanding the behavior of fire, involves an understanding of physics and engineering. In this chapter, we expand upon these applications and the role engineers play in accident reconstruction, failure analysis, and related topics. At its heart, forensic engineering is all about energy—tracing it, calculating it, and determining what effects energy has on objects, structures, and people. The fundamental underlying relationship is the **first law of thermodynamics**, that energy is conserved. Energy may be converted to several types of energy (heat, kinetic energy, friction, etc.), but it is neither created nor destroyed. Forensic engineers work with these and other fundamental relationships to understand and recreate accidents as well as develop ways to avoid them or minimize destruction when they do occur. The study of collisions, be it between two cars or an airplane with a building, is based on a study of the energy involved and the consequences of this energy being moved and changed. The **second laws of thermodynamics**, which states that when energy is converted, the process is never 100% efficient, is also key to forensic engineering.

In Chapter 15, we discussed firearms evidence. Think about how a gun works from the perspective of energy: There is **chemical energy** (CE) stored in the propellant that fills the cartridge. When the propellant is ignited, the chemical energy is released, creating hot expanding gases that force the bullet forward. This is an example of the conversion of chemical energy (propellant) to **kinetic energy** (KE; movement of the bullet), but not all the chemical energy is converted into kinetic energy. Some is lost as heat (the barrel gets hot), light (muzzle flash), and friction (as the bullet moves down the barrel). All the energy released by burning of the propellant is accounted for, but not all of it goes into moving the bullet. We can write a simple equation to describe this:

$$CE = KE + Other\ ``lost"\ energy$$

The kinetic energy imparted to the bullet is delivered to the target, usually with devastating consequences. This idea can be expanded to collisions between cars in which the kinetic energy of one vehicle combines with the kinetic energy of another. Similarly, we can estimate the energy transferred from a jet liner full of fuel to a stationary building and begin to understand why the building collapsed.

19.2 AUTOMOBILE ACCIDENTS

Vehicular accidents continue to have a major economic impact in the United States. The economic losses due to vehicular accidents were more than $4 billion in 2016, the latest date for which final figures are available. For many years, traffic fatalities decreased; improving safety features (e.g., airbags) and decreases in DUIs are among the factors credited with this trend. However, in 2016 and continuing into 2017, the rate of fatalities on the road increased. In part this could be due to more people driving more miles, but other factors appear to be playing a role, such as distracted driving (use of cell phones, texting, etc.), drowsiness, and less stringent enforcement of seat belt use, speeding, and DUI laws, often due to limits on personnel. Sadly, the death rates for motorcyclists, bicyclists, and pedestrians are also increasing (Figure 19.1).

From the forensic engineering perspective, accident investigations and reconstructions can involve determining the speed of the vehicles at the time of the accident, the relative positions of the vehicles with respect to yielding or right-of-way requirements, adherence by the drivers and pedestrians to traffic signs or controls around the accident, whether the driver was impaired, in some cases who was driving, and if there were factors related to weather or equipment failures.

19.2.1 ANALYTICAL TOOLS USED TO EVALUATE ACCIDENTS

The two most important analytical tools used by engineers in vehicular accident evaluations are the laws of conservation of momentum and conservation of energy. For this brief overview, we focus on conservation of energy in accident investigations. Physical evidence in vehicular accidents usually includes tire and skid marks, a representational plan of the accident area, the location and depth of impact damage on the vehicles and other damaged items observed at the scene, the mechanical condition of the vehicles, the type and location of impact debris noted at the scene, paint transfer marks and other vehicle-to-vehicle and vehicle-to-object contact marks, road conditions, weather conditions, and the physical and mental conditions of the drivers. Other information that could bear upon the causation of the accident includes specific vehicle model

FIGURE 19.1   A typical car accident scene. Notice the damage features and the physical evidence available. Even minor accidents can have a significant economic impact.

FIGURE 19.2   Examples of skid marks.

performance, roadway specifications, type and placement of traffic control devices and signs, and date and time of the accident (Figure 19.2).

Although eyewitness accounts of the accident are important sources of information, they must be scrutinized. Distance, lapsed time, and speed—the key elements of most vehicular accident evaluations—are difficult to judge accurately even for an experienced observer. Further, drivers involved in the accident are notorious for underestimating their own speed and overestimating the speed of the other driver.

AN UNUSUAL ACCIDENT

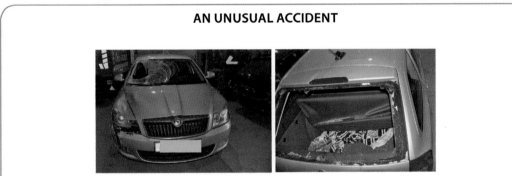

The front and rear of the car. The body was wedged into the trunk area (right). (Reproduced from Nerger, E., et al., Traffic Accident or Dumping? Striking Results of a Traffic Accident Reconstruction, Legal Medicine 24 [Jan 2017]: 63–66. With permission from Elsevier.)

A journal article from 2017 demonstrated the value of accident reconstruction, including computer-based techniques, in solving a puzzling accident. The accident occurred at night when a driver noted an impact on the front of his car. He stopped and looked for an injured animal but did not find one. Intending to put out markers to alert drivers to the accident scene, he went to the back of his car to retrieve signs. He found the rear window of the hatch-back broken and the mangled body of a man stuffed into the trunk area. The man called the police who initially feared that the car was being used to transfer and dump a body. The autopsy showed death was caused by extensive trauma, with injuries consistent with being hit by a car; the question was how the body ended up in the trunk. The accident occurred on dry pavement and skid marks showed that the car stopped in ~210 yards beyond the initial impact point. The low beams of the lights were on at the time of the accident and no mechanical problems were found with the car or tires. Damage was significant in the right front of the vehicle, and the rear hatchback was severely damaged as well. The speed limit on the road was 130 km/h or ~81 miles/h (mph). Reconstruction showed that the driver was traveling at about 120 km/h or ~75 mph. The victim was hit about a second after he stepped onto the road.

Simulation of sequence of events. (Reproduced from Nerger, E., et al., Traffic Accident or Dumping? Striking Results of a Traffic Accident Reconstruction, Legal Medicine 24 [Jan 2017]: 63–66. With permission from Elsevier.)

A computer simulation was used to recreate the sequence of events that led to the body being in the trunk. The simulation was created using data from the scene and estimates of speed. As shown in the figure, the impact launched the man into the air while the car was moving forward but was in the braking process. Due to the unique timing, the body struck the trunk with sufficient force to break the hatchback and wedge into the trunk area. Had the car not braked, the body would have fallen to the road behind the car. It turned out that the victim (age 79) was a resident of a local senior home who had been missing for almost a day and had been previously diagnosed with early dementia.

19.2.2 CONVERTING SCENE DATA INTO AN EVENT SEQUENCE

An engineer or investigator often does not have the opportunity to examine an accident scene before it is cleaned up and the vehicles are towed away. In fact, it may be weeks or months after the event before an engineer is involved. Thus, as for any crime scene, detailed scene documentation is critical, as it is this data that the investigator must rely upon for analysis. The goal of the investigation will depend on the case, but causes and factors related to the accident are the focus, and in some cases, an accident reconstruction is conducted, either physically or using software. One important outcome is a timeline or sequence of events for the accident.

19.2.3 ACCIDENT RECONSTRUCTION: ENERGY

As we discussed earlier, the conservation of energy law states that, in any physical process, the total energy of the system at the beginning of the process is equal to the total energy of the system at the end of the process. The total energy at the end of the process includes any **irreversible work** done during the accident event. In an irreversible process, the change from one energy type to another is a one-way conversion. In our gun example in the first part of this chapter, the conversion of chemical energy in the propellant to friction and heating of the barrel by the moving bullet (kinetic energy) cannot be reversed. **Skidding** is another example of irreversible work. The initial kinetic energy of the vehicle is converted to irreversible work as tires rub against pavement and brake pads rub against brake disks.

A moving car possesses kinetic energy. The amount of kinetic energy it has depends on its mass and how fast it is going; the faster it moves, the more kinetic energy it has. Bringing a vehicle to a stop requires that a vehicle's kinetic energy be reduced to zero, which requires that its kinetic energy be dissipated by converting it to another type of energy. In most accidents, this is done by using the kinetic energy to do some type of irreversible work, such as braking, skidding, or various types of crushing, twisting, or bending.

The kinetic energy of a vehicle is the same as for a bullet, with different assignments of variables:

$$KE = 1/2 \times m \times v^2$$

where:
  KE = Kinetic energy
  m = Mass of the vehicle
  v = Velocity of the vehicle

Because velocity is squared, doubling the velocity increases the kinetic energy by a factor of 4. Accordingly, the braking distance for a vehicle traveling at 60 mph is approximately four times more than the braking distance of the same vehicle traveling at 30 mph. Damage severity and injury severity, like braking distance, also increase in direct proportion to the square of the velocity.

When a vehicle comes to a complete stop in the normal way, its kinetic energy is reduced to zero by applying the vehicle's brakes. The brake pads rub against a disk or drum, and the friction between the two surfaces converts the kinetic energy of the vehicle into irreversible work. Some of the energy is used to abrade away material on the two rubbing surfaces; some of it is converted into heat. As the vehicle's kinetic energy is dissipated, the vehicle slows. When all the energy has been dissipated, the vehicle stops.

In an accident, the initial kinetic energy of a vehicle can be dissipated in many ways. One of the most common ways is skidding. To see how this works, consider the following example of a simple skid: A car skids 100 feet to a stop on dry concrete pavement. Halfway through the skid, the car strikes a pedestrian. How fast was the car going just before the driver applied his brakes and initiated skidding, and how fast was the car going when the pedestrian was struck?

Because the definition of work is force applied through a distance, the irreversible frictional work done in skidding 100 feet on dry concrete pavement is given by the following formula:

$$E_{\text{work}} = (mg) \times f \times d$$

where:
  E_{work} = Work done by skidding
  m = Mass of the vehicle
  g = Acceleration of gravity (32.17 ft/s^2)
  (mg) = Weight of the vehicle and its contents (mass × acceleration of gravity)
  f = Frictional coefficient between the tires and the pavement, which is about 0.75 in this
       case
  d = Distance skidded

If the kinetic energy of the car before skidding is set equal to the energy dissipated by skidding and then we apply algebra to simplify the results, the following is obtained:

$$\text{Energy start} = \text{Energy end} \left(\text{energy is conserved}\right)$$

KE   = E_{work}
$(1/2)mv^2$ = $(mg) \times f \times d$
v = $[2gfd]^{1/2}$

Thus, the initial speed of the vehicle just before skidding began can be computed by measuring the skid mark. The other values in the formula—g for the acceleration of gravity and f for the tire-to-pavement frictional coefficient—are already known. This formula, or variations of it, is widely known as the **skid formula**. Most police officers who have had some formal accident reconstruction training at an academy are familiar with the use of this formula. Examples of how skid marks can provide additional information are given in Figure 19.3.

In our example, where the skid mark was measured to be 100 feet long, the initial speed of the vehicle is calculated to have been 69 ft/s [fps], or approximately 47 mph, just before the skid began. To determine the speed of the vehicle when the pedestrian was struck, a person can simply work backward from where the vehicle stopped to where the pedestrian was struck and then reapply the skid formula. In this case, the car skidded 50 feet after striking the pedestrian. Thus, the speed of the car when it struck the pedestrian is equivalent to a skid of 50 feet. This computes to a speed of 49 fps, or approximately 33 mph.

FIGURE 19.3   Information from skid marks. Left: A "barber pole" pattern indicates that the tire was rolling forward and skidding sideways simultaneously. Right: Straight marks indicate that the tire is not rotating much.

If the driver had been traveling at the posted speed limit of 35 mph or less, it is possible that he might have been able to come to a stop, or almost come to a stop, in 50 feet. Fifty feet is the distance in which the driver responded after recognizing the situation. Because the driver was not traveling at or less than the posted speed limit, but was traveling at 47 mph or perhaps more, he could not stop within 50 feet. Consequently, excessive speed is a factor in both the cause and severity of the accident. Being struck by a vehicle at 33 mph is certainly very different in severity from being struck at 2 mph.

Another way that kinetic energy can be dissipated in an accident is by crushing the front end. It takes work to push in the front end of a vehicle. Most people know by experience that if a car impacts an unforgiving, massive brick wall, the faster the car impacts the wall, the deeper the front end is crushed. Passengers in the vehicle are also subjected to the effects of the kinetic energy involved in the accident. Without a seat belt or airbag to provide cushion and absorb the kinetic energy of the driver's body, the driver would be thrown into the steering wheel, dashboard, and windshield area at a speed of 19 mph. This is the equivalent of a fall from about 19 feet. Such an impact would result in serious injury or death.

19.3 STRUCTURAL COLLAPSES

We can extend our understanding of energy to a different type of engineering challenge—understanding why structures such as bridges and buildings fail. Structures such as the Great Pyramid of Cheops, the Taj Mahal, or the Golden Gate Bridge symbolize aspects of a culture. Building methods, materials, and architectural style can even be used to broadly characterize a civilization. Distinguishable architectural details among buildings and structures can then be used to separate eras within a civilization. For example, the New York skyscrapers built in the 1930s are readily distinguishable from the New York skyscrapers built in the 1970s, even though both groups characterize the skyline of modern New York. Once a structure is erected, factors such as corrosion, weather, various aging effects inherent in the choice of materials, original design mistakes, abuse, unexpected loads, and external forces all work together to bring a building down. These items can be divided into two fundamental categories: static load support deficiencies and dynamic load deficiencies.

19.3.1 STATIC LOADS

Static loads include the basic weight of the building itself and its contents. A building must be strong enough to resist gravity and hold itself up. It should do so without excessive deflections and movements that might scare the occupants, make the occupants uncomfortable, or make the building difficult to use. For example, a large foundation settlement in a frame building may allow the floors to sag significantly, which then allows wheeled furniture to slowly slide to the low point, causes doors and windows to jam, and makes walking around an uncomfortable experience.

The static loads of a building are often subdivided into two categories: dead loads and live loads. **Dead loads** are loads that never seem to change in a building, such as the weight of the floors, walls, supports, and roof. **Live loads** are loads that can sometimes change due to weather, occupancy, or building use. They include such things as the temporary weight of snow or ice on the roof, the weight of the people in the building and where they are congregated at various times of the day, and the weight of furniture, machinery, and equipment in the building and how they are distributed.

A building can collapse when its primary structural components do not have sufficient strength to support the applied static loads. This can happen because of an error in original design; an omission or mistake during construction; abuse or neglect of the building; sabotage; external forces such as earthquakes, storms, or floods; use of the building for unintended purposes; or perhaps because of degradation over time by corrosion, wear, or weathering. To compensate for degradation over time, wear, possible minor design mistakes, minor construction mistakes, and certain types of abuse or neglect that can be reasonably anticipated, buildings are designed to support static loads that are several times stronger than what the designer anticipates would typically be needed. This is the building's margin of safety.

19.3.2 DYNAMIC LOADS

Dynamic loads are loads on a building that change during a relatively brief period. They are repeatedly applied and released. Dynamic loads add to the static loads that a building must be able to handle, which means that a building, or perhaps certain parts of a building, must be made even stronger than is required to handle its static loads. Unexpected dynamic loads eat into a building's margin of safety.

Dynamic loads include forces due to strong winds, gusting, or winds from varying directions; machinery inside the building or nearby that pounds or shakes the floors and walls; and ground motion, such as earthquakes, heavy traffic, or nearby construction work. Dynamic loads, when sufficiently strong and when applied often enough, can cause some materials to fail due to material fatigue. Fatigue is the premature fracture and failure of material due to the repeated application and release of loads. Fatigue can occur even when there is margin between the intrinsic static strength of the material and the sum of the applied forces. In other words, a varying load can sometimes cause failure even when the varying load is less than the static load strength of the material.

19.3.3 COLLAPSE OF THE TWIN TOWERS IN 2001

The terrorist attacks of 2001 on New York and Washington killed thousands and, in the case of the World Trade Center (WTC), led to a far-ranging investigation of what caused the towers to collapse relatively soon after the impacts of the passenger jets. Although several buildings were associated with the WTC, it conspicuously consisted of two very large buildings: One World Trade Center (North Tower) and Two World Trade Center (South Tower). One World Trade Center, completed in 1972, was 1368 feet tall. Two World Trade Center was completed a year later and was 6 feet shorter. Both buildings were 110 stories high and structurally similar.

Structurally, each building was a vertical, hollow, rectangular tube within another vertical, hollow, rectangular tube. The outer rectangular tube consisted of two hundred forty-four 14-inch steel box columns spaced 39 inches apart. The inner rectangular tube was 90 feet long and was composed of tightly spaced steel girders around a central core of elevator shafts and stairways. The inner tube supported much of the weight of the building. The inner tube and the outer tube were connected by steel spandrel members overlaid with steel decking and 4 inches of concrete. The floor decking system supported the 40,000-square-foot floor and acted as a structural stiffener between the inner and outer rectangular tubes. Each floor system by itself weighed perhaps 3–3.5 million pounds. At the foundation of each building, the total bearing load was approximately 1 billion pounds.

At the time the building complex was designed, it occurred to the designers that accidental impact by an airplane was a possibility. The Empire State Building, which is 102 stories, 1250 feet high, and in the same neighborhood as the WTC, was struck by an errant U.S. Army B-25 bomber during a fog in 1945, just 14 years after the building was completed. In that accident 14 people were killed. Although designers realized that an impact from a larger aircraft was possible, purposeful high-speed impacts of fully fueled passenger jets was understandably not anticipated.

At 8:46 a.m. local New York time on September 11, 2001, the North Tower of the WTC was struck by a Boeing 767 that was deliberately steered into the building. A second Boeing 767 struck the South Tower at 9:03 a.m.; it also was deliberately steered into the building (Figure 19.4). Both aircraft apparently impacted the towers at cruising speed to maximize damage to the buildings.

The impact occurred on the North Tower between the 90th and 96th floors. Seismometers located in Palisades, New York, about 21 miles north of the building, recorded a 12-second ground shock with a 0.8-second dominant period that had an equivalent earthquake magnitude of 0.9 on the Richter scale. The impact to the South Tower occurred between the 75th and 84th floors. It generated the equivalent of a 0.7-magnitude earthquake and had a 6-second ground shock with a 0.7-second dominant period.

In the North Tower, the initial impact severed about two-thirds of the steel supports on the tower's north side. Despite this severe structural damage, however, the floors above the impact area did not collapse. After impact, fire immediately broke out in the affected floors and rapidly spread through the crash-affected area, feeding upon the spilled fuel from the decimated aircraft. The North Tower eventually succumbed to the fire and collapsed after 102 minutes at 10:28 a.m. When it fell, it generated the equivalent of a 2.3-magnitude earthquake; the South Tower's impact was the equivalent of a 2.1-magnitude quake.

The kinetic energy at impact of each aircraft that struck the towers can be estimated by the same formula we used with vehicles:

$$KE = 1/2 \times m \times v^2$$

where:

  KE = Kinetic energy
  m = Mass of the aircraft
  v = Velocity of the aircraft (530 mph, or 777 fps)

FIGURE 19.4 Impact of the second jet with the South Tower. The impact site of the first jet on the North Tower (at right) is clearly visible.

$$KE=(1/2)(395,000 \text{ pounds force}/32.17 \text{ fps}^2)(777 \text{ fps})^2$$

$$=3,710,000,000 \text{ foot pounds force}$$

By way of comparison, a compact car that weighs 2500 pounds and is traveling at 100 mph has a kinetic energy of 836,000 foot-pounds of force. Thus, the impact energy of one of the aircraft was equivalent to about 4400 compact cars, all traveling at 100 mph, hitting the same location.

The aircraft did not penetrate through the building and come out the other side, although some parts and some fuel did. Both aircraft buried themselves into the buildings' interiors. Based on this fact, it is estimated that the impact time was about 0.4 seconds. This amount of time is consistent with the length of the aircraft (159 feet), the average speed during impact (~389 fps), and the depth dimensions of the building.

With respect to the aircraft impacts themselves, both the North and South Towers performed admirably. Both towers absorbed remarkable amounts of impact energy, sustained significant structural damage, and endured significant applied forces and moments and still stood upright and did not collapse. What the two towers could not endure was the ensuing fire associated with the jet fuel.

Upon impact, both aircraft were destroyed as they penetrated the towers and released all of their fuel into the interior of the buildings. Since both aircraft had just taken off and both were bound for the West Coast, both had nearly full fuel tanks. The impact to the North Tower was spread over six floors, and the impact to the South Tower was spread over nine floors. If 19,000 gallons of jet fuel are spread evenly over nine floors, each with an area of about 40,000 square feet, this amounts to 0.067 gallons for every square foot, or 9 ounces of jet airplane fuel for every square foot of space. Jet fuel can theoretically create temperatures greater than 3000°F. Actual flame temperatures in an uncontrolled building interior environment, such as the interior of one of the WTC towers, were likely several hundred degrees less than this value, but still sufficiently high to compromise the steel skeleton of the buildings.

Structural steel loses approximately half of its tensile strength at 1000°F. At 1300°F and higher, it loses most of its strength and stiffness and ceases to be a viable structural component. Likewise, steel-reinforced concrete degrades and cracks at temperatures of 1200°F or more. After the fire initiated, temperatures built up significantly within the building interiors. With jet fuel as the initial primary fire load, temperatures exceeding 1300°F certainly occurred within the impact areas. As the temperature increased, the strength of structural components within the fire-affected areas diminished. When the temperature of the structural components approached and perhaps exceeded 1200°F, the components could no longer carry their loads and failed.

Because the second impact was spread between the 75th and 84th floors of the South Tower, the remaining columns, connections, and supports near the second impact were supporting about 19 floors' worth of weight above the damaged area. Similarly, because the first impact was spread between the 90th and 96th floors of the North Tower, the remaining columns, connections, and supports near the first impact were supporting only about 14 floors' worth of weight above the damaged area. As fire near the impact increased the temperature of the structural components located there, those components that carried proportionally more load failed first. Therefore, the South Tower collapsed after 57 minutes of burn time and the North Tower collapsed after 102 minutes of burn time.

What occurred on September 11, 2001, caused many engineers, architects, and planners to consider whether building codes for skyscrapers should incorporate defensive measures and, some have even suggested, offensive measures to deal with the possibility of airplane crashes and other direct acts of sabotage. Some defensive architectural measures were evaluated and incorporated into some types of buildings after the April 19, 1995, Murrah Federal Building bombing event in Oklahoma City where 168 people were killed. Likewise, the U.S. Embassy bombing attack in Lebanon in 1983 that killed 60 people and the U.S. Embassy bombing in Kenya in 1998 where 212 were killed and perhaps 4000 were injured have caused people to think more defensively about building design. The topic of deliberate or accidental airplane impact, however, is still being debated. As the public memory of the event has somewhat faded, the construction of a new structure, One World Trade Center ("Freedom Tower"), was completed in 2014 at the site of collapse (Figure 19.5).

FIGURE 19.5 The Freedom Tower (center) stands where the Twin Towers once did.

THE *COLUMBIA* DISASTER, 2003

Forensic engineering was a critical part of the reconstruction of the Columbia *shuttle breakup in 2003. Here, debris recovered is placed on a grid corresponding to where the part came from on the shuttle.*

In 2003, the space shuttle *Columbia* disintegrated upon reentry, killing all seven astronauts aboard. Debris was scattered over several states, principally Texas and Louisiana. Forensic engineering was central to understanding what happened and why it happened and for generating a sequence of events. Eventually it was shown that a piece of foam fell off the external tank and impacted the left wing and left a hole that allowed hot gases to enter the body of the spacecraft and compromise the structure.

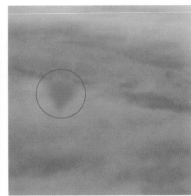

Figure 6. In-flight conditions 26 minutes after launch (about 0724).
Note: The ground is again visible through thin clouds.

Figure 8. In-flight conditions 40 minutes after launch and 4 minutes before the accident (about 0738).
Note: This photograph was taken facing east; the balloon was tracking north. The cloud layer appears to extend to the horizon.

Figure 9. In-flight conditions 42 minutes after launch and 2 minutes before the accident (about 0740).
Note: This photograph was taken facing west. The balloon's shadow (circled) can be seen just below the break in the clouds. A power line tower is visible through the break in the clouds.

FIGURE 19.6   Sequence of photos showing visibility during the flight. (Images and captions from the NTSB report.)

19.4 INTEGRATED FORENSIC EXAMPLE

Forensic engineering involves the study of mechanical and material failure, but human failures and factors are also part of the investigation. When fatalities occur, the tools of death investigation (Chapter 5) and postmortem toxicology (Chapter 6) are also involved. A recent example is a hot-air balloon accident that occurred in Texas in July 2016 that was investigated by the National Transportation Safety Board (NTSB). A report was issued in October 2017.

Around dawn on July 30, 2016, a hot-air balloon took off with the pilot and 15 sightseeing passengers aboard. At the time of the launch (~6:58 a.m.), fog was closing in on the field and weather conditions were deteriorating. A sequence of photos taken by passengers prior to the accident demonstrate the deterioration in the weather as the flight progressed (Figure 19.6). For a balloon, clear visibility is vital to safe operations, but on this morning, the balloon passed into and eventually above a cloud layer a few minutes after launch. The ground crew eventually lost sight of the craft. About 44 minutes after launch, the balloon struck a power line as it was descending; one of the towers associated with this power line can be seen in the far right frame of Figure 19.6.

Engineering analysis indicated that cables supporting the balloon's basket were cut when they collided with the power lines. Evidence of electrical arcing (Figure 19.7) caused by the collision was seen on the metal frame of the basket. Once the cables were cut, the basket separated from the balloon and fell ~100 feet to ground impact (Figure 19.8). This impact also cut fuel lines and caused a fire. All aboard were killed. Autopsies showed that the causes of death included combinations of blunt force trauma, thermal burns, and inhalation injuries. The focus of the forensic investigation was on the cause of the deaths and to what extent the operator of the balloon was at fault for the deaths of his passengers.

In this type of accident, postmortem toxicology plays an important part of the investigation. In this case, no alcohol or illegal drugs were detected, but other intoxicating drugs were found, including oxycodone (an opiate), antidepression medication, diazepam (a sedative and muscle relaxant), cyclobenzaprine (effects similar to diazepam), and methylphenidate (used to treat attention deficit hyperactivity disorder [ADHD]). In addition, diphenhydramine (used as an over-the-counter sleep aid and antihistamine), dextromethorphan (a cough suppressant), and acetaminophen were detected. Although legally prescribed and present in levels consistent with prescribed amounts, the presence of multiple impairing drugs was considered a factor in the poor decisions made by the pilot. The NTSB listed poor decision making as the probable cause of the accident and the pilot's medical conditions and medications as contributing factors.

FIGURE 19.7 The round hole in the metal frame shows where electrical arcing occurred as the balloon hit the power lines. (Image from the NTSB report.)

FIGURE 19.8 Impact site of the basket. (Image from the NTSB report.)

Interestingly, the medical requirements and restrictions for balloon pilots are not the same as those for aircraft pilots; commercial balloon pilots are not required to have a Federal Aviation Administration medical certificate. If that had been required, the balloon pilot would have been disqualified based on his diagnosed ADHD and depression. In addition, some of the drugs he was taking (oxycodone, diazepam, and diphenhydramine) are considered "do not fly" substances, meaning that licensed pilots are not allowed to take these medications when they are flying. The report recommended that commercial balloon pilots be required to get the same medical certification as aircraft pilots, although to date this change has not been made.

19.5 REVIEW MATERIAL: KEY CONCEPTS AND QUESTIONS

19.5.1 KEY TERMS AND CONCEPTS

Chemical energy
Dead load
Dynamic load
First law of thermodynamics
Irreversible work
Kinetic energy

Live load
Second law of thermodynamics
Skidding
Static load

19.5.2 REVIEW QUESTIONS

1. In vehicular accident cases, eyewitness testimony is usually gathered and recorded. Explain why this type of evidence should be evaluated with caution.
2. What is the difference between the dead load and the live load of a building?
3. If you visit high floors of skyscrapers, you may feel the building sway. What is this an example of?
4. Did the impacts of the aircraft into the two towers of the WTC directly cause the towers to collapse? Why or why not?

19.5.3 ADVANCED QUESTIONS AND EXERCISES

1. Alcohol decreases in a man's blood at a rate of about 0.017% per hour. Bill has an accident, and his blood alcohol level is checked about 2 hours after the accident at a local hospital. During that time, he is in police custody and does not drink anything. The analysis determines that his blood alcohol at the time of testing is 0.05%. If the legal limit is 0.08%, was he legally drunk at the time of the accident?
2. Alcohol decreases in a woman's blood at a rate of about 0.015% per hour. Susan has been drinking and has a blood alcohol level of 0.10%, as determined by a breathalyzer administered by a barkeeper. Being a responsible driver, Susan does not want to drive while legally drunk. How long does she have to wait to have a blood alcohol of less than 0.08%?
3. The NTSB maintains a database of accident reports at https://www.ntsb.gov/_layouts/ntsb.aviation/index.aspx. Select a recent accident that obtained national attention in aviation or railroads. Find the report and analyze it as was done for the balloon accident.

BIBLIOGRAPHY AND FURTHER READING

BOOKS

Bohan, T. L. *Crashes and Collapses*. New York: Checkmark Books, 2009.
Franck, H., and Franck, D. *Forensic Engineering Fundamentals*. Boca Raton, FL: CRC Press, 2013.

ARTICLES AND REPORTS

National Transportation Safety Board. Aviation Accident Report: Impact with Power Lines: Heart of Texas Hot Air Balloon Rides: Balóny Kubíček BB85Z, N2469L: Lockhart, Texas: July 30, 2016. NTSB/AAR-17/03 PB2018-100161. https://www.ntsb.gov/investigations/AccidentReports/Reports/AAR1703.pdf.

WEBSITES

Link	Description
http://www.nafe.org/	National Academy of Forensic Engineers
https://www.ntsb.gov/Pages/default.aspx	National Transportation Safety Board

Computer Forensics

20

20.1 A NEW TYPE OF EVIDENCE

We have seen many examples of how computers are applied in forensic science, such as in crime scene scanning (Chapter 3) and accident reconstruction. In this chapter, we discuss examples of a relatively new discipline in which forensic science is applied to computers and digital evidence. One of the unique features of digital evidence compared with physical evidence is the ability to make copies, but that introduces a new set of issues that must be addressed. ***Digital evidence*** is a term that covers a wide range of information stored electronically, such as digital images, videos, audio recordings, digital files, and software that is found on devices such as computers, mobile phones, and smart devices that can store and recording data, like smart watches. The advent of the **Internet of things** (IoT) continues to expand the range of devices that can become or contain evidence. Examples of an IoT device are home assistant systems, refrigerators that keep track of inventory, and even lamps that turn on with a voice command; what these devices have in common is the ability to send and receive data through the Internet.

As with all evidence, it is critical to collect, protect, and preserve the digital evidence in its original condition. If the data is within a device, that can be transported to a laboratory. However, often the device is not the critical evidence; rather, it is the data encoded within it. In a computer, it is usually the hard drive (physical evidence) that contains the digital evidence. That drive must be maintained and protected, but the data it contains can be copied and evaluated. As another example, consider a digital image taken on your phone. You might download that image to your computer and then utilize software such as Adobe Photoshop® to adjust color and add effects. If this image were to become evidence, it would be critical to either find the original file or have a record of how it was changed. Many modern software packages keep such a record. Thus, just as the integrity of physical evidence is documented and ensured by a chain of custody document, so also must digital evidence be. If a copy is made for study, that copy must be identical to the original or traceable directly back to it. The ability to copy digital evidence is unique; you can't "copy" a postmortem blood sample. The key is to ensure that the copy is exact. We discuss this in detail in a later section.

FIGURE 20.1 Types of media on which information and evidence might be stored.

Digital evidence can be stored or recorded on or in several types of media (Figure 20.1), some of which you have probably never seen or heard of. Magnetic tape was one of the earliest forms of storage, as was film used in cameras and still used for some movies. Memory keys and cards are likely familiar to you, but floppy disks many not be. The form of the media matters because not all electronic evidence is stored on hard drives or newer types of media. For example, a cold case from the 1990s might have digital evidence stored on floppy disks (Figure 20.1); thus, a laboratory would need access to a device capable of reading these disks and making copies of the data.

BTK KILLER AND DIGITAL EVIDENCE

A notorious serial killer of the late 20th century was known as "BTK" (bind-torture-kill) for the method he used in his murders. His first known killings were in 1974, a family of four in Wichita, Kansas. In subsequent years, six additional killings were assumed to have been committed by the same person. By 1991, the murders appeared to stop, and little progress was made in the investigation. In 2004, the police and media outlets began receiving communications from someone claiming to be BTK; in one envelope, the killer included evidence from one of the killings that confirmed that BTK was the source. Police continued to correspond with the individual in hopes of finding him. The sender asked police if he could be traced if he sent data to them on a floppy disk (Figure 20.1); the police intimated that they could not trace him. Incredibly, the killer sent a floppy disk to a local media source. This digital evidence broke the case open. When investigators studied information about where the digital contents had been created, they learned that the file had been saved by "Dennis" and that the disk had been used at a local church and library. The police quickly identified Rader as a suspect and located his home. However, before arresting him, they performed a DNA test on samples obtained from Rader's daughter during a routine medical examination. The results were compared to case samples, and it was clear that she was the daughter of BTK and that the killer was Dennis Rader. He was arrested in 2005 and convicted of 10 murder counts.

20.2 CYBERCRIME

Computers and related devices can be **ancillary** to a crime or the means by which a crime is committed. For example, a cell phone could be useful for tracking a suspect's movements during a period of time by determining which cell phone towers the phone contacted. The phone was not used to commit a crime, but it plays a role in the investigation; therefore, it is ancillary. However, computers and the Internet have evolved to the point that crimes can be committed using computers and over the Internet. These are classified as **cybercrime**. A perpetrator still commits a crime, but the technology is the means by which the crime is committed. As an example, if someone breaks into your online bank account and empties it, the crime is just as much as a theft as would be robbing a bank in person, but the former is a cybercrime. One of the most difficult aspects of cybercrimes is that the perpetrator does not have to ever come into direct contact with the victim or victims of the crime. The victim and perpetrator often live in different countries with different law enforcement and justice systems and will never come face-to-face. Cybercime is borderless, which means that many law enforcement agencies across different locations and countries can be involved. As such, solving cybercrimes and bringing justice to the perpetrator can be more challenging than noncybercrimes (which are difficult enough already). Two example categories of cybercrime are briefly discussed in the following sections, but the types of cybercrime expand as fast as the technology behind it.

20.2.1 HACKING, SECURITY BREACHES, AND EXTORTION

The term ***hacking*** can be used to describe some of these crimes, particularly when the perpetrator breaks into a secure or private computer or network to commit a crime. This is a form of trespassing, and it is often the first step in perpetrating much more serious crimes. Currently, two of the most common types of cybercrimes are theft of personal data and extortion. The first step in both instances is installation of malicious software that will be activated to commit the crime. Such software can be referred to as a virus or worm, or generically as **malware** (malicious software). The differences between these types of software are not critical for our discussion.

There are numerous ways for malware to be installed on a computer, including methods that do not require the user to do anything. Another way is through **phishing** (pronounced *fishing*), in which a target is sent an email or text that appears to come from a legitimate source, such as a company or someone that they know or someone in their work organization. If the victim opens the email and follows embedded links, this provides the hacker with a way into the person's computer, where data such as passwords, bank account information, social security numbers, credit card data, and medical data can be copied and stolen. This data may be used to make fraudulent purchases, get loans, or steal money; this is called **identify theft**. Alternatively, the stolen data can be collated with others and sold to other criminals. The theft from the computer may never be detected, and it is only after the data is gone and has been used illegally that the person becomes aware of the theft. Identity theft is currently one of the most common types of cybercrime. Data can also be stolen from large companies and used for criminal purposes. In 2017, the credit bureau Equifax and Yahoo were breached; the retail store chain Target was hacked in 2013, Sony in 2015, and the Office of Personnel Management of the U.S. federal government in 2015. As a result, millions of people had critical personal data stolen.

Extortion via computers is typically carried out using malware called **ransomeware**. Once the software gets into the target system, it locks the user out and threatens to permanently delete all the files unless the user pays a ransom, often using an untraceble form of payment. Ransomware can hit individual computers or large networked systems, including healthcare facilities. From January 2017 to February 2018 (latest available), more than 300 cases of data breaches were reported for healthcare systems in the United States as reported by the Department of Health and Human Services. The number of attacks are likely underreported, and it is unknown how many victims pay the ransom.

20.2.2 PIRACY

On the Internet, piracy refers to the theft of creative works, such as books, music, and movies. These works are referred to as **intellectual property**, and in these types of crimes, the property is stored in digital format. Prior to widespread use of computers, these works were physical things, like books, paintings, records, or movies on film, and stealing them required a physical act of theft. The ability to copy files has changed the nature of the crime and made it a greater problem. Certainly books or film could be stolen, but making copies was not easy, which discouraged the crime. However, now intellectual property is easily copied and thus easily stolen. Adding to the problem is the widespread perception that content on the Internet should be "free," and as a result, a significant number of people do not consider taking intellectual property to be a crime.

The way someone is paid for intellectual property is different than a direct payment for service. If you venture to a fast-food place for lunch, you pay cash (or the equivalent) for the food. The same holds true for purchasing gas or buying something at the grocery store. You are purchasing physical things. Now consider writing and performing music that is digitally recorded in a studio. If you like the song and purchase it on a platform such as iTunes®, the artist receives a small percentage of what you pay; this is called a **royalty**. Royalties are paid to artists, actors, writers, and people who hold patents and is the means by which they receive payment for service. When works are obtained without payment, there is no royalty and no payment for services to the person who created it.

One of the milestones in the history of digital piracy was the development of a method of storing large files such as movies, videos, and music in a compact digital format. Ideas for this began taking shape in the early 1980s, and by 1988, the International Organization for Standardization joined the call for developing a standard digital file format. By 1995, the format that is now known as MP3 was available, and it is still widely used for music. Having the format allowed for the creation of music CDs and players such as the iPod®, and now directly on cell phones. Unfortunately, having a standard format that created compact files also made it easier to share and ultimately steal music.

The best-known entity of that era (late 1990s) was a service called Napster, which used a protocol called **peer-to-peer** (P2P), a file-sharing protocol. With P2P, the connection between computers does not go through a central website or service but links users directly. As a result, detecting the trafficking of the files became much more difficult. The controversy over Napster and similar services drove legislation and discussions between content creators and users that eventually led to the demise of Napster and the current model of content purchasing and sharing. However, from the forensic and law enforcement point of view, P2P protocols became an important factor for investigation. Using P2P requires an Internet connection, but it bypasses centralized websites and servers; this can make detection and tracking more challenging for forensic investigators.

MOST PIRATED MOVIES

2017	2016
Thor: Ragnarok	Deadpool
The Shape of Water	Batman vs. Superman
Jumanji: Welcome to the Jungle	Captain America: Civil War
Bright	Star Wars: The Force Awakens
Blade Runner 2049	X-Men Apocalypse
Murder on the Orient Express	Warcraft
Downsizing	Independence Day: Resurgence
Geostorm	Suicide Squad
Justice League	Finding Dory
Coco	The Revenant

20.3 PROCEDURES FOR DIGITAL EVIDENCE AND STORAGE DEVICES

The steps taken with digital evidence mimic those taken with physical evidence, and often, both physical and digital evidence exist at a crime scene. The first step for the first responder is to recognize that an object contains potential evidence. Common examples of this type of evidence include cell phones, global positioning system (GPS) devices, computers, tablet computers, and memory storage devices. Once recognized and the finding documented, a chain of custody is started for the physical item. For example, suppose a cell phone is found during a crime scene search. A chain of custody would be started on the phone, which is the physical evidence; the digital evidence would be photos, numbers, messages, and so forth, that are stored electronically within the phone. Both the phone and the data within must be preserved.

The examination of the data begins by generating a copy of the data for further testing, preserving the original data stored on the phone. The copy must be an *exact* duplicate (authentic) of the original, and that copy must be made in such a way that there is no chance that the original data is altered. The copying process is typically referred to as **imaging**. An image is a bit-by-bit replicate of the original device, and it cannot be created by simple copy and paste operations that most users are familiar with. It is possible to copy images and messages from the phone to a computer (files), but that does not duplicate the entire contents as stored on the phone. This arises because of how storage media such as hard drives and chips inside mobile phones store data. The process of imaging and verifying the copy is shown in Figure 20.2.

Consider a brand new tablet computer fresh out of the box. The manufacturer of the computer has already installed the software needed to operate the device (the **operating system** [OS]). Examples of OSs include Apple Mac OS, Unix, Linux, Windows, and Android. As you install software programs or apps (applications), you are writing information to a storage device. Suppose you store a new podcast in the device, listen to it, and then delete it. This operation does not restore the disk to the original condition. Typically, what the "delete" operation does is remove the information that the OS uses to find the file on the storage device. It appears to you as if the file is gone, but it is still stored there, just not easily accessible. The next file you save might overwrite all or part of this region, but the remnants of the file you deleted are still present on the storage media. Given that even a simple computer device with an OS is constantly creating, storing, deleting, and overwriting the storage media, the contents of the disk at any one location are difficult to predict. In addition, a file is not necessarily stored in one piece; it may be broken up into many smaller portions and stored wherever space permits. When you recall that data (say a photo), the OS reassembles the data to display the image. All these operations are invisible to the user, but all leave "footprints" on the storage media. Where and how the data is stored on one device differs from how and where it is stored on the other.

To be admissible, the copy has to be proven as exact. This is usually verified by a technique called **hashing**. A hashing algorithm uses the original information as input and uses an algorithm to create an output string that is unique to the original at the bit-level of detail. Currently, the MD5 (Message Digest 5) algorithm is commonly used, but there are many others. Consider the text sentence *Take a left at Albuquerque*, which is used in the example in Figure 20.2. The MD5 hash for this string is **e9618b637bb5d4250e740f212e62630f**. If the sentence is changed slightly by

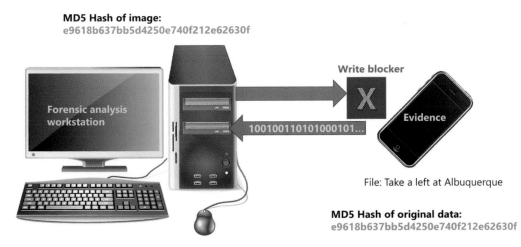

FIGURE 20.2 How storage devices are imaged. The digital evidence on the phone is protected by a write blocking device so that the data can only flow one way to the analysis station. The hash values match, proving that the copy is exact and valid.

misspelling the city name as *Albuquerque*, the hash string changes to **4f3ed76bdf3dc60d282e-ab241f31ea07**. Even a nearly invisible change, such as inserting an extra space between the word *at* and *Albuquerque*, changes the hash to **1a2126defe80b0dc0f8751023a1c5b11**. Thus, if the hash of the copy is the same as the hash of the original, the original has been successfully imaged. The original would be set aside and forensic analysis undertaken on the copy or copies. As long as a copy is shown to be authentic, it can be analyzed. Note in Figure 20.2 that a **write blocker** is shown. This is a hardware device that is placed between the evidence and the device that the data is being copied to. The blocker prevents any data from being written to the original device and ensures that the original digital data has not been altered as part of the copying process.

Another use of hash values is as a means to represent images. Sadly, computers and the Internet have become the ancillary in crimes related to sexual exploitation and child pornography. Sharing images and videos is one of the ways these crimes are enabled and committed. Hash values of an image file can be generated and added to lists that can be monitored and tracked. The FBI has established a task force dedicated to finding and prosecuting such crimes against children: Violent Crimes Against Children/Online Predators (VCAC; www.fbi.gov/investigate/violent-crime/cac).

There are numerous software packages available to conduct forensic analysis. Examples of analyses include recovering erased or deleted files and images, reconstructing web search histories, and searching a storage device for a specific text string, such as a name. This software may be purchased from companies, but there are also open-source tools available. Because law enforcement and digital forensics experts rely on these tools, it was recognized that independent testing of these tools was vital. The task fell to the National Institute of Standards and Technology (NIST), which in collaboration with other agencies, established the Computer Forensics Tool Testing (CFTT) program and website. NIST also established a Computer Forensics Tool Catalog that helps investigators find the right software tool for a specific task, ranging from cloud services tools to social media to **voice over Internet protocol** (VoIP) or computer calling. A recent addition to the site is tools for the analysis of digital evidence related to drones.

20.4 CELL PHONES AND GPS

As cell phones have become ubiquitous, they have become a common form of physical and digital evidence. The increasing capabilities of these devices means that phones can contain photos, videos, and voice recordings in addition to data about where the phone has been. The evolution of cell phones from bulky "car phones" to smart phones is shown in Figure 20.3. Cell phones are called "cells" or "cellular" because of the way the network coverage is designed (Figure 20.4). At a basic level, a mobile phone can be thought of as a computer with internal storage and a radio transmitter. To greatly simplify, when you place a call, the phone transmits a signal that is picked up by towers that are close enough to detect it. The strength of the radio signal gets weaker over distance and eventually becomes too weak to be detected. The tower relays the signal and the call along until it reaches the phone that you dialed. The area around a tower is a cell and the network of these cells creates the **cellular network.**

It is sometimes possible to determine the location of a phone at any given time based on the strength of the signal received at a tower. The closer a phone is to the tower, the stronger the signal. If more than one tower detects a signal, the location of the source can be located using a technique called **triangulation** (Figure 20.5). The strength of the signal received at the tower to the left (purple) is measured and corresponds to a distance of 2 miles. If that were the only data available, all that could be said for certain is that the phone is somewhere 2 miles from the tower, but where on that circle can't be determined. Now add the data from the middle tower (yellow); the signal strength indicates that the phone is somewhere along the circle 10 miles away. Adding the data from the third tower (right, green) adds the 20-mile circle; where those three circles intersect is where the signal originated.

This concept can be expanded to reconstruct the route a phone traveled over time (Figure 20.6). Here, the same set of towers has been overlaid on a map to show a simplified example. If a phone was recovered and the movement of the phone and its owner is critical to a case, records of contacts with towers can help reconstruct the route taken. During stages 1–3 of the drive, only

FIGURE 20.3 Evolution of cell phones from the late 1980s to the present. Notice how the early models had external antennas; modern phones have antennas as well, but they are internal.

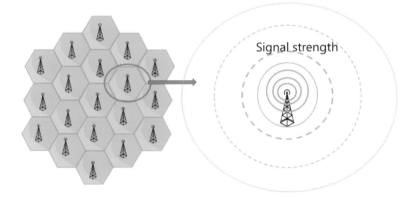

FIGURE 20.4 Simplified diagram of a cellular network. Signals are passed from tower to tower to reach their destination. The closer a phone is to a tower, the stronger the signal, if there are no obstructions between the phone and the tower.

tower 1 (T1) would detect signal from the phone. Once the car is on the interstate and heading west, T2 would detect a signal that would start out relatively weak and get stronger, reaching maximum strength near Sedillo. This is the closest the phone got to T2. Moving farther west, the T2 (and T1) signals become weaker. To the west of Sedillo, T3 will start detecting a weak signal. Near point 5, the T2 signal again strengthened and then decreased, and by the time the phone is at point 6, only T3 will be detecting signal. Although greatly simplified, this is the process used to recreate the movement of the phone.

Under ideal conditions, records of signals received at towers may be useful in determining where a phone was at a given time, but interpretation must be undertaken with caution. The examples shown here are meant to be used for illustration purposes. Suppose you live in a city along with thousands of others packed into a small area. Many people are using their phones at the same time, which can exceed the capacity of towers. In that case, signals can be routed

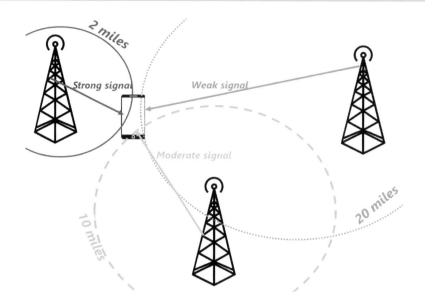

FIGURE 20.5   The process of triangulation.

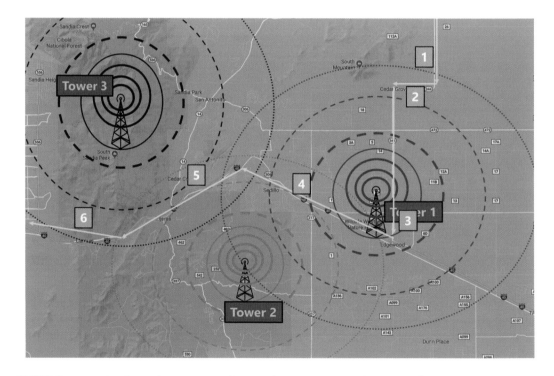

FIGURE 20.6   Applying the concept of triangulation to retrace the route of a moving car.

through other towers. This and many other, often complex, factors determine the route that a cell phone call actually takes when being transmitted from the caller to the receiver. As shown in Figure 20.4, the network consists of many towers, but that does not mean that a call will always pass from one tower to another using the shortest distance between your phone and the person you are calling. Even in simple cases, such as shown in Figure 20.5, triangulation does not pin-point a position of a phone exactly; rather, it defines an area of various sizes. It is important to keep this in mind.

If signals sent to cell towers are usually not reliable for pinpointing a location, how can a smart phone be used to provide reliable directions when walking or driving? To accomplish that, location data has to be accurate down to a few feet and adjust in real time. The reason is that smart phones have GPS chips that perform a different function than the hardware responsible for using the phone. GPS, first introduced to the public in the 1990s, utilizes multiple satellites to provide reliable information on your location, including elevation. As an example of that complexity, the satellites used contain clocks that are accurate to a tiny fraction of a second and that take into account factors that arise from Einstein's theory of relativity. However, concepts of triangulation are still used to find location, but the system is much more advanced than radio triangulation.

MH370 AND SATELLITE LOCATION

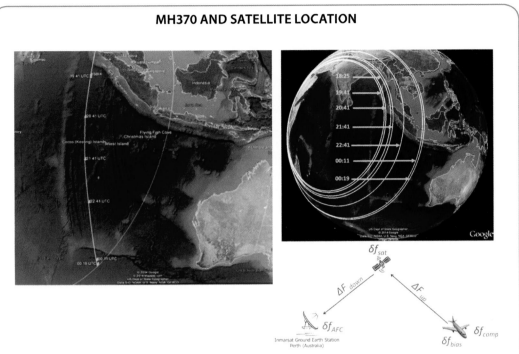

Malaysian Airlines flight 370 with 293 people aboard took off from Kuala Lumpur in the early morning hours of March 7, 2014. The flight was headed to Beijing but went missing, and as of this writing, the plane has not been found, although a few pieces of wreckage have been located in the area of the southern Indian Ocean.

Satellite data was used to determine the likely flight path and impact area of the airliner. Unfortunately, these areas are remote and the water is very deep, making the search extremely difficult. Modern airliners routinely communicate (ping) satellites as they fly to provide information on performance, but this type of communication is not meant to identify specific locations like GPS. However, as with cell phones, it is possible to utilize these signals to help locate areas in which the plane likely flew and ultimately crashed. The procedures used are conceptually similar to triangulation. In the lower right side of the figure, signals are shown moving between the plane and the ground station via a satellite. However, in this case, the plane and the satellite are both moving quickly, unlike our example with a cell tower and a phone, which if moving is moving relatively slowly. As a result, the frequency of the transmission (F) is altered by the motion itself. This is an example of the Doppler effect; when you stand still and a train approaches and then passes you, the frequency increases and the pitch of the noise increases as the train approaches and decreases as it moves past and away. In this case, the change in frequency was used to help estimate distances between the plane, satellite, and receiver. The arcs of space that the satellite monitors vary as the earth and satellite move, resulting in arcs on the map. If the plane is somewhere along the arc, it can communicate with the satellite. However, as noted with cell phones (Figure 20.4), with only one piece of data, all that can be said for certain is that the plane is somewhere on that arc. By combining data from several communications, it was possible to estimate the plane's likely path and where it crashed. However, as with cell phone location, these are probable areas or zones, not pinpoint locations.

Source: Ashton, C., et al., The Search for MH370, *Journal of Navigation* 68, no. 1 (Jan 2015): 1–22. Open-access publication. DOI: 10.1017/S037346331400068X.

20.5 CYBERCRIME AND THE INTERNET OF THINGS

Each year, more devices are connected to the Internet, expanding the IoT with additional smart devices. Law enforcement and digital evidence specialists are understandably concerned about the vulnerability of the IoT to attacks such as ransomware. An overview of the IoT and threats is shown in Figure 20.7. The IoT works over the Internet, which in the figure is shown as a cloud. In the context of computers and computer forensics, the **cloud** describes storage locations that are not local to a personal or individual device. If you store a file or photo "in the cloud," the file is stored in a location that is maintained by a company or other entity. Current examples of cloud-based storage are Microsoft's OneDrive®, Apple's iCloud®, and Dropbox®. Cloud storage still requires a physical storage media, such as a hard drive, but that hard drive is not located on your device. Many software applications are cloud-based as well, which means the program runs on a remote computer that you access, but the entirety of the program or app is not stored on the local device. The digital evidence is still created, but not exclusively on a single device.

Many devices on the IoT do not have features that make them vulnerable to ransomware attacks given that their primary purpose is data transmission and receipt. Smart microphones/speakers (current examples include Alexa® by Amazon and Echo® by Google) are voice-activated devices that communicate through the Internet but do not do a significant amount of internal computing. However, as these devices add capability, the threat increases. Hospitals (Figure 20.7 upper left) have already become frequent targets of cyberattacks. The challenge presented to computer forensics will grow along with the sophistication of the IoT.

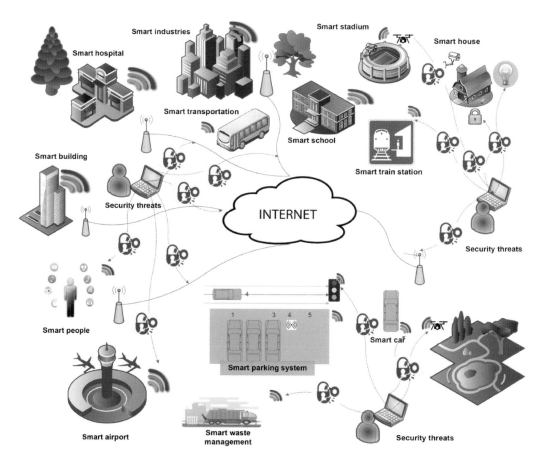

FIGURE 20.7 The IoT environment. Red locks indicate vulnerable locations. (Reproduced from Yaqoob, I., et al., The Rise of Ransomware and Emerging Security Challenges in the Internet of Things, *Computer Networks* 129 [Dec 2017]: 444–458. With permission.)

20.6 REVIEW MATERIAL: KEY CONCEPTS AND QUESTIONS

20.6.1 KEY TERMS AND CONCEPTS

Ancillary
Cellular network
Cloud
Cybercrime
Digital evidence
Hacking
Hashing
Identity theft
Imaging
Intellectual property
Internet of things
Malware
Operating system
Peer-to-peer
Phishing
Ransomware
Royalty
Triangulation
Voice over Internet protocol
Write blocker

20.6.2 REVIEW QUESTIONS

1. Suppose a person murders another and attempts to conceal the crime by writing a suicide note on the victim's tablet computer. Is this a cybercrime or is the digital evidence ancillary?
2. Search online for an MD-5 hash generator. What is the hash for your name? Try it without using capital letters. Does it change?
3. When you take a picture on your phone and later delete it, is it "gone"? Why or why not? What about a text message?
4. Using the diagram below, describe the relative signal strengths at each tower for all three locations based on the ideal of triangulation.

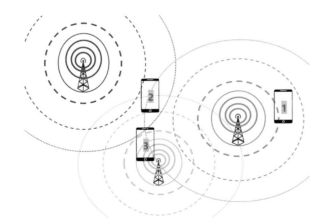

5. Why must triangulation of cell phone tower signals be interpreted with caution?
6. Suppose that a suspect is arrested and found to be in possession of one the newest models of smart phones with all features on. What types of digital evidence might be found on this device?

7. Are digital images and videos taken at a crime scene by crime scene investigators considered to be digital evidence? Why or why not?
8. Why is the process of imaging a disk different than copying and pasting?
9. A person manages to hack into a film studio computer and download a copy of a summer blockbuster before it is released. He sends this to 10 other hackers using a P2P protocol. If 1 of the 10 is arrested and found to have a copy on her computer, where would investigators search for digital evidence?
10. Why are write blockers critical to ensuring the integrity of digital evidence?
11. In Figure 20.7, an icon for "Smart People" is shown. What do you think this means in terms of the IoT? What devices might be included?

BIBLIOGRAPHY AND FURTHER READING

BOOKS

Holt, T. J., Sossler, A. M., Seigfried-Spellar, K. C. *Cybercrime and Digital Forensics*. 2nd ed. New York: Routledge, 2018. ISBN: 978-1138238732.

Middleton, B. *A History of Cyber Security Attacks: 1980 to Present*. Boca Raton, FL: CRC Press, 2017. ISBN: 978-1498785860.

Steinmetz, K. F., and Nobles, M. R. *Technocrime and Criminological Theory*. Boca Raton, FL: CRC Press, 2017. ISBN: 978-1138305205.

ARTICLES

Chung, H., Park, J., and Lee, S. Digital Forensic Approaches for Amazon Alexa Ecosystem. *Digital Investigation* 22 (Aug 2017): S15–S25.

Yaqoob, I., Ahmed, E., Rehman, M. H. U., Ahmed, A. I. A., Al-Garadi, M. A., Imran, M., and Guizani, M. The Rise of Ransomware and Emerging Security Challenges in the Internet of Things. *Computer Networks* 129 (Dec 2017): 444–458.

WEBSITES

Link	Description
https://www.nist.gov/topics/digital-multimedia-evidence	NIST website on digital evidence and links to tools
https://toolcatalog.nist.gov/index.php	NIST Computer Forensics Tool Catalog
https://www.cftt.nist.gov/index.html	NIST Computer Forensics Tool Testing website
https://www.npr.org/sections/therecord/2011/03/23/134622940/the-mp3-a-history-of-innovation-and-betrayal	Information on the history of the MP3 file format
https://www.swgde.org/	Scientific Working Group on Digital Evidence; the page describing "myths" is particularly interesting
http://forensicswiki.org/wiki/Main_Page	Website with open-source tools and other information regarding digital forensics

Behavioral Science

21

21.1 ROLE OF BEHAVIORAL SCIENCE IN THE JUSTICE SYSTEM

We have spent most of our time discussing things associated with forensic science, such as physical evidence, laboratories, procedures, and practices. We end the book with an overview of the ultimate reason for the existence of forensic science—human beings and their behavior. We have devoted our time to discussing crime scenes from the forensic perspective where the goal is to answer questions such as who and how. Now we delve into why, but with the recognition that the answer to the question why is often unknowable. However, there are patterns and consistencies that can be useful to law enforcement, forensic science, and the judicial system.

The study of human behavior is addressed through the disciplines of psychology and psychiatry. Both disciplines deal with human behavior; psychiatrists are trained as medical doctors, while psychologists are trained in universities. As physicians, psychiatrists can work with medical diagnostic equipment and testing and prescribe medication, while psychologists primarily work through counseling and interviews. For our purposes here, the differences are not critical, so the term *behavioral science* will be used to include both disciplines as applied to forensic science.

21.2 BEHAVIORAL SCIENCE AND THE JUSTICE SYSTEM

There are many junctures in the justice system where the expertise of behavioral science is needed and called upon; we touch upon a few here as examples of a much larger and diverse field. Once a person becomes involved in the justice system (charged, arrested, etc.), behavioral scientists can be called in to assist. For example, one of the first questions may be regarding the current mental health of the person and if they are able to understand what is happening to them and why. Often, the psychologist or psychiatrist is called upon to assess the mental state of an individual as part of determining risk to the community and sentencing. The goal of these interventions is to see a just outcome while weighing the risk to the public. For example, if someone commits a crime while mentally ill, is imprisonment or treatment the best option? What is the risk that a person convicted of stalking will repeat the behavior? How can you tell if someone is truly mentally ill or faking it? Conversely, can you tell if a mentally ill person is pretending not to be? What can you learn about a perpetrator based on how he or she commits a crime? These are the types of questions that behavioral scientists address.

21.2.1 DECEPTIVE BEHAVIOR

In forensic settings, distortion and exaggeration of symptoms commonly occur for a variety of reasons. One common type of deception is **malingering** (also referred to as **simulation**), a conscious attempt to fake a mental illness; another is **dissimulation**, a conscious and deliberate attempt to minimize or deny symptoms of a mental disorder. Criminal offenders might feign mental illness to avoid criminal responsibility, or they might also deny mental illness to get released from prison or a hospital. Interestingly, offenders who simulate symptoms of an illness often select a currently "trendy" illness, such as multiple personality disorder; moreover, they often greatly exaggerate the clinical picture. Not surprisingly, it is very difficult to maintain consistent malingered psychotic symptoms for extended periods.

21.2.2 COMPETENCY TO STAND TRIAL

When considering the **competency** of an individual, the question is really, "Competency to do what?" In criminal matters, the primary concern is that of competency to stand trial, but another important competency is that of making a statement to the police or giving a confession. Was the individual competent to make the statement? Did he or she understand his or her *Miranda* warnings, and could he or she voluntarily proceed after having been given the warnings? Was the confession made at the time valid and trustworthy? Was it coerced or freely given? This is a challenging task as the psychologist or psychiatrist must make a judgment of what happened at the time in the context of that particular situation and the person's mental state at that time.

The issue of competency to stand trial is perhaps the most important one facing many defendants. Most defendants consider themselves competent to stand trial; however, to be found competent, the defendant must be free of mental illness that impacts his or her ability to know what is happening in his or her case. In other words, someone who is not competent to stand trial probably is not competent to decide if they are competent!

Several factors are considered when competency for trial is evaluated. For example, does the defendant truly understand what the consequences of these charges are? If not, then he or she may be found incompetent to stand trial. The defendant should also understand the basics of the courtroom processes and procedures and the roles of the judge, lawyers, and jury (if present). Some defendants may have an intellectual understanding of the nature of the courtroom proceedings but may have delusions about the principals and their roles. Others may want to act as their own attorney even when they are clearly not qualified or well enough to do so.

If a defendant is found incompetent to stand trial, he or she will be hospitalized, usually at a forensic psychiatric hospital, where he or she will be treated (or perhaps educated) until he or she is competent. In some rare cases, the defendant, after a period of treatment, will be deemed incompetent to stand trial for the foreseeable future.

CASE THAT CHANGED THE INSANITY PLEA

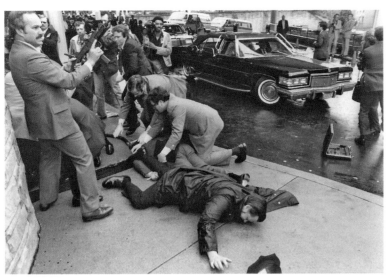

Seconds after the shooting. The president was in the car; the police officer and James Brady were down. Hinckley is surrounded by agents at the left.

Left: James Brady. Right: Hinckley at his arrest in 1981.

On March 30, 1981, John Hinckley Jr. shot and wounded President Ronald Reagan, his press secretary James Brady, and two others. Reagan and Brady were seriously wounded; the president's life was saved because a Secret Service agent insisted he be taken directly to a nearby hospital. Brady was shot in the head and suffered significant impairment for life. Hinckley was found not guilty by reason of insanity the following year, which ignited a public outcry regarding the insanity defense and how it was used in this case. Rather than prison, Hinckley was sent to a mental institution and resided there while receiving treatment. The case resulted in significant changes to the law regarding the insanity plea. Hinckley was released in 2016 (when he was 61) to live with his elderly mother. James Brady died in 2014, and the cause of death was listed as homicide because his death was the result of the injuries he received in 1981.

21.2.3 LEGAL INSANITY

The definition of **insanity** varies in different jurisdictions. Most states follow the **McNaughten rules**, which were promulgated in England in 1843. McNaughten rules are primarily a cognitive test of insanity. Most jurisdictions indicate that a person would be found not guilty due to insanity if, at the time of the commission of the crime, the person was suffering from such mental infirmity or disease that he or she did not know the nature and quality of his or her action. This is purely cognitive in that the individual could not or did not know what he or she was doing or that what he or she was doing was wrong. Over the years, various criteria were developed to assist in determining if the insanity defense is viable. The year 1982 was pivotal for the insanity defense; in that year, John Hinckley Jr., who admitted to shooting and wounding President Reagan in 1981, was found not guilty by reason of insanity. The case is discussed in the insert.

Follow this controversial decision, changes were made, including the 1984 passage of the Omnibus Crime Code for Insanity by the U.S. Congress, which stated that a person would be found not guilty by reason of insanity if at the time of the commission of the crime the defendant could not appreciate the criminality of his behavior. Also following the *Hinckley* decision was the concept of "guilty but mentally ill." This designation is important in determining how the courts hand out appropriate sentences after a guilty verdict is reached.

21.3 INTERACTIONS WITH FORENSIC SCIENCE AND ACTIVE INVESTIGATIONS

21.3.1 MANNER OF DEATH

In Chapter 5, we discussed the procedures associated with medicolegal death investigation. The medical examiner or coroner has the responsibility of determining the manner of death as natural, accidental, suicide, homicide (NASH), or undetermined. There are many cases in which the differentiation of accidental death and suicide benefits from the assistance of behavioral science. With the rapid increase in overdose deaths that was noted in earlier chapters, the need has become critical. As an example, consider a typical scenario: A young man is found unresponsive in his apartment. First responders arrive and declare him dead. They suspect an overdose, which initiates the death investigation system. An autopsy is performed along with postmortem toxicology. The results of the toxicology show concentrations of a synthetic fentanyl analog and metabolites that were in the low range of a fatal concentration. Alcohol was also found at a concentration of 0.12% (recall that the legal limit for driving is 0.08%), along with low amounts of an antianxiety medication.

The death investigation team learns that the deceased had a prescription for this last medication. The cause of death is listed as cardiac arrest due to an overdose of the illegal drug. However, the *manner* of death is more difficult to answer; the only manner of death that can be immediately discounted is natural, which leaves homicide, suicide, and accidental. Sorting through those options requires further investigation, evaluation of the physical evidence, and ideally the perspective of a behavioral scientist. The term ***psychological autopsy*** is sometimes used to describe what the behavioral scientist is tasked with producing.

In our example, law enforcement will want to know if the case is a homicide. Assume that investigators found no evidence of a struggle at the apartment, along with drug paraphernalia, including a syringe and a packet, that are found to contain residues of the same drug that was found in the deceased's blood. Interviews with friends and family reveal that he had a history of drug use and had been in a drug rehabilitation program up until recently. He did not have any significant financial assets and no life insurance, and nothing had been taken from his apartment. As a result, police are satisfied that this case is not a homicide and close the criminal investigation. Homicide as a manner of death is eliminated, leaving accidental, suicide, or undetermined.

Sorting this out requires more than physical evidence and toxicological data, and this is where behavioral scientists can be of help. The goal of their investigation is to uncover the person's state of mind and intentions at the time when the fatal dose was self-administered. The example of an overdose was selected purposely, as the rate of overdose deaths has soared in the past 5 years (Figure 21.1). Notice that the number of deaths attributed to accidental overdose dramatically increased, while the number of suicidal and undetermined cases remained

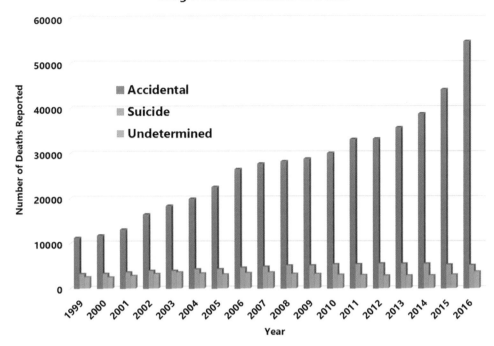

FIGURE 21.1 Compiled death rates due to drug overdose, 2009–2016. (Data obtained from the WONDER database, Centers for Disease Control and Prevention [cdc.gov].)

relatively unchanged; questions have been raised as to whether the number of suicides is being underestimated.

Why would this be important? In some circumstances, if the decedent committed suicide, this would have financial implications (e.g., life insurance) or implications based on religious beliefs of the family. The stigma of suicide is also a concern. Conversely, knowing that the death was a suicide is important in determining if signs were missed or if something can be learned to prevent others from committing suicides. As a result, a key goal of a psychological autopsy is to determine the presence or absence of **suicidal intent**.

Returning to our example, suppose that no suicide note was found in the apartment or on the young man's computer and that he had attempted suicide twice before. This is evidence of suicidal intent but not conclusive. Only about one in four people who commit suicide leave a note. On the other hand, suppose no note was found and despite past suicide attempts, his friends thought he was clean and trying to stay that way. Investigators learn that the victim had enrolled in online classes to start in a few weeks and had started attending support group meetings. A current grocery list was found on a pad in the kitchen. These findings argue against suicidal intent, at least in the recent past. The investigators will have to sort through all this type of evidence before a reasonable conclusion can be drawn. These are just examples; ultimately, the collection and analysis of a large body of evidence is used in the psychological autopsy. As with the actual physical autopsy, that information will be integrated with all other investigative information to draw a conclusion as to the manner of death.

21.3.2 CRIME SCENE STAGING

On occasion, investigators will find a crime scene that has been tampered with before the arrival of authorities. This is referred to as **crime scene staging**, which involves acts that are committed to conceal evidence or to change the perception regarding the manner of death. A homicide might be staged so that it appears that the death was natural, accidental, or a suicide, for example.

In other cases, there may not be criminal intent. A family member may move, dress, or clean a body of a loved one who committed suicide before calling first responders. Similarly, someone who commits suicide may try to stage the death as an accident to ensure that his relatives are not denied payment of a life insurance benefit from a policy that would otherwise be voided. There are also cases of suicides staged as homicides.

The key to identifying a staged scene is to look for elements that are internally inconsistent, which means they will not fit together logically, nor will they be properly supported by forensic evidence. This is the same concept that we discussed regarding crime scenes and forensic investigations in general. A correct hypothesis will be supported by evidence (physical and psychological) that is consistent across all findings. Consider a scene in which a man is found on the floor, a gunshot wound to the head and a gun nearby. It could have been an accident, a suicide, or a homicide, but whatever it was, all the evidence would have to be consistent with that hypothesis. From the behavioral science perspective, suicidal intent would be a key piece of evidence, while the key evidence from the scene would be the position of the body, livor mortis (Chapter 3), bloodstain patterns (Chapter 4), the nature of the wound (contact or distance), and all the other kinds of factors that we have noted in previous chapters. Suppose that a suicide note is found on a computer but patterns of livor mortis show that the body was moved and placed into a sitting position. This inconsistency is evidence of staging, which in turn is evidence of an attempt to conceal the true nature of the event.

As an example, suppose a husband murders his wife in a fit of rage, realizes what he has done, and stages the scene to look as if a burglar broke into the home, encountered the wife, and killed her. The crime is violent by nature and the injuries inflicted on the wife's body may reflect overkill. This is not consistent with an encounter with a typical burglar, who would tend to want to avoid confrontation with an owner and would have no underlying reason for overkill. Such a crime is not driven by rage as is the husband's act. These inconsistencies will alert the investigator to look for other inconsistencies in the physical and behavioral evidence.

MYTHS OF FORENSIC SCIENCE

Profiling is an art.

Profiling as a forensic and criminal investigative technique has reached an almost mythical status in the public mind through television shows, books, and movies. It is commonly thought that investigators that use profiling (a term that doesn't effectively capture all that goes into the process) do so based more on gut instinct and intuition than data. The hero of the book or movie catches the killer because he or she can "think like" the criminal. This concept works fine for entertainment but sells the discipline short. Databases and past cases are used to generate statistics, frequencies, and common elements of crimes based on many factors, such as time of day of killing, the type of victims, how the crimes are carried out, and features of the crime scene. Psychology and behavioral analysis are integral to criminal investigative analysis, but profiling has deep roots in data.

21.4 CRIMINAL INVESTIGATIVE ANALYSIS

The term *profiling* has become popularized in the media and fiction, but the term doesn't capture all the ways that behavioral science can be utilized during an investigation. Better descriptors are behavioral analysis or **criminal investigative analysis**. A significant part of any investigation of this type is based on accumulated data, statistics, and probability. This is contrary to the popular notion that behavioral science investigators use advanced psychology and uncanny insights to identify suspects and predict their behavior. Criminal investigative analysis integrates many techniques, including statistical analysis, a range of behavioral analysis tools, and crime scene data and characteristics, to provide information to investigators.

One example of such information would be to allow investigators to determine the likelihood that separate crimes are linked to a single offender (**linkage analysis**). For example, assume that police find three murder victims in their city over the course of a few weeks. They could be isolated incidents committed by different people, or they could be the work of a single individual. Physical evidence and forensic analysis would be part of such a determination, but behavioral information could significantly increase the chances of identifying linked offenses. If linkage proves likely, criminal investigative techniques might assist in narrowing the list of suspects to consider. Among these are the methods used to commit the crime, characteristics of the victims and how they were selected, and unique features of the crime scene. Common elements suggest the possibility of a single perpetrator. In 1985, the Federal Bureau of Investigation (FBI) established a database of unsolved homicides that incorporates information related to behaviors useful in linkage analysis. This database is referred to as the Violent Criminal Apprehension Program (ViCAP; www.fbi.gov/wanted/vicap), and it provides a repository of information regarding homicides, sexual assaults, and missing persons.

21.4.1 MODUS OPERANDI AND SIGNATURE ANALYSIS

A criminal's **modus operandi** (MO) is the methods that he or she uses to commit his or her crimes. If a car thief always breaks into parked cars in darkened garages while another breaks the windows of cars parked on the street, the MOs are different even though the crime (car theft) is the same. The MO encompasses the acts that are necessary to committing a crime, and it may change over time. The Unabomber is a serial bomber whose first known bomb was set in 1978. From then until his arrest in 1996, his bombs killed three people and injured many more. He constructed his bombs from simple and widely available materials and delivered the bombs mostly through the mail; this pattern describes his MO. Other characteristics of an MO are the types of victims, when and where they strike, and other approaches that are part of committing a crime.

A **signature** is a related concept describing characteristics found at the crime scene that are evidence of acts that are not essential to the crime itself. Elements of a signature provide insight into behaviors that are reflective of the offender's personality and motivations, but these elements are beyond the acts that are necessary to commit a crime (the MO). Some murderers pose the body of their victim; this is an element of a signature and not the MO because rearranging the body in a pose is not essential to committing the murder itself. As with the MO, a signature can be useful in linking crimes and limiting the suspect pool.

21.4.2 VICTIM SELECTION AND VICTIMOLOGY

The psychological processes involved in the selection of victims range from simplistic to complex, depending on the offender's motivation. The complexity is somewhat commensurate with the richness of the offender's fantasy, criminal sophistication, and constraints of the moment. When a victim is targeted, it is an indication of some level of association between the victim and the offender. Other victims may have merely had the misfortune of being "at the wrong place at the wrong time." Some offenders have idealized their victim type and involved them in their mental play acting; conversely, offenders may act with impulsivity and choose a victim simply based on their availability.

Some of the more elusive serial killers have established selection rules for their victims. Such rules are unknown to the victims, and how they respond to gambits thrown their way by the offender often determines their fate. In these situations, the offender creates scenarios and measures the victims' attitudes and responses to determine if they measure up to the standards he has established. The infamous Ted Bundy created an idealized standard for his victims wherein they had to be "worthy" of his selection.

One source of information for investigators trying to understand the crime scene involves detailed knowledge of the crime victim's life and lifestyle. The history of a victim in some ways defines the analysis of the crime. From one perspective, unsolved violent crime frequently results from an incomplete or faulty understanding of the victim. So important is this phase of investigation that the study of the victim must be thorough and undertaken immediately. People have a variety of personality factors, even quirks, which help define them and can contribute to their being targeted. The victim's characteristics may also offer insight regarding the personality of the attacker. Why was that person attacked instead of some other person? The goal is to develop sufficient **victimology** to be able to answer the question, "Was the victim's lifestyle a contributing factor toward victimization?" As the investigator initiates the search for the perpetrator, a full understanding of the crime scene and interactions between the victim and the offender may be elusive.

21.4.3 BODY DISPOSAL

Offenders have choices to make when confronted with the reality that they have murdered another person and now have a body to dispose of. The body disposal location and methodology are often revealing as to a prior victim and offender relationship, the offender's criminal sophistication level, the degree of planning by the offender in committing the crime, the attitude of the offender toward the victim as an individual, whether the victim represented a class of people, and the offender's knowledge of the body disposal site. The location may indicate whether the offender resides close to the crime scene and body disposal site. The choices facing the offender are limited. He may simply walk away from the body, abandoning it where it fell, or he may choose to spend a brief or considerable time concealing the body. The offender may arrange the body in a manner that, in his thinking, makes a statement about the victim. Furthermore, he may also move the body to another location and dump the remains at the second location.

To walk away from the body, abandoning it where it fell, may signal a lack of planning or forethought. It may also suggest a certain pattern of "disorder" that might lead to the conclusion that the crime was more spontaneous and impulsive, rather than controlled and thought out. The offender who invests time and effort in concealment of the body suggests that he has thought about the crime and is aware of the need to delay discovery of the body; the offender may need time to establish an alibi because his name will likely surface in the investigation. The concealment of the body also serves the purpose of concealing any linking evidence, thereby denying the investigator vital information on which to establish an investigation based on linkages to known associates or activities. A highly organized offender may choose the disposal site prior to selecting the victim.

By moving the body to another site, the offender creates multiple crime scenes (primary, secondary, etc., as we discussed in Section 2). The vehicle used for transportation becomes a crime scene, as does the new location. The significance of the acts is related directly to the thought process of the offender. If the movement is part of the process of concealing the body, those acts are likely from an organized offender, who has specific goals in mind to protect his identity or delay the discovery of the crime. The movement of the body may be to place it in an area where it will not be found or, if found, the area itself would be noted for specific activity, such as a lovers' lane or a trash dump site. The movement of the body may also accomplish the purpose of allowing the offender to simply dump the body in a more remote area. Such actions often provide critical hair, fiber, and DNA forensic evidence that can be found in vehicles; such evidence often links the victim to the specific offender.

MASS SHOOTINGS AND BEHAVIORAL SCIENCE

One of the most tragic events in any community is a mass shooting. Recent examples include the Las Vegas mass shooting in the summer of 2017, where 58 people were killed, and the one in February 2018, where a 19-year-old former student killed 12 at his former high school in Florida. In the Las Vegas case, the shooter committed suicide, while the 19-year-old walked away from the scene and was later captured.

Locations of mass shootings in the United States in 2015. (Image from http://www.gunviolencearc hive.org/reports/mass-shooting. This site maintains current listings of shooting incidents of all types across many years.)

Behavioral science is a key component of these tragic events, from trying to understand motives to attempting to develop strategies to prevent future occurrences. One aspect that has been studied is the presence or absence of suicidal intent in the shooters and how that might play into the type of crime committed. These individuals, rather than killing just those people they feel have aggrieved them, purposely attempt to kill as many people as possible regardless of whether the victims include people that have been purposely targeted. Recent data shows that only about 4% of murderers commit suicide and only about 12% of terrorist attacks involve suicide. In contrast, almost 40% of mass shooters kill themselves and another 10% are killed by police in a situation called 'suicide by cop.' In those instances, the suspects ignore commands or otherwise provoke police into killing them. It is known that mass shooters suffer from mental illness at a much higher rate than other murderers, but this is thought to be a contributing factor to the crime rather than the sole cause. Researchers found that killers who commit suicide tend to come to the scene with more weapons and ammunition, an observation that could reflect underlying suicidal intent. Men who commit mass shootings in former workplaces also seem more prone to suicide by cop behavior. Researchers suggested that those who feel the angriest and abused are the ones who try to kill the most people but are also the ones who are likely to feel the most self-hatred and are the most likely to kill themselves as part of the crime. The value in these types of studies is that they can provide information for first responders as well as hopefully provide tools that can be used to prevent future incidents.

Source: Lankford, A., Mass Shooters in the USA, 1966–2010: Differences between Attackers Who Live and Die, *Justice Quarterly* 32, no. 2 (2015): 360–379.

21.5 REVIEW MATERIAL: KEY CONCEPTS AND QUESTIONS

21.5.1 KEY TERMS AND CONCEPTS

Competency
Crime scene staging
Criminal investigative analysis
Dissimulation
Insanity plea
Linkage analysis
Malingering
McNaughten rules
Modus operandi
Psychological autopsy
Signature
Simulation
Suicidal intent
Victimology

21.5.2 REVIEW QUESTIONS

1. What is the difference between a criminal's MO and his or her signature?
2. How does behavioral evidence differ from physical evidence?
3. How does physical evidence relate to behavior?

21.5.3 ADVANCED QUESTIONS AND EXERCISES

1. Look at the case study in Chapter 6 regarding cyanide poisoning. How would you describe this man's suicidal intent?
2. How would behavioral science be applied to the crime scene scenario described in Chapter 3? List at least three specifics based on what you have learned in this chapter.

BIBLIOGRAPHY AND FURTHER READING

BOOKS

Huss, M. T. *Forensic Psychology*. 2nd ed. Hoboken, NJ: John Wiley & Sons, 2014.
Ramsland, K. *The Psychology of Death Investigations: Behavioral Analysis for Psychological Autopsy and Criminal Profiling*. Boca Raton, FL: CRC Press, 2018.

ARTICLES

Edmond, G., Towler, A., Growns, B., Ribeiro, G., Found, B., White, D., Ballantyne, K., et al. Thinking Forensics: Cognitive Science for Forensic Practitioners. *Science and Justice* 57, no. 2 (Mar 2017): 144–154.
Franchi, A., Bagur, J., Lemoine, P., Maucort-Boulch, D., Malicier, D., and Maujean, G. Forensic Autopsy of People Having Committed Suicide in 2002 and in 2012: Comparison of Epidemiological and Sociological Data. *Journal of Forensic Sciences* 61, no. 1 (Jan 2016): 109–115.
Kaltiala-Heino, R., and Eronen, M. Ethical Issues in Child and Adolescent Forensic Psychiatry: A Review. *Journal of Forensic Psychiatry & Psychology* 26, no. 6 (Nov 2015): 759–780.
Shapiro, S., and Rotter, M. Graphic Depictions: Portrayals of Mental Illness in Video Games. *Journal of Forensic Sciences* 61, no. 6 (Nov 2016): 1592–1595.

WEBSITES

Link	Description
https://leb.fbi.gov/articles/featured-articles/criminal-investigative-analysis-practitioner-perspectives-part-one-of-four	Articles from the *FBI Law Enforcement Bulletin* describing criminal investigative analysis
https://www2.fbi.gov/hq/isd/cirg/ncavc.htm	National Center for the Analysis of Violent Crime
https://www.fbi.gov/history/famous-cases/unabomber	Compilation of well-known FBI cases with video; includes the Unabomber case

Appendix A: Abbreviations

A	Adenosine
AAFS	American Academy of Forensic Sciences
ABC	American Board of Criminalistics
ABP	American Board of Pathology
ACE-V	Analysis, comparison, evaluation, and verification
ADME	Absorption, distribution, metabolism, and excretion
AFIS	Automated Fingerprint Identification System
ALS	Alternate light source
ASCLD	American Association of Crime Laboratory Directors
ASCLD/LAB	American Association of Crime Laboratory Directors/Laboratory Accreditation Board
BAC	Blood alcohol concentration
bp	Base pair
BPA	Bloodstain pattern analysis
BrAC	Breath alcohol concentration
BSE	Backscattered electron
C	Cytosine
CE	Capillary electrophoresis
CE	Chemical energy
CIA	Criminal investigative analysis
CODIS	Combined DNA Index System
Cp	Peak plasma concentration
CSA	Controlled Substances Act
CSI	Crime scene investigator
CT	Computer-aided tomography
DEA	Drug Enforcement Administration
DFO	1,8-Diazafluoren-9-one
DUI	Driving under the influence
ED	Effective dose
EDS	Energy-dispersive x-ray spectroscopy
EP	Eyepiece lens in a microscope
ESDA	Electrostatic detection apparatus
F/A ratio	Fuel-to-air ratio
FBI	Federal Bureau of Investigation
G	Guanine
GC/MS	Gas chromatography–mass spectrometry
GPR	Ground-penetrating radar
GPS	Global positioning system
GRIM	Glass refractive index measurement instrument
GSR	Gunshot residue
IAFIS	Integrated Automated Fingerprint Identification System
IBIS	Integrated Ballistic Identification System
IED	Improvised explosive device
IoT	Internet of things
IR	Infrared radiation
K	Known or control sample
KE	Kinetic energy
LCM	Laser capture microdissection
LD	Lethal dose

LFL	Lower flammability limit
MDI	Medicolegal death investigator
ME	Medical examiner
MO	Modus operandi
MRI	Magnetic resonance imaging
MS	Mass spectrometry
MSP	Microspectrophotometry
mtDNA	Mitochondrial DNA
NAME	National Association of Medical Examiners
NASH	Natural, accidental, suicidal, or homicidal
NG	Nitroglycerin
NIST	National Institute of Standards and Technology
NPS	Novel psychoactive substance
OBJ	Objective lens in a microscope
OS	Operating system
OSAC	Organization of Scientific Area Committees
OTC	Over the counter
P2P	Peer to peer
PCR	Polymerase chain reaction
PLM	Polarized light microscopy
PMI	Postmortem interval
PMR	Postmortem distribution
PSA	Prostate-specific antigen
Q	Questioned sample
QA/QC	Quality assurance/quality control
R	Retardation
RBC	Red blood cell
RI	Refractive index
SAP	Seminal acid phosphatase
SE	Secondary electron
SEM	Scanning electron microscopy
SOFT	Society of Forensic Toxicology
SOP	Standard operating procedure
SPR	Small particle reagent
STR	Short tandem repeat
SWGSTAIN	Scientific Working Group on Bloodstain Pattern Analysis
T	Thymine
THC	Tetrahydrocannabinol
TM	Total magnification
UFL	Upper flammability limit
ViCAP	Violent Criminals Apprehension Program
VoIP	Voice over Internet Protocol
VSC	Video spectral comparison
WBC	White blood cell

Appendix B: Glossary

Abductive reasoning: The process of reasoning and thinking in which known pieces of evidence are considered to produce a theory that represents the most likely and usually the simplest explanation.

Accelerant: Substance used to ignite or support a fire.

Accidental marks: Marks that arise from wear or random interaction, such as on shoe soles.

Accidental whorl: A type of whorl pattern in fingerprints.

Accreditation: The process that ensures that laboratories are following standard protocols and procedures set forth by the accrediting body.

Acid phosphatase: A group of enzymes. Seminal acid phosphatase is used in screening tests for seminal fluid.

Adjudicated: The settlement of a legal matter.

Admissibility: Allowing something (physical or testimonial) to be admitted as evidence in a legal proceeding.

Adversarial system: The process in which the merit of opposing sides is evaluated and settled by argument.

Algor mortis: Cooling of the body that occurs after death.

Alkaloids: A group of plant-based drugs such as cocaine, morphine, and caffeine.

Allelic ladder: A set of reference base pairs that are used to calibrate DNA analyses.

Allometry: Patterns of relationships between bone lengths that can be used to estimate stature.

Amorphous: Unordered, without organization. Glass is an amorphous substance.

Amplicons: The amplified DNA that results from the polymerase chain reaction process.

Amylase: A family of enzymes that break down starches and sugars. They are targeted in tests for saliva.

Analogs: Compounds purposely synthesized to have similar effects as controlled or illegal drugs.

Analyzer: In polarizing light microscopy, a second polarizing filter.

Ancillary: Associated with; for example, digital evidence is often part of a criminal investigation even if the data itself is not central to the commission of a crime.

Angle of impact: Acute or internal angle formed by the direction of a blood drop and the plane of the surface it strikes.

Anisotropic: A substance that does not have the same refractive index throughout.

Annealing: In the polymerase chain reaction cycle, binding of free nucleotides with short tandem repeat loci on open DNA strands.

Anorexic: Drugs that suppress appetite.

Antemortem: Before death.

Anti-Drug Abuse Act: An extension of the Controlled Substances Act meant to control analogs.

Apocrine gland: A type of sweat gland.

Aqueous: A system or environment that is based on water.

Arch: A type of pattern in fingerprints.

Area of convergence: A point or area to which a bloodstain pattern can be projected on a two-dimensional surface. It is determined by tracing the long axes of well-defined bloodstains within the pattern back to a common point or area.

Area of origin: A three-dimensional point or area from which the blood that produced a bloodstain originated. Determined by projecting angles of impact of well-defined bloodstains back to an axis constructed through the point or area of convergence.

Arson: Intentional setting of a fire.

Arterial spurt: Bloodstain patterns resulting from blood exiting the body under pressure from a breached artery.

Asphyxia: Interruption of the oxygen flow to the brain.

Autopsy: Postmortem dissection of a human body.

Azoospermia: Condition in which seminal fluid does not contain sperm cells.

Backdraft: An eruption of flame when oxygen is suddenly reintroduced to an oxygen-depleted zone.

Backscattered electrons: In scanning electron microscopy, electrons scattered off the surface and used to generate an image.

Back spatter: Blood directed back toward the source of energy or force that caused the spatter. Back spatter is often associated with entrance gunshot wounds.

Ball powder process: A method of making gunpowder propellants by dropping a heating mixture into water.

Base pairs: In DNA, A-T or C-G nucleotides.

Bath salts: A term for synthetic stimulants.

Becke line: A line visible under a microscope between a sample and the immersion liquid. How it moves as a function of defocusing and focusing is useful for refractive index measurements.

Bias: An inclination, conscious or unconscious, to interpret evidence and draw conclusions that favor one result over another.

Bifurcation: A fingerprint ridge feature in which ridges divide.

Biological profile: A collection of data from skeletal remains, such as size, shapes, estimation of stature, and unique features of the bones.

Biometric identifiers: Physical features of a body considered unique to an individual.

Birefringence: The difference in the refractive indices of a material that has two refractive indices.

Black box studies: Studies that use blind samples to determine how a methodology performs across many examiners and conditions.

Blank: A sample that does not contain any of the compounds or materials associated with a specific analysis.

Blasting cap: A device used to initiate a detonation.

Blast wave: Zone(s) of compressed air and hot expanding gases created by an explosion.

Blind sample: A sample provided to an analyst with a composition that is known to the submitter but not to the analyst.

Blood alcohol concentration: The concentration of alcohol in blood, usually expressed as a percent and used to determine if a person is legally intoxicated.

Breath alcohol: The concentration of ethanol in exhaled breath that can be used to estimate the blood alcohol concentration.

Breathalyzer: A device used to measure the amount of alcohol in exhaled breath.

Breech face marks: Marks on the base of a cartridge of ammunition created when the cartridge is driven backwards into the back of the chamber.

Buccal swab: A cotton swab swiped along the inside of the cheek to collect DNA.

Burden of proof: The responsibility of establishing proof within a legal setting.

Caliber: The dimension of the barrel of a gun (nominally). Also refers to the size of a bullet.

Cannabinoids: Substances that produce physiological effects similar to the active ingredients in marijuana (tetrahydrocannabinol).

Carboxyhemoglobin: Hemoglobin that is bound to carbon monoxide instead of oxygen.

Carfentanil: An extremely potent synthetic opioid.

Carrion: The decaying flesh of an animal.

Cartridge: A self-contained unit of ammunition.

Case manager: The person at a forensic laboratory who oversees and organizes different types of forensic testing on evidence from one scene or one crime.

Castoff patterns: Bloodstain pattern created when blood is released or thrown from a blood-bearing object in motion.

Cathinone: The active ingredient in Khat leaves; stimulant.

Cause of death: The disease or injury that started an irreversible chain of events leading to death.

Cellular network: A network of zones associated with towers that create the basis of a phone network.

Centerfire: An ammunition cartridge in which the primer is mounted in the center of the base.

Central blood: Blood sample taken from the heart during autopsy.

Central pocket whorl: A type of whorl pattern in fingerprints.

Certification: A process that includes written tests and sometimes laboratory work that if passed ensures that the analyst can perform work at the level specified by the certification body.

Chain of custody: The documented process the evidence goes through from the point of gathering to the final presentation in court. Intended to ensure that there has been no tampering with or altering of the evidence.

Chemical energy: A form of energy that is stored in chemical bonds.

Chemical heat: Heat released by chemical reaction.

Christmas tree stain: A staining procedure used to highlight sperm cells for visualization under a microscope.

Chromogenic substance: A compound that can react to form a colored substance.

Chromosomes: Found in the nucleus of cells and composed of DNA. They carry genetic traits.

Circumstantial evidence: Evidence that indirectly attests to a fact.

Civil law: The legal system that deals with noncriminal matters or disputes between individuals.

Clandestine grave: A grave that is meant to remain hidden.

Class characteristic: A characteristic that allows items to be sorted into different classes. Sex is a class characteristic of people that allows them to be sorted into categories of men and women.

Clearance rate: The rate at which a drug or metabolite is removed from general circulation in the bloodstream.

Cloud: Collectively describes storage locations for data that are not localized to a specific personal device.

Comparison microscope: An instrument made of two microscopes linked by an optical bridge.

Competency: Measure of ability to stand trial, understand the proceedings, and so forth.

Comprehensive Crime Control Act: Federal legislation that extended the Controlled Substances Act.

Compression heat: Heat generated when a gas is compressed.

Concentric cracks: In broken glass, cracks that are arranged in a roughly circular pattern around the impact site.

Conchoidal lines: Lines formed in glass as a result of an impact.

Conduction: Heat transfer through direct contact.

Connecting strokes: In handwriting, the strokes that connect one letter to the next.

Contact wound: Gunshot wound caused by contact of the weapon with skin.

Contextual information: Information about how evidence was found, collected, or evaluated; the story surrounding a piece of evidence.

Controlled Substances Act: Federal legislation that defines what drugs are illegal. Drugs are classified by schedule.

Control samples: Samples collected at a crime scene for comparison purposes.

Convection: Heat transfer by a circulating substance such as air.

Core: A type of minutia feature in fingerprints.

Coroner: A government official responsible for death investigations and issuing death certificates. In the United States, this position is elected or appointed.

Cortical bone: The dense outer layer of bone.

Crime scene map: A map of a crime scene that results from the use of GPS instrumentation.

Crime scene reconstruction: A reenactment of a crime undertaken based on evidence collected and what hypotheses have been proposed to explain it.

Crime scene staging: Purposeful manipulation of a crime scene designed to confuse or mislead investigators.

Criminal investigative analysis: A suite of investigative and statistical techniques used in criminal investigation that focus on behavior and behavioral evidence.

Criminal law: The legal system that deals with criminal acts.

Cross-reactivity: In an immunoassay, a reaction that is not the same as the intended one.

Cyanoacrylate: Informally, superglue; used to develop latent fingerprints.

Cybercrime: Crimes committed with computers.

***Daubert* standard:** Name for a 1993 court decision that set forth the concept of the judge as a gatekeeper for determining the admissibility of evidence.

***Daubert* trilogy:** Three case rulings in the 1990s that set the concept of the judge as a gatekeeper for determining the admissibility of evidence and that added relevancy to the criteria as well as allowing a range of experts to testify.

Dead load: The total loads in a structure; divided into live load and dead load.

Deciduous dentition: A person's baby teeth.

Deductive reasoning: Reasoning that is based on known facts that can be logically applied.

Defendant: The entity or person who has been charged with a crime or the subject of a dispute filed by a plaintiff.

Deflagration: A burning or combustion reaction in which the speed of propagation is slower than the speed of sound in that medium (usually air).

Degree of intoxication: The extent to which a person is impaired or unable to function normally.

Delta: A type of pattern in fingerprints.

Delusterant: Substance in paint or other materials placed there to reduce the shine.

Dental stone: A plaster-like material that when mixed has a consistency like pancake batter and that hardens to a mold. It is used to cast impressions such as shoeprints in mud.

Depressants: Drugs that cause suppression of the central nervous system processes; cause sleepiness.

Designer drugs: Another term for novel psychoactive substances.

Detonation: Reaction in which the speed of the reaction exceeds the speed of sound.

Detonation cord: A ropelike material designed to be part of the initiation system of an explosive.

Detonation velocity: How fast a detonation front propagates through the media it is in.

Differential extractions: In DNA analysis, extractions designed to separate male from female DNA contributions.

Digital evidence: A category of evidence that is stored electronically, such as digital images, videos, and PDF files.

Direct evidence: Evidence that directly determines a fact.

Directionality: Parameter that indicates the direction the blood was traveling when it impacted the target surface. Directionality of flight can usually be established from the geometric shape of the bloodstain.

Disarticulated: Separation of the parts of the body; skeleton and joints disconnected.

Disguised writing: Forgery in which there is a deliberate attempt to alter writing, such as using the opposite hand to write than usually used.

Dissimulation: Denying the presence of mental illness.

Distance determination/estimation: A series of experiments used to estimate the distance of the muzzle of a weapon from a target.

DNA polymerase: An enzyme that facilitates the recombination of separated strands of DNA.

Dot: A type of minutia feature in fingerprints.

Double action: In a firearm, a design in which pulling the trigger cocks and fires the weapon.

Double base: A propellant in which nitrocellulose and nitroglycerin are the energetic materials.

Double-blind sample: A sample submitted to a laboratory that has a composition known to the submitter but not to anyone in the laboratory or to the analyst performing the test.

Double-loop whorl: A type of whorl fingerprint pattern.

Drip patterns: Bloodstain pattern resulting from blood dripping into blood.

Drug: Substance that when ingested causes a physiological response.

Drug paraphernalia: Items used as part of the ingestion of illicit drugs.

Dry origin impression: Impressions such as from footwear in which the impression is made in a dry material such as dust.

Dynamic load: A load in a structure that can change rapidly, such as when a gust of wind hits it.

Eccrine gland: A type of sweat gland.

Effective dose: The dose of a drug that generates the desired physiological response.

Ejector: A mechanism in a semiautomatic weapon that throws a cartridge clear of the weapon after discharge.

Ejector marks: Marks on a cartridge of ammunition created by the ejector mechanism.

Electrical heat: Heat produced by electrical conditions, such as shorts or resistance.

Electropherogram: The chart output of short tandem repeat DNA typing.

Elimination samples: Samples collected from individuals authorized to be present at a scene to allow materials arising from their presence, such as fingerprints or DNA, from being confused with samples relevant to the case.

Entomology: The study of insects.

Equifinality: Occurs when different processes produce similar types of bone damage.

Equivocal death: A death in which there is some question as to why and how it occurred; a questioned death.

Erythrocytes: Red blood cells.

Exclusionary evidence: Evidence that excludes the possibility of a common source with Q versus K.

Exclusive evidence: Evidence that tends to exclude a person or a possibility.

Exculpatory: A type of evidence that tends to exclude a suspect as a possible source.

Exemplars: Examples of writing.

Expired bloodstain pattern: Blood propelled from the nose, mouth, or a wound as a result of air pressure and/or airflow.

Explosion: A rapid and often violent release and expansion of hot gases.

Exsanguination: Death after a significant amount (usually half or more) of blood is lost. Bleeding to death.

Extruded powder process: A method of making gunpowder propellants by forcing them through a mold and cutting.

False negative: Occurs when a test yields a negative result when the target substance is actually present.

False positive: Occurs when a test yields a positive result when the target substance is not present.

Falsifiability: A method of presenting results such that others have sufficient information to repeat the test and demonstrate that the original results were true or false.

Federal Rules of Evidence: The rules that govern the admissibility of evidence in federal cases.

Felony: A serious crime, the conviction of which can result in lengthy prison time or death.

Fentanyl: A synthetic opioid.

Fire point: Temperature at which a liquid produces sufficient vapors to support combustion.

Fire tetrahedron: A version of the fire triangle that shows the role of complex chemical chain reactions.

Firing pin: The part of a firearm that strikes the primer.

First law of thermodynamics: Paraphrased as energy is neither created nor destroyed, and in any process all that can happen is a conversion.

First-pass metabolism: A process that occurs in the liver after a drug is ingested orally.

First responder: The first person to arrive at a crime scene in an official capacity, such as police, fire, or rescue personnel.

Flame front: The leading edge of a deflagration.

Flammability range: The concentration range (fuel to air) that supports combustion.

Flanking region: The region adjacent to a short tandem repeat locus. Target of the primer in a polymerase chain reaction.

Flash point: Temperature at which a liquid emits enough vapor to support a combustible mixture at the surface.

Fluorescence: Emission of light due to illumination with another source, such as a laser.

Forensic anthropology: The application of the theory and methods of anthropology to forensic problems.

Forensic archaeology: The application of archaeological methods to the recovery of forensically relevant human remains.

Forensic mapping: The process of using GPS and related scanning technology to create a digital map of a scene.

Forensic pathologist: A doctor who specializes in the diagnosis of disease and determination of cause of death.

Forward spatter: Blood that travels in the same direction as the source of energy or force; often associated with exit gunshot wounds.

Freehand simulations: A method of attempting to emulate another person's writing.

Friction ridge skin: Outer skin layer of humans and primates.

***Frye* standard:** Named for a 1923 ruling that set forth the standard of general acceptance by the relevant scientific community as a requirement for admissibility of scientific evidence.

Fuel-to-air ratio: The volume ratio of fuel to air used to evaluate the potential for combustion to occur.

Gastric contents: Stomach contents collected at autopsy.

Gauge: The size of a shotgun barrel. A smaller gauge corresponds to a larger barrel.

Gene: A portion on a chromosome that codes for a protein.

Grabbers: Devices that grip and pull paper through copiers and printers.

Grooves: The cutaway portion of the rifling of a firearm barrel.

Ground-penetrating radar: A type of radar that can be used to visual and find objects buried in soil.

Gunshot residue: Particulate residues produced from the primer when a weapon is fired.

Hacking: Digitally breaking into a website or database. Evading digital security measures to obtain access to information.

Hammer: The part of a firearm that moves backward and then drives the firing pin into the primer.

Hashing: A digital computation that is used to ensure that a digital copy is an exact duplicate of the original.

Headstamp: Manufacturer markings on the base of a cartridge of ammunition.

Hemastix®: A screening test for blood that uses a commercial test strip.

Heme group: The portion of the hemoglobin molecule that complexes with oxygen.

Hemoglobin: The substance in red blood cells that carries oxygen.

Henry system: A system used to classify fingerprints in the West.

Heterozygous: In genetics, two different genes or inherited sequences.

Hinge lifters: A hinged pocket with tape used to lift fingerprints.

Homozygous: The inherited sequence or gene is the same from the mother and the father.

Hydrocarbon: *See* petroleum distillates.

Hyoid bone: Small delicate bone in the throat that is sometimes broken during manual strangulation.

Hypervariable regions: Regions in mitochondrial DNA that have genetic variation.

Identity theft: Use of digital technologies to assume another's identifying information for malicious or criminal purposes.

Ignitable liquid: A type of accelerant, such as charcoal lighter fluid or gasoline.

Ignition temperature: The minimum temperature at which a substance can ignite.

Ignition time: The time of direct contact between an ignition source and a substance required to cause ignition.

Imaging: A process used to create exact duplicates of computer files.

Immersion method: A method of determining the refractive index of a substance by immersing it in a liquid of a known refractive index.

Immunoassay: A method of analysis that relies on reactions between antigens and antibodies.

Immunoglobulin: A molecule created by joining a drug to a protein.

Impact spatter: Stain produced when liquid blood strikes a surface.

Incendiary device: Device used to ignite an intentionally set fire.

Incendiary fire: Intentionally set fire; arson.

Incised wound: Cutting or slicing wound.

Inclusionary evidence: Evidence that includes the possibility of a common source with Q versus K.

Inclusive evidence: Evidence that tends to include a person or possibility.

Inculpatory evidence: A type of evidence that tends to include a suspect as a possible source.

Indented writing: An image of writing transferred by pen pressure from pages above the page where indented writing is found.

Indictive reasoning: Reasoning that is based on generalizations based on data.

Individualization: An outdated term meant to link an item of evidence to one and only one possible source.

Ingestion: Taking a substance into the body.

Inorganic substance: A compound that does not contain carbon. Exceptions do exist, such as carbon dioxide (CO_2).

Insanity plea: A plea in judicial proceedings that invokes mental illness as the driver of criminal behavior.

Instar: Stages of development in a maggot.

Intellectual property: A creative work or idea rather than a tangible thing that can be stolen digitally.
Interference colors: Colors that result when light waves interfere. Used in the context of polarizing light microscopy.
Internally consistent: Occurs when all evidence is supporting one and only one explanation; no contradictory information.
Internet of things: Devices that are linked to the Internet, such as smart speakers, phones, and televisions.
Interoperability: The ability of different Automated Fingerprint Identification System computer systems to link up and communicate among themselves.
Iodine fuming: A method of visualizing latent fingerprints.
Irreversible work: A conversion of energy that cannot be reversed.
Island: A minutia feature of fingerprints.
Junk science: A practice or process that is represented as scientific but that does not use science or scientific methods.
Jurisdiction: A region or geographical area over which law enforcement or a legal entity can excise authority.
Kastle–Meyer test: A screening test for blood that uses phenolphthalein.
Khat: The leaves of this plant are chewed to release cathinone, a stimulant.
Kinetic energy: The energy associated with movement. Skidding is an example.
Known: Same as a positive control.
Laceration: Tearing wound.
Lacquer: Paint in which the vehicle evaporates to form the surface.
Lands: The elevated portion of rifling.
Lands and grooves: The pattern of rifling of a firearm barrel.
Latent print: A fingerprint left on a surface such that is is invisible or nearly so.
Lead snowstorm: X-ray pattern seen in a firearm wound when bullets fragment.
Lethal dose: The dose of a drug that caused death.
Leukocytes: White blood cells.
Ligature: Something used to tie or bind, like a rope or string.
Linear burn rate: How fast a burning reaction moves when set up in a straight-line type of experiment.
Linkage analysis: Use of evidence to determine if separate crimes could have been committed by the same person.
Live load: The portion of static load that changes with environmental conditions.
Livor mortis/lividity: Pooling of blood driven by gravity that occurs after death.
Locard's exchange principle: Paraphrased as "every contact leaves a trace," a foundational concept of forensic science and trace evidence attributed to Edmund Locard.
Long guns: Rifles and shotguns; firearms too large to be handheld.
Loop: A type of pattern in fingerprints.
Lower flammability limit: The lowest concentration of fuel-to-air ratio that supports combustion.
Luminol: A compound that reacts in the presence of blood and emits light.
Macroscopic crime scene: Description based on the size of a crime scene. The overall or "big picture" crime scene.
Magna Brush: A device used to deliver magnet fingerprint powder.
Malingering: Faking of mental illness.
Malware: Malicious software. Software designed to be secretly loaded into digital devices for criminal or malicious purposes.
Manual strangulation: Causing asphyxia using the hands pressed around the throat.
McNaughten rules: Early formulation of rules and tests to determining if a person is mentally ill.
Mechanical heat: Heat produced mechanically, such as through friction.
Mechanical trauma: The results of an applied force exceeding the strength of skin to resist it.
Mechanism of death: The medical/biochemical/physiological abnormality leading to a death.
Medicolegal death investigator: A person who responds to deaths and death scenes and is part of a death investigation system.
Medicolegal investigation of death: Procedures and practices used to investigate deaths in which the cause is not immediately obvious.
Michel–Lévy chart: A chart that uses the observed interference colors and sample thickness to determine birefringence.
Microanalysis: The use of microscopic analysis in the characterization of evidence.
Micrometry: Measurements on a microscopic scale.
Microscopic crime scene/microscene: Crime scene description based on the type of physical evidence present.
Microspectrophotometry: A combination of a microscope with a spectrometer.
Minutia: A fingerprint visible to the naked eye.
Misdemeanor: A less serious crime that can be punished by sentences including fines and confinement.
Misting: Blood reduced to a fine spray as the result of the application of energy or force.

Mitochondria: A small organelle in cells that contains DNA.

Mode of ingestion/mode of administration: The method or route by which a substance is taken into the body.

Modus operandi: The specific methods, procedures, or techniques used by an offender in the commission of a crime.

Molecular explosive: A molecule that contains fuel and oxygen.

Morphine: The first opiate extracted from opium.

Necropsy: Postmortem dissection of an animal.

Negative control: Same as a blank.

Ninhydrin: A chemical used to visualize latent fingerprints.

Nitrocellulose: A key ingredient in propellants, made by treating cotton with acid. Also called guncotton.

Nitroglycerin: A key ingredient in propellants, made by treating glycerin with acid.

Nonrequest writing: Writing samples collected from a person's past writing.

Normal hand forgery: Forgery of writing using the same hand as usually used to write.

Novel psychoactive substances: Synthetic compounds specifically designed to mimic the effects of existing illegal drugs to skirt around regulation and control.

Nuclear DNA: DNA found in the nucleus of the cell.

Nucleotide: Chemical unit in DNA connected to a sugar–phosphate backbone; here one of A, T, C, or G.

Oblique lighting: Illumination from a small angle; sideways illumination.

Odontology: The study of teeth.

Operating system: The software that is used to control a computer or device, such as Microsoft Windows or Apple iOS.

Opiates: Substances derived from the opium plant or that are chemically similar to morphine.

Opinion evidence: Testimonial evidence in which an expert offers an opinion based on analysis or data.

Opioids: Synthetic opiates.

Optical bridge: A method of linking two microscopes into a single comparison microscope system.

Organic substance: A compound that contains carbon.

Ossification centers: Locations where bone growth begins.

Osteoarthritis: A degenerative disease of the joints that is usually the result of aging. The wearing down of cartilage over time.

Osteology: The study of bones.

Osteometry: The technique of bone measurement used by forensic anthropologists.

Osteoporosis: A loss of bone, usually due to age.

Overpressure: *See* blast wave.

Over-the-counter drugs: Drugs available without a prescription, such as aspirin or cold medicines.

Oxidation: A type of chemical reaction.

Parent drop: Drop of blood from which a wave castoff or satellite spatter originates.

Passive drop: Blood that drops as a result of gravity and not additional force, such as blood dripping from a bloody nose.

Patent print: A visible fingerprint.

Pattern matching: Forensic disciplines that involve comparison of patterns such as fingerprints, firearms, and handwriting analysis.

Peer review: The process in science in which results are presented to other scientists and reviewed before publication.

Peer-to-peer: A protocol used by computers and devices to communicate over the Internet.

Pen lift: A characteristic of handwriting.

Perimortem: Events that occurred very near to the time of death.

Peripheral blood: Blood sample taken from the femoral vein in the leg at autopsy.

Petroleum distillates: Compounds and mixtures derived from oil.

Petroleum hydrocarbons: *See* petroleum distillates.

Phadebas®: A reagent used in screening tests for saliva.

Phenolphthalein: A compound that changes color in the presence of blood in the Kastle–Meyer test; a weak acid/base indicator.

Phishing: A type of cybercrime in which an email is disguised as representing a legitimate sender or source.

Physical developer: A type of chemical used to visualize latent fingerprints.

Physical evidence: Tangible materials and evidence analyzed by forensic scientists.

Pipe bomb: An improvised explosive device made from pipe (metal or plastic) and an explosive such as gunpowder.

Plaintiff: In civil law, the party that brings the suit against another entity.

Plane polarized light: Light that is polarized so that it vibrates in only one plane.

Plasma: The fluid portion of blood.

Plastic deformation: A warping in bone fragments that is caused by blunt force trauma.

Plastic print: A fingerprint impression in a soft surface.

Platelets: Cells in blood involved in clotting.

Plea bargain: An agreement negotiated between a prosecutor and someone charged with a crime. The accused typically agrees to plead guilty to a lesser charge in exchange for a less harsh sentence.

Point of origin: Location where a fire begins.

Poison: Toxin. A substance that when ingested in a sufficient amount will generate a toxic response.

Polarizer: A filter that polarizes light into one direction of vibration.

Polygenetic: A trait that is controlled by more than one gene.

Polymerase chain reaction: A process used to replicate DNA at targeted loci.

Polymorphic: More than one form.

Positive control: A sample with a known composition.

Postmortem: After death.

Postmortem interval: Time since death.

Postmortem toxicology: The study of drugs and poisons in the context of medicolegal death investigation.

Powder dusting: A method of visualizing latent fingerprints.

Precedent: Previous court ruling and opinions.

Precursor: A compound that is the starting point of the synthesis of another drug. Pseudoephedrine is a precursor of methamphetamine, for example.

Press test: A test in which an absorbent paper is pressed against a sample to collect sample. The paper is then tested.

Pressure wave: *See* blast wave.

Presumptive test: A simple test designed to determine if a substance is likely present; indicative of presence, not definitive.

Primary container: The first layer or container used to enclose evidence collected at a scene.

Primary crime scene: Description of a crime scene based on the location of the original criminal activity; the original scene.

Primary transfer: A transfer directly from the source of interest.

Primer: A small section of a cartridge of ammunition that contains a shock-sensitive explosive that ignites the propellant.

Primers: In DNA typing, the compounds used to locate and bind to flanking regions of short tandem repeat loci.

Privileged direction: The one direction of vibration of light that passes through a polarizing filter.

Probable cause: A reasonable basis of belief or suspicion of a crime being committed.

Propagation reaction: A reaction or process that moves in a direction like a flame front.

Propellant: Gunpowder. Material that forms hot expanding gases used to propel a bullet forward.

Prosecution: In criminal law, the government representative.

Prosecutorial bias: Bias that favors the prosecution's position.

Prostate-specific antigen: A substance that is used to confirm the presence of seminal fluid.

Pseudoscience: Another term for junk science.

Psychological autopsy: A postmortem evaluation of a person's behavior and state of mind at and near the time of death.

Pubic symphysis: A portion of the pelvis useful in determining the biological sex of an individual.

Puparial: A stage in the life cycle of an insect that follows the larval stage (here, maggot).

Pyrolysis zone: Area in a combustion where thermal degradation is occurring.

Quality assurance/quality control: Procedures and practices used to ensure that data and results reported are reliable.

Questioned death: A death in which there is some question as to why and how it occurred. An equivocal death.

Radial cracks: In broken glass, cracks that originate at the impact and radiate away.

Radial loop: A fingerprint loop that flows toward the radius bone in the arm.

Radiation: Heat in the form of infrared radiation.

Random match probability: The probability that another sample collected at random will share the exact same characteristics as the item or evidence in question.

Ransomware: Software secretly installed on a computer that denies user access until a ransom is paid.

Refractive index: The degree to which a sample bends or refracts light.

Remodeling: The process of bone shape changes.

Reproducibility: The ability to recreate results of an analysis and obtain comparable or identical results.

Requested writing: Writing collected at request as part of an investigation.

Retardation: The difference in refractive indices multiplied by a sample's thickness.

Retention time: In gas chromatography linked to mass spectrometry, the time from injection to emergence of a compound from the column and into the detector.

Revolver: A handgun that uses a revolving cylinder to deliver cartridges to the chamber.

Rifling: Machining of the barrel of a firearm that causes the bullet to spin when ejected from the muzzle.

Rigor mortis: Stiffening of muscles that occurs after death.

Rimfire cartridge: An ammunition cartridge in which the primer is wrapped around the base of the cartridge.

Royalty: A percentage payment to the originator of intellectual property; for example, payment to a musician for each use of a song they composed and performed.

Ruhemann's purple: *See* ninhydrin.

Rules of evidence: Rules put in place to assist judges in determining admissibility of evidence.

Satellite spatter: Small droplets of blood projected around a drop of blood upon impact with a surface. A wave castoff is considered a form of satellite spatter.

Sciatic notch: A portion of the pelvis useful in determining the biological sex of an individual.

Screening test: *See* presumptive test.

Secondary container: The second level of protection or container used for evidence collected at a scene.

Secondary electrons: In scanning electron microscopy, electrons that are emitted from the sample as a result of electron bombardment.

Secondary scene: Crime scene location after the original or primary crime scene.

Secondary transfer: A transfer that is one or more removed from the primary transfer.

Second law of thermodynamics: Paraphrased as the conversion of energy is never 100% efficient.

Seized drugs: Substances submitted as physical evidence and suspected to contain illegal drugs.

Self-correction: A process that is designed to favor improvements and advancements. A system that can find problems and address them.

Semen/seminal fluid: The fluid ejaculated by males that contains sperm cells.

Sequential unmasking: The process of revealing contextual information about a case after analyses have taken place. A practice used to limit bias.

Serial number restoration: Processes used to treat surfaces from which a serial number has been removed or obliterated.

Serology: The analysis of blood and body fluids.

Sexual dimorphism: Differences in skeletal features associated with biological sex.

Shock wave: A zone of extremely compressed material.

Shooting reconstruction: An experiment or experiments used to try and recreate a shooting event based on evidence recovered.

Shrapnel: Debris, fragments, or other objects ejected from an explosion.

Signature: Additional characteristics or acts performed at a crime scene by the offender. Unlike the methods of the modus operandi, signature actions are not necessary for the crime to occur.

Simulation: *See* malingering.

Single action: In a handgun or long gun, a design in which the weapon must first be manually cocked to deliver a cartridge to the chamber.

Skeletonized stain: Bloodstain consisting only of an outer periphery after the central area is removed by wiping when the liquid was partially dried. Can also be produced by flaking away of the center of a completely dried stain.

Skidding: Occurs when the wheels of a vehicle stop turning but motion does not.

Skid formula: Mathematical expressions used in auto accidents to estimate speed based on skid marks.

Slide: A part of a semiautomatic weapon that moves backwards when the weapon is fired.

Smoke: Visible by-products of fire, to include particulates and droplets.

Soot: Tiny particles of carbon.

Spermatozoa: Sperm cells.

Stance: The track width between two tires, such as between the two front tires.

Standard operating procedure: A set and validated written procedure that dictates how a given analysis is to be performed.

Static load: In a building, the load on structural elements that is based only on the weight of portions above. For example, in a 10-story building, the static load on the first floor is the combined weight of the floors above.

Stature: A person's height while living.

Stereo binocular microscope: A microscope with two eyepieces that allow for stereo vision.

Stimulants: Drugs that stimulate the central nervous system, such as amphetamine.

Stippling: Burn wounds caused by hot propellant impacting the skin; associated with gunshot wounds. Also called tattooing.

Striations: Parallel or roughly parallel lines associated with firearms and tool mark examinations.

Stringing: A technique that uses string, the measurement of blood drop dimensions, and a protractor to estimate the angle of impact and potential areas of origin of a stain pattern.

Subpubic angle: A portion of the pelvis useful in determining the biological sex of an individual.

Successional colonization: The process by which a body is colonized by a series of insects spaced out over time.

Successive classification: Placing an item into several categories in order. For example, a fiber can be successively classified as nylon, blue, and shiny.

Suicidal intent: The degree of seriousness of a person who attempts or commits suicide; degree of motivation to purposefully end life.

Superglue: *See* cyanoacrylate.

Surface tension: A force that draws the surface of a liquid inward.

Taphonomic context: The area around where human remains have been found.

Taphonomy: The study of the process that cause changes to occur in human remains after death.

Tattooing: Burn wounds caused by hot propellant impacting the skin; associated with gunshot wounds. Also called stippling.

Tented arch: A type of arch pattern in fingerprints.

Testimonial evidence: Evidence transmitted via written or spoken testimony.

Tetrahydrocannabinol: The primary active ingredient in marijuana.

Therapeutic dose: The dose of a drug that must be taken to generate a therapeutic response in most people.

Three-dimensional impressions: Impressions (footwear, tires, etc.) that have depth, such as a tire track in mud.

Tire tracks: The distance between two or more tires on a car or truck; usually refers to the width.

Total magnification: In microscopy, the product of the magnification by the eyepiece lens and objective lens.

Touch DNA: Tiny amounts of DNA deposited by a touching; typically, a trace amount.

Toxin: Poison; substance that causes a toxic response.

Trabecular bone: The spongy inner portion of the bone.

Transfer evidence: Evidence that is created by contact or interaction.

Trash marks: Marks made on copies because of defects in a copier or printer.

Triangulation: Estimating the location of a device based on the relative signal strength received at three or more receivers.

Trier of fact: The person or party responsible for establishing the facts in a legal proceeding.

Turning diameter: The diameter of a circle that a car will make when the steering wheel is fully turned.

Ulnar loop: A fingerprint loop that flows toward the ulnar bone in the arm.

Upper flammability limit: The highest concentration of fuel-to-air ratio that supports combustion.

Vacuoles: Bubbles in blood that leave circular areas when dry; usually associated with expired blood.

Vehicle: The primary solvent in paint.

Victimology: An evaluation of the characteristics of victims of crime assessed to determine if a pattern is evident.

Videospectral comparator: A device used in questioned document analysis; works with different types of illumination and filters.

Virtual autopsy: Medical imaging techniques used to image a body in three dimensions.

Viscosity: Resistance to flow; informally "thickness" of a liquid.

Vitreous humor: Gelatinous filling of the eyeball.

Voice over Internet protocol: A method by which voice communication can occur over the Internet; calling by computer.

Void areas: Absence of bloodstain in an otherwise continuous bloodstain pattern. The geometry of the void may suggest an outline of the object that intercepted the blood, for example, furniture, a shoe, or a person.

Volatile/volatility: Compounds that evaporate easily and quickly; a relative term.

Walk-through: Preliminary crime scene survey performed to orient the crime scene investigator to the scene and the physical evidence at the scene.

Wear characteristics: Characteristics that are acquired by an item, such as a tool or shoe, over time by use.

Weather/weathering: The natural degradation through the evaporation of accelerants.

Weathering: The natural degradation through evaporation of accelerants.

Wheelbase: Distance between the front and rear tires on a vehicle measured center to center.

Whorl: A type of pattern in fingerprints.

Write blocker: A device used in the imaging of digital storage devices to ensure that data can only flow from the original to the copy.

Xenobiotic: A substance that is foreign to the body.

Index